de Gruyter Expositions in Mathematics 40

de Gruyter Expositions in Mathematics

Embedding Problems in Symplectic Geometry

by

Felix Schlenk

Walter de Gruyter · Berlin · New York

Author

Felix Schlenk
Mathematisches Institut
Universität Leipzig
Augustusplatz 10/11
04109 Leipzig
Germany
e-mail: schlenk@math.uni-leipzig.de

Mathematics Subject Classification 2000: 53-02; 53C15, 53D35, 37J05, 51M15, 52C17, 57R17, 57R40, 58F05, 70Hxx

Key words: symplectic embeddings, symplectic geometry, symplectic packings, symplectic capacities, geometric constructions, Hamiltonian systems, rigidity and flexibility

Library of Congress Cataloging-in-Publication Data

Schlenk, Felix, 1970−
  Embedding problems in symplectic geometry / by Felix Schlenk.
    p.   cm − (De Gruyter expositions in mathematics ; 40)
  Includes bibliographical references and index.
  ISBN 3-11-017876-1 (cloth : acid-free paper)
  1. Symplectic geometry.   2. Embeddings (Mathematics)   I. Title.
  II. Series.
  QA665.S35   2005
  516.3'6−dc22

                                                        2005000895

ISBN 3-11-017876-1

Bibliographic information published by Die Deutsche Bibliothek

Die Deutsche Bibliothek lists this publication in the Deutsche Nationalbibliografie;
detailed bibliographic data is available in the Internet at <http://dnb.ddb.de>.

Typesetting using the author's TEX files: I. Zimmermann, Freiburg.
Printing and binding: Hubert & Co. GmbH & Co. KG, Göttingen.
Cover design: Thomas Bonnie, Hamburg.

To my parents

Preface

Symplectic geometry is the geometry underlying Hamiltonian dynamics, and symplectic mappings arise as time-1-maps of Hamiltonian flows. The spectacular rigidity phenomena for symplectic mappings discovered in the last two decades demonstrate that the nature of symplectic mappings is very different from that of volume preserving mappings. The most geometric expression of symplectic rigidity are obstructions to certain symplectic embeddings. For instance, Gromov's Nonsqueezing Theorem states that there does not exist a symplectic embedding of the $2n$-dimensional ball $B^{2n}(r)$ of radius r into the infinite cylinder $B^2(1) \times \mathbb{R}^{2n-2}$ if $r > 1$. On the other hand, not much was known about the existence of interesting symplectic embeddings. The aim of this book is to describe several elementary and explicit symplectic embedding constructions, such as "symplectic folding", "symplectic wrapping" and "symplectic lifting". These constructions are used to solve some specific symplectic embedding problems, and they prompt many new questions on symplectic embeddings.

We feel that the embedding constructions described in this book are more important than the results we prove by them. Hopefully, they shall prove useful for solving other problems in symplectic geometry and will lead to further understanding of the still mysterious nature of symplectic mappings.

The exposition is self-contained, and the only prerequisites are a basic knowledge of differential forms and smooth manifolds. The book is addressed to mathematicians interested in geometry or dynamics. Maybe, it will also be useful to physicists working in a field related to symplectic geometry.

Acknowledgements. This book grew out of my PhD thesis written at ETH Zürich from 1996 to 2000. I am very grateful to my advisor Edi Zehnder for his support, his patience, and his continuous interest in my work. His insight in mathematics and his criticism prevented me more than once from further pursuing a wrong idea. He always found an encouraging word, and he never lost his humour, even not in bad times. Last but not least, his great skill in presenting mathematical results has finally influenced, I hope, my own style.

Many ideas of this book grew out of discussions with Paul Biran, David Hermann, Helmut Hofer, Daniel Hug, Tom Ilmanen, Wlodek Kuperberg, Urs Lang, François Laudenbach, Thomas Mautsch, Dusa McDuff and Leonid Polterovich. I am in particular indebted to Dusa McDuff who explained to me symplectic folding, a technique basic for the whole book.

Doing symplectic geometry at ETH has been greatly facilitated through the existence of the symplectic group. When I started my thesis in 1996, this group consisted of Casim Abbas, Michel Andenmatten, Kai Cieliebak, Hansjörg Geiges, Helmut Hofer, Markus Kriener, Torsten Linnemann, Laurent Moatty, Matthias Schwarz, Karl Friedrich Siburg, Edi Zehnder and myself, and when I finished my thesis, the group consisted of Meike Akveld, Urs Frauenfelder, Ralph Gautschi, Janko Latschev, Thomas Mautsch, Dietmar Salamon, Joa Weber, Katrin Wehrheim, Edi Zehnder and myself. I in particular wish to thank Dietmar Salamon, who helped creating a great atmosphere in the symplectic group; his enthusiasm for mathematics has been a continuous source of motivation for me.

This book is the visible fruit of my early studies in mathematics. A more important fruit are the friendships with Rolf Heeb, Laurent Lazzarini, Christian Rüede and Ivo Stalder.

Sana, Selin kedim, o zamanki sonsuz sabır ve sevgin için teşekkür ederim.

Some final work on this book has been done in autumn 2004 at Leipzig University. I wish to thank the Mathematisches Institut for its hospitality, and Anna Wienhard and Peter Albers for carefully reading the introduction. Last but not least, I thank Jutta Mann, Irene Zimmermann and Manfred Karbe for editing my book with so much patience and care.

Leipzig, December 2004   *Felix Schlenk*

Contents

Chapter 1

Introduction

In the first section of this introduction we recall how symplectomorphisms of $\mathbb{R}^{2n}$ arise in classical mechanics, and introduce such notions as "Hamiltonian", "symplectic" and "volume preserving". In the second section we briefly tell how symplectic rigidity phenomena were discovered, and then state two paradigms of symplectic non-embedding theorems as well as a symplectic non-embedding result proved in Chapter 2. From Section 1.3 on we describe various symplectic embedding theorems and, in particular, the results proved in this book.

1.1 From classical mechanics to symplectic geometry

Consider a particle of mass 1 in some $\mathbb{R}^n$ subject to a force field F. According to Newton's second law of motion, the acceleration of the particle is equal to the force acting upon it, $\ddot{x} = F$. In many classical problems, such as those in celestial mechanics, the force field F is a potential field which depends only on the position of the particle and on time, so that

$$\ddot{x}(t) = \nabla U(x(t), t).$$

Introducing the auxiliary variables $y = \dot{x}$, this second order system becomes the first order system of twice as many equations

$$\left.\begin{aligned} \dot{x}(t) &= y(t), \\ \dot{y}(t) &= \nabla U(x(t), t). \end{aligned}\right\} \tag{1.1.1}$$

Besides for special potentials U, it is a hopeless task to solve (1.1.1) explicitly. One can, however, obtain some quantitative insight as follows. The structure of the system (1.1.1) is not very beautiful. Notice, though, that (1.1.1) is "skew-coupled" in the sense that the derivative of x depends on y only and vice versa. We capitalize on this by considering the function

$$H(x, y, t) = \frac{y^2}{2} - U(x, t), \tag{1.1.2}$$

which represents the total energy (i.e., the sum of kinetic and potential energy) of our particle. With this notation, the Newtonian system (1.1.1) becomes the *Hamiltonian system*

$$\left.\begin{aligned} \dot{x}(t) &= \frac{\partial H}{\partial y}(x, y, t), \\ \dot{y}(t) &= -\frac{\partial H}{\partial x}(x, y, t). \end{aligned}\right\} \tag{1.1.3}$$

Notice the beautiful skew-symmetry of this system. In order to write it in a more compact form, we consider the constant exact differential 2-form

$$\omega_0 = \sum_{i=1}^{n} dx_i \wedge dy_i \tag{1.1.4}$$

on $\mathbb{R}^{2n}$. It is called the *standard symplectic form* on $\mathbb{R}^{2n}$. It's n'th exterior product is $\omega_0^n = \omega_0 \wedge \cdots \wedge \omega_0 = n! \, \Omega_0$, where the volume form

$$\Omega_0 = dx_1 \wedge dy_1 \wedge \cdots \wedge dx_n \wedge dy_n$$

agrees with the Euclidean volume form $dx_1 \wedge \cdots \wedge dx_n \wedge dy_1 \wedge \cdots \wedge dy_n$ up to the factor $(-1)^{\frac{n(n-1)}{2}}$. It follows that ω_0 is a non-degenerate 2-form, so that the equation

$$\omega_0 \left(X_H(z, t), \cdot \right) = dH(z, t) \tag{1.1.5}$$

of 1-forms has a unique solution $X_H(z, t)$ for each $z = (x, y) \in \mathbb{R}^{2n}$ and each $t \in \mathbb{R}$. The time-dependent vector field X_H is called *Hamiltonian vector field* of H. Notice now that $X_H = \left(\frac{\partial H}{\partial y}, -\frac{\partial H}{\partial x} \right)$, so that the Hamiltonian system (1.1.3) takes the compact form

$$\dot{z}(t) = X_H(z(t), t). \tag{1.1.6}$$

Under suitable assumptions on the potential U this ordinary differential equation can be solved for all initial values $z(0) = z \in \mathbb{R}^{2n}$ and for all times t. The resulting flow $\{\varphi_H^t\}$ defined by

$$\left.\begin{aligned} \frac{d}{dt} \varphi_H^t(z) &= X_H \left(\varphi_H^t(z), t \right), \\ \varphi_H^0(z) &= z, \quad z \in \mathbb{R}^{2n}, \end{aligned}\right\} \tag{1.1.7}$$

is called the *Hamiltonian flow* of H. Each diffeomorphism φ_H^t is called a *Hamiltonian diffeomorphism*. More generally, any time-t-map φ_H^t obtained in this way via a smooth function $H \colon \mathbb{R}^{2n} \times \mathbb{R} \to \mathbb{R}$, not necessarily of the form (1.1.2), is called a Hamiltonian diffeomorphism.

The above formal manipulations were primarily motivated by aesthetic considerations. As the following facts show, there are important pay-offs, however.

Fact 1. *If H is time-independent, then H is preserved by the flow φ_H^t.*

Proof. Since H is time-independent and in view of definitions (1.1.7) and (1.1.5),

$$\tfrac{d}{dt} H(\varphi_H^t) = dH(\varphi_H^t) \tfrac{d}{dt} \varphi_H^t = dH(\varphi_H^t) X_H(\varphi_H^t) = \omega_0(X_H, X_H) \circ \varphi_H^t,$$

which vanishes because ω_0 is skew-symmetric. $\qquad\square$

Historically, this fact was the main reason for working in the Hamiltonian formalism. For us, two other pay-offs will be more important.

Fact 2. *Hamiltonian diffeomorphisms preserve the symplectic form ω_0.*

Proof. Using Cartan's formula $\mathcal{L}_{X_H} = d\iota_{X_H} + \iota_{X_H} d$, definition (1.1.5) and $d\omega_0 = 0$, we compute

$$\mathcal{L}_{X_H}\omega_0 = d\iota_{X_H}\omega_0 + \iota_{X_H} d\omega_0 = ddH + 0 = 0 \quad \text{for all } t.$$

Therefore, $\tfrac{d}{dt}\left(\varphi_H^t\right)^* \omega_0 = \left(\varphi_H^t\right)^* \mathcal{L}_{X_H}\omega_0 = 0$ for all t, so that $\left(\varphi_H^t\right)^* \omega_0 = \omega_0$. $\qquad\square$

A diffeomorphism φ of $\mathbb{R}^{2n}$ is called *symplectic diffeomorphism* or *symplectomorphism* if it preserves the symplectic form ω_0,

$$\varphi^* \omega_0 = \omega_0.$$

In classical mechanics, symplectomorphisms play the role of those coordinate transformations which preserve the class of Hamiltonian vector fields, and are thus called *canonical transformations*. Symplectic geometry of $(\mathbb{R}^{2n}, \omega_0)$ is the study of its automorphisms, which are the symplectomorphisms. By Fact 2, the set of Hamiltonian diffeomorphisms is embedded in this geometry.

A diffeomorphism φ of $\mathbb{R}^{2n}$ is called *volume preserving* if it preserves the volume form Ω_0,

$$\varphi^* \Omega_0 = \Omega_0.$$

Of course, a diffeomorphism is volume preserving if and only if it preserves the Euclidean volume form.

Fact 3. *Symplectomorphisms preserve the volume form Ω_0.*

Proof. $\varphi^* \Omega_0 = \varphi^* \left(\tfrac{1}{n!}\omega_0^n\right) = \tfrac{1}{n!}\left(\varphi^* \omega_0\right)^n = \tfrac{1}{n!}\omega_0^n = \Omega_0.$ $\qquad\square$

These facts go back to the 19th century. Putting Fact 2 and Fact 3 together we find Liouville's Theorem stating that Hamiltonian diffeomorphisms preserve the volume in phase space. There is no analogue of Liouville's Theorem for the flow in $\mathbb{R}^n$ generated by the Newtonian system (1.1.1).

Summarizing, we have

$$\text{Hamiltonian} \implies \text{symplectic} \implies \text{volume preserving}.$$

It is well-known that symplectomorphisms of $\mathbb{R}^{2n}$ are Hamiltonian diffeomorphisms (see Appendix A), so that "Hamiltonian" and "symplectic" is the same. In dimension 2, "symplectic" and "volume preserving" is also the same. In higher dimensions, however, the difference between "symplectic" and "volume preserving" turns out to be huge and lies at the heart of symplectic geometry.

1.2   Symplectic embedding obstructions

The most striking examples for the difference between "symplectic" and "volume preserving" are obstructions to certain symplectic embeddings. Before describing such obstructions, we briefly tell the story of

1.2.1   The discovery of symplectic rigidity phenomena.   We only describe the discovery of the first rigidity phenomena found. For the theories invented during these discoveries and for preceding and subsequent developments and further results we refer to the papers [18], [31] and to the books [32], [39], [62], [63].

Local considerations show that there are much less symplectomorphisms than volume preserving diffeomorphisms of $\mathbb{R}^{2n}$ if $n \geq 2$: The linear symplectic group has dimension $2n^2 + n$, while the group of matrices with determinant 1 has dimension $4n^2 - 1$. Moreover, locally any symplectic map can be represented in terms of a single function, a so-called generating function, see [39, Appendix 1], while one needs $2n - 1$ functions to describe a volume preserving diffeomorphism locally. Also notice that the Lie algebra of the group of symplectomorphisms can be identified with the set of time-independent Hamiltonian functions, while the Lie algebra of the group of volume preserving diffeomorphisms consists of divergence-free vector fields, which depend on $2n - 1$ functions. These local differences do not imply, however, that the set of symplectomorphisms of $\mathbb{R}^{2n}$ is also much smaller than the set of volume preserving diffeomorphisms from a global point of view: Every time-dependent compactly supported function on $\mathbb{R}^{2n}$ generates a Hamiltonian diffeomorphism, and so one could well believe that whatever can be done by a volume preserving diffeomorphism can "approximately" also be done by a Hamiltonian or symplectic diffeomorphism. This opinion was indeed shared by many physicists until the mid 1980s.

Global properties distinguishing Hamiltonian or symplectic diffeomorphisms from volume preserving diffeomorphisms were discovered only around 1980. There are various reasons for this "delay" in the discovery of symplectic rigidity. One reason is that the preservation of volume of Hamiltonian diffeomorphisms and stability problems in celestial mechanics had attracted and absorbed much attention, leading to ergodic theory and KAM theory. Another reason is that many interesting questions in classical mechanics (such as the restricted 3-body problem) lead to problems in dimension 2, where "symplectic" and "volume preserving" is the same. But the main reason for this delay was undoubtedly the difficulty in establishing global symplectic rigidity phenomena. No such phenomenon known today admits an easy proof.

In the 1960s, Arnold pointed out the special role played by the 2-form ω_0 in Fact 2 and made several seminal and fruitful conjectures in symplectic geometry, whose proofs in particular would have demonstrated that both Hamiltonian and symplectic diffeomorphisms are distinguished from volume preserving diffeomorphisms by global properties.

A breakthrough came in 1983 when Conley and Zehnder, [18], proved one of Arnold's conjectures. Denote the standard $2n$-dimensional torus $\mathbb{R}^{2n}/\mathbb{Z}^{2n}$ endowed with the induced symplectic form ω_0 by (T^{2n}, ω_0).

Arnold conjecture for the torus. *Every Hamiltonian diffeomorphism of the standard torus (T^{2n}, ω_0) must have at least $2n + 1$ fixed points.*

It in particular follows that volume preserving (or symplectic) diffeomorphisms of (T^{2n}, ω_0) cannot be C^0-approximated by Hamiltonian diffeomorphisms in general. Indeed, translations demonstrate that this global fixed point theorem is a truly Hamiltonian result, which does not hold for all volume preserving or symplectic diffeomorphisms. For a Hamiltonian diffeomorphism which is generated by a time-independent Hamiltonian or which is C^1-close to the identity, the theorem follows from classical Lusternik–Schnirelmann theory. The point of this theorem is that it holds for arbitrary Hamiltonian diffeomorphisms, also for those far from the identity.

The first rigidity phenomenon for symplectomorphisms of $\mathbb{R}^{2n}$ was found by Gromov and Eliashberg. In the early 1970s, Gromov proved the following alternative.

Gromov's Alternative. *The group of symplectomorphisms of $\mathbb{R}^{2n}$ is either C^0-closed in the group of all diffeomorphisms (hardness), or its C^0-closure is the group of volume preserving diffeomorphisms (softness).*

Notice that "symplectic" is a C^1-condition, so that there is no obvious reason for hardness. The soft alternative would have meant that there are no interesting global invariants in symplectic geometry. In the late 1970s, Eliashberg decided Gromov's Alternative in favour of hardness.

C^0-stability for symplectomorphisms. *The group of symplectomorphisms of $\mathbb{R}^{2n}$ is C^0-closed in the group of all diffeomorphisms.*

It follows that volume preserving diffeomorphisms of $\mathbb{R}^{2n}$ cannot be C^0-approximated by symplectomorphisms in general. For references and an elegant proof we refer to [39, Section 2.2]. An important ingredient of each known proof is a symplectic non-embedding result.

1.2.2 Symplectic non-embedding theorems. A smooth map $\varphi \colon U \to \mathbb{R}^{2n}$ defined on an open (not necessarily connected) subset U of $\mathbb{R}^{2n}$ is called *symplectic* if $\varphi^*\omega_0 = \omega_0$. Locally, symplectic maps are embeddings. Indeed, symplectic maps preserve the volume form $\Omega_0 = \frac{1}{n!}\omega_0^n$ and are thus immersions.

Flexibility for symplectic immersions. *For every open set V in $\mathbb{R}^{2n}$ there exists a symplectic immersion of $\mathbb{R}^{2n}$ into V.*

Proof. Since translations are symplectic, we can assume that V contains the origin. Let D be an open disc in $\mathbb{R}^2$ centred at the origin whose radius r is so small that $D \times \cdots \times D \subset V$. We shall construct a symplectic immersion φ of $\mathbb{R}^2$ into D. The product $\varphi \times \cdots \times \varphi$ will then symplectically immerse $\mathbb{R}^{2n}$ into V.

Step 1. Choose a diffeomorphism $f \colon \mathbb{R} \to]0, r^2/2[$. Then the map

$$\mathbb{R}^2 \to]0, r^2/2[\times \mathbb{R}, \quad (x, y) \mapsto \left(f(x), \frac{y}{f'(x)} \right), \tag{1.2.1}$$

is a symplectomorphism.

Step 2. The map

$$]0, r^2/2[\times \mathbb{R} \to D, \quad (x, y) \mapsto \left(\sqrt{2x}\cos y, \sqrt{2x}\sin y \right),$$

is a symplectic immersion.   □

A symplectic map $\varphi \colon U \to \mathbb{R}^{2n}$ is called a *symplectic embedding* if it is injective. In view of the above result, we shall only consider symplectic embeddings from now on.

A *domain* V in $\mathbb{R}^{2n}$ is a non-empty connected open subset of $\mathbb{R}^{2n}$. The basic problem addressed in this book is

Basic Problem. *Consider an open set U in $\mathbb{R}^{2n}$ and a domain V in $\mathbb{R}^{2n}$. Does there exist a symplectic embedding of U into V?*

A reader with a more physical or dynamical background may rather ask for embeddings of U into V induced by *Hamiltonian* diffeomorphisms of $\mathbb{R}^{2n}$. This is not the same problem in general:

Example. Let U be the annulus

$$U = \left\{ (x, y) \in \mathbb{R}^2 \mid 1 < x^2 + y^2 < 2 \right\}$$

of area π and let V be the disc of area a. As is easy to see (or by Proposition 1 below), U symplectically embeds into V if and only if $a \geq \pi$. On the other hand, U symplectically embeds into V via a Hamiltonian diffeomorphism of $\mathbb{R}^2$ only if $a \geq 2\pi$ since such an embedding must map the whole disc of radius $\sqrt{2}$ into V.   ◇

For a large class of domains U in $\mathbb{R}^{2n}$, however, finding a symplectic or a Hamiltonian embedding is almost the same problem. A domain U in $\mathbb{R}^{2n}$ is called *starshaped* if U contains a point p such that for every point $z \in U$ the straight line between p and z is contained in U.

Extension after Restriction Principle. Assume that $\varphi \colon U \hookrightarrow \mathbb{R}^{2n}$ is a symplectic embedding of a bounded starshaped domain $U \subset \mathbb{R}^{2n}$. Then for any subset $A \subset U$ whose closure in $\mathbb{R}^{2n}$ is contained in U there exists a Hamiltonian diffeomorphism Φ_A of $\mathbb{R}^{2n}$ such that $\Phi_A|_A = \varphi|_A$.

A proof of this well-known fact can be found in Appendix A. In most of the results discussed or proved in this book, U will be a bounded starshaped (and in fact convex) domain – or a union of finitely many balls, for which the Extension after Restriction Principle also applies, see Appendix E.

Since symplectic embeddings preserve the volume form Ω_0 and are injective, they preserve the total volume $\mathrm{Vol}(U) = \int_U \Omega_0$. A necessary condition for the existence of a symplectic embedding of U into V is therefore

$$\mathrm{Vol}(U) \leq \mathrm{Vol}(V). \tag{1.2.2}$$

For *volume preserving embeddings*, this necessary condition is also sufficient.

Proposition 1. *An open set U in $\mathbb{R}^{2n}$ embeds into a domain V in $\mathbb{R}^{2n}$ by a volume preserving embedding if and only if $\mathrm{Vol}(U) \leq \mathrm{Vol}(V)$.*

Notice that we did not assume that $\mathrm{Vol}(U)$ is finite. A proof of Proposition 1 can be found in Appendix B. Since in dimension 2 an embedding is symplectic if and only if it is volume preserving, the Basic Problem is completely solved in this dimension by Proposition 1. In higher dimensions, however, strong obstructions to symplectic embeddings which are different from the volume condition (1.2.2) appear. Consider the open $2n$-dimensional ball of radius r

$$B^{2n}(\pi r^2) = \left\{ (x, y) \in \mathbb{R}^{2n} \mid \sum\nolimits_{i=1}^{n} x_i^2 + y_i^2 < r^2 \right\}$$

and the open $2n$-dimensional *symplectic cylinder*

$$Z^{2n}(\pi) = \left\{ (x, y) \in \mathbb{R}^{2n} \mid x_1^2 + y_1^2 < 1 \right\}.$$

While the ball $B^{2n}(a)$ has finite volume for each a, the symplectic cylinder $Z^{2n}(\pi)$ has infinite volume, of course. The following theorem proved by Gromov in his seminal work [31] is the most geometric expression of symplectic rigidity.

Gromov's Nonsqueezing Theorem. *The ball $B^{2n}(a)$ symplectically embeds into the cylinder $Z^{2n}(\pi)$ if and only if $a \leq \pi$.*

Remarks. 1. Proposition 1 shows that for $n \geq 2$ the whole $\mathbb{R}^{2n}$ embeds into $Z^{2n}(\pi)$ by a volume preserving embedding. Explicit such embeddings are obtained by making use of maps of the form (1.2.1). The linear volume preserving diffeomorphism

$$(x, y) \mapsto (\epsilon x_1, \epsilon^{-1} x_2, x_3, \ldots, x_n, \epsilon y_1, \epsilon^{-1} y_2, y_3, \ldots, y_n)$$

of $\mathbb{R}^{2n}$ embeds the ball of radius ϵ^{-1} into $Z^{2n}(\pi)$.

2. The "symplectic cylinder" $Z^{2n}(\pi)$ in the Nonsqueezing Theorem cannot be replaced by the "Lagrangian cylinder"

$$\{(x, y) \in \mathbb{R}^{2n} \mid x_1^2 + x_2^2 < 1\}.$$

Indeed, for $a = \sqrt{2}/2$ the n-fold product of the map (1.2.1) symplectically embeds the whole $\mathbb{R}^{2n}$ into this cylinder. The linear symplectomorphism

$$(x, y) \mapsto (\epsilon x, \epsilon^{-1} y) \tag{1.2.3}$$

of $\mathbb{R}^{2n}$ embeds the ball of radius ϵ^{-1} into this cylinder.

3. Combined with Gromov's Alternative, the Nonsqueezing Theorem implies the C^0-Stability Theorem at once. In [20], Ekeland and Hofer observed that the C^0-Stability Theorem easily follows from the Nonsqueezing Theorem alone, see also [39, Section 2.2].

4. For far reaching generalizations of Gromov's Nonsqueezing Theorem we refer to Remark 9.3.7 in Chapter 9.   ◇

Gromov deduced his Nonsqueezing Theorem from his compactness theorem for pseudo-holomorphic spheres. In his proof, the obstruction to a symplectic embedding of $B^{2n}(a)$ into $Z^{2n}(\pi)$ for $a > \pi$ is the symplectic area $\int_S \omega_0$ of a holomorphic curve S in $B^{2n}(a)$ passing through the centre, which is at least $a > \pi$.

Using Gromov's compactness theorem for pseudo-holomorphic discs with Lagrangian boundary conditions, Sikorav found another amazing Nonsqueezing Theorem. Let S^1 be the unit circle in $\mathbb{R}^2(x, y)$, and consider the torus $T^n = S^1 \times \cdots \times S^1$ in $\mathbb{R}^{2n}$. Sikorav proved in [79] that *there does not exist a symplectomorphism of $\mathbb{R}^{2n}$ which maps T^n into $Z^{2n}(\pi)$*. Notice that the volume of T^n in $\mathbb{R}^{2n}$ vanishes, and that T^n does not bound any open set! In Sikorav's proof, the obstruction is the symplectic area $\int_D \omega_0$ of a closed holomorphic disc $D \subset \mathbb{R}^{2n}$ with boundary on T^n, which is at least π.

Sikorav's result combined with the Extension after Restriction Principle implies the following remarkable version of the Nonsqueezing Theorem, which is due to Hermann, [37].

Symplectic Hedgehog Theorem. *For $n \geq 2$, no starshaped domain in $\mathbb{R}^{2n}$ containing the torus T^n symplectically embeds into the cylinder $Z^{2n}(\pi)$.*

A more martial reader may prefer calling it Symplectic Flail Theorem. Notice that there are starshaped domains containing T^n of arbitrarily small volume!

We finally describe a symplectic non-embedding result which will lead to the search for interesting symplectic embedding constructions. Given an open subset U of $\mathbb{R}^{2n}$ and a number $\lambda > 0$ we set

$$\lambda U = \{\lambda z \mid z \in U\}.$$

Our Basic Problem can be reformulated as

Problem UV. *Consider an open set U in $\mathbb{R}^{2n}$ and a domain V in $\mathbb{R}^{2n}$. What is the smallest λ such that U symplectically embeds into λV?*

In the two theorems above, the symplectic embedding realizing the smallest λ was simply the identity embedding. In these theorems, the set U was a round ball $B^{2n}(a)$ with $a > \pi$ and a starshaped domain containing the "round" torus T^n, the set V was the "long and thin" cylinder $Z^{2n}(\pi)$, and the outcome was that these "round" sets U cannot be symplectically "squeezed" into the "long but thinner" set V. In order to formulate a symplectic embedding problem in which we have a chance to find interesting symplectic embeddings, we therefore take now U "long and thin" and V "round". Our hope is then that U can be symplectically "folded" or "wrapped" into V. To fix the ideas, we take V to be a ball and U to be an ellipsoid. Using complex notation $z_i = (x_i, y_i)$, we define the open *symplectic ellipsoid* with radii $\sqrt{a_i/\pi}$ as

$$E(a_1, \ldots, a_n) = \left\{ (z_1, \ldots, z_n) \in \mathbb{C}^n \mid \sum_{i=1}^{n} \frac{\pi |z_i|^2}{a_i} < 1 \right\}.$$

Here, $|\cdot|$ denotes the Euclidean norm in $\mathbb{R}^2$. Notice that $E(a, \ldots, a) = B^{2n}(a)$. Since a permutation of the symplectic coordinate planes is a (linear) symplectic map, we may assume $a_1 \leq a_2 \leq \cdots \leq a_n$. If, for instance, $a_1 = \cdots = a_{n-1}$ and a_n is much larger than a_1, then $E(a_1, \ldots, a_n)$ is indeed "long and thin". With these choices for U and V, Problem UV specializes to

Problem EB. *What is the smallest ball $B^{2n}(A)$ into which $E(a_1, \ldots, a_n)$ symplectically embeds?*

Of course, the inclusion symplectically embeds $E(a_1, \ldots, a_n)$ into $B^{2n}(A)$ if $A \geq a_n$. The following rigidity result shows that one cannot do better if the ellipsoid is still "quite round".

Theorem 1. *Assume $a_n \leq 2a_1$. Then the ellipsoid $E(a_1, \ldots, a_n)$ does not symplectically embed into the ball $B^{2n}(A)$ if $A < a_n$.*

In the case $n = 2$, Theorem 1 was proved in [26] as an application of symplectic homology. Our proof is simpler and works in all dimensions. It uses the n'th Ekeland–Hofer capacity. Symplectic capacities are special symplectic invariants prompted by Gromov's work [31] and introduced by Ekeland and Hofer in [20]. Definitions and a discussion of properties relevant for this book can be found in Chapter 2 and Appendix C, and a thorough exposition of symplectic capacities is given in the book [39]. For now, it suffices to know that with starshaped domains U and V in $\mathbb{R}^{2n}$ a symplectic capacity c associates numbers $c(U)$ and $c(V)$ in $[0, \infty]$ in such a way that

A1. Monotonicity: $c(U) \le c(V)$ if U symplectically embeds into V.

A2. Conformality: $c(\lambda U) = \lambda^2 c(U)$ for all $\lambda \in \mathbb{R} \setminus \{0\}$.

A3. Nontriviality: $0 < c\big(B^{2n}(\pi)\big)$ and $c\big(Z^{2n}(\pi)\big) < \infty$.

A symplectic capacity c is *normalized* if

A3′. Normalization: $c\big(B^{2n}(\pi)\big) = c\big(Z^{2n}(\pi)\big) = \pi$.

In view of the monotonicity axiom, symplectic capacities can be used to detect symplectic embedding obstructions. Indeed, the existence of any normalized symplectic capacity implies Gromov's Nonsqueezing Theorem at once. It therefore cannot be easy to construct a symplectic capacity.

From Gromov's work on pseudo-holomorphic curves one can extract normalized symplectic capacities, and the afore mentioned proofs of the Nonsqueezing Theorem and the Symplectic Hedgehog Theorem can be formulated in terms of these capacities. These normalized symplectic capacities are useless for Problem EB, however. Indeed,

$$B^{2n}(a_1) \subset E(a_1, \ldots, a_n) \subset Z^{2n}(a_1) := B^2(a_1) \times \mathbb{R}^{2n-2},$$

so that $c\,(E(a_1, \ldots, a_n)) = a_1$ for any normalized symplectic capacity. Given a symplectic embedding $E(a_1, \ldots, a_n) \hookrightarrow B^{2n}(A)$, such a symplectic capacity therefore only yields $a_1 \le A$, an information already covered by the volume condition (1.2.2).

Shortly after the appearance of Gromov's work, Ekeland and Hofer found a way to construct symplectic capacities via Hamiltonian dynamics. In order to give the idea of their approach, we consider a bounded starshaped domain $U \subset \mathbb{R}^{2n}$ with smooth boundary ∂U. A *closed characteristic* on ∂U is an embedded circle in ∂U tangent to the *characteristic line bundle*

$$\mathcal{L}_U = \{(x, \xi) \in T \partial U \mid \omega_0(\xi, \eta) = 0 \text{ for all } \eta \in T_x \partial U\}.$$

If ∂U is represented as a regular energy surface $\{x \in \mathbb{R}^{2n} \mid H(x) = \text{const}\}$ of a smooth function H on $\mathbb{R}^{2n}$, then the Hamiltonian vector field X_H restricted to ∂U is a section of $\mathcal{L}_U$ in view of its definition (1.1.5), and so the traces of the periodic orbits of X_H on ∂U are the closed characteristics on ∂U. The *action* $A\,(\gamma)$ of a closed characteristic γ on ∂U is defined as

$$A\,(\gamma) = \left| \int_\gamma \lambda \right|,$$

where λ is any primitive of ω_0. In view of Stokes' Theorem, $A(\gamma)$ is the symplectic area $\left| \int_D \omega_0 \right|$ of a closed disc $D \subset \mathbb{R}^{2n}$ with boundary γ. The set

$$\Sigma\,(U) = \big\{kA\,(\gamma) \mid k = 1, 2, \ldots; \ \gamma \text{ is a closed characteristic on } \partial U\big\}$$

is called the *action spectrum* of U. In [20], Ekeland and Hofer associated with U a number $c_1(U)$ defined via critical point theory applied to the classical action functional of Hamiltonian dynamics, and in this way obtained a symplectic capacity c_1 satisfying $c_1(U) \in \Sigma(U)$. If U is convex, then $c_1(U)$ is the smallest number in $\Sigma(U)$. Therefore, $c_1(E(a_1, \ldots, a_n)) = a_1$. From this, the Nonsqueezing Theorem follows at once, and also the Symplectic Hedgehog Theorem can be proved by using the symplectic capacity c_1, see [79] and [83]. But again, c_1 is useless for Problem EB. In [21], then, Ekeland and Hofer repeated their construction from [20] in an S^1-equivariant setting and used S^1-equivariant cohomology to obtain a whole family $c_1 \leq c_2 \leq \cdots$ of symplectic capacities satisfying $c_j(U) \in \Sigma(U)$. Besides c_1, these capacities are not normalized, and for an ellipsoid $E = E(a_1, \ldots, a_n)$ they are given by

$$\{c_1(E) \leq c_2(E) \leq \cdots\} = \{ka_i \mid k = 1, 2, \ldots; \ i = 1, \ldots, n\}.$$

For an ellipsoid $E(a_1, \ldots, a_n)$ with $a_n \leq 2a_1$ and for a ball $B^{2n}(A)$ we therefore find

$$c_n(E(a_1, \ldots, a_n)) = a_n \quad \text{and} \quad c_n(B^{2n}(A)) = A,$$

so that Theorem 1 follows in view of the monotonicity of c_n. In Chapter 2 we shall prove a stronger result. For further symplectic non-embedding results we refer to Chapters 4 and 9, Appendix E, and the references given therein.

Summarizing, we have seen that there are various symplectic non-embedding theorems, and that the methods used in their proofs are quite different. The obstructions found, though, have a common feature: Once it is the symplectic area of a holomorphic curve through the centre of a ball, once it is the symplectic area of a holomorphic disc with Lagrangian boundary conditions, and once it is the symplectic area of a disc whose boundary is a closed characteristic.

1.3 Symplectic embedding constructions

What, then, can be done by symplectic embeddings? The main characters of this book are various symplectic embedding constructions. They are best motivated, described and understood when applied to specific symplectic embedding problems. In this section we describe the results thus obtained and only give a vague idea of the constructions. They will be carried out in detail in Chapters 3 to 9. These symplectic embedding constructions are all elementary and explicit. The need for explicit symplectic embedding constructions could be sufficiently motivated by purely mathematical curiosity alone. More importantly, these constructions will shed some light on the nature of symplectic rigidity. Sometimes, they show that the known symplectic embedding obstructions are sharp. More often, they yield symplectic embedding results which are not known to be optimal or known to be not optimal, and thereby prompt new questions on symplectic embeddings. Certain non-explicit symplectic embeddings can be obtained via the so-called h-principle for symplectic embeddings of codimension at least 2 and via the symplectic blow-up operation. While the former method is addressed only briefly right below, the latter will be important in Chapter 9, which is devoted to symplectic packings by balls.

1.3.1 From rigidity to flexibility.

The following result, which is due to Gromov [32, p. 335] and is taken from [26, p. 579], gives a partial answer to Problem EB and shows that the assumption $a_n \leq 2a_1$ in Theorem 1 cannot be omitted.

Symplectic embeddings via the h-principle. *For any $a > 0$ there exists an $\epsilon > 0$ such that the $2n$-dimensional ellipsoid $E(\epsilon, \ldots, \epsilon, a)$ symplectically embeds into $B^{2n}(\pi)$.*

Proof. This is an immediate consequence of Gromov's h-principle for symplectic embeddings of codimension at least 2. Indeed, choose a *smooth* embedding φ_0 of the closed disc $\overline{B^2(a)}$ into $B^{2n}(\pi)$. According to [32, p. 335] or [24, Theorem 12.1.1], arbitrarily C^0-close to φ_0 there exists a symplectic embedding $\varphi_1 \colon \overline{B^2(a)} \hookrightarrow B^{2n}(\pi)$, meaning that $\varphi_1^* \omega_0 = \omega_0$. Using the Symplectic Neighbourhood Theorem, we find $\epsilon > 0$ such that φ_1 extends to a symplectic embedding of $B^{2n-2}(\epsilon) \times \overline{B^2(a)}$ and in particular to a symplectic embedding of $E(\epsilon, \ldots, \epsilon, a)$. $\square$

Since the C^0-small perturbation φ_1 of the smooth embedding φ_0 provided by the h-principle is not explicit, this embedding method gives no quantitative information on the number $\epsilon > 0$. In a large part of this book we shall be concerned with providing quantitative information on ϵ. We first investigate the zone of transition between rigidity and flexibility in Problem EB. Our hope is still to "fold" or "wrap" a "long and thin" ellipsoid into a smaller ball in a symplectic way. To find such constructions, we start with giving a list of

Elementary symplectic embeddings

1. Linear symplectomorphisms. The group $\mathrm{Sp}(n; \mathbb{R})$ of linear symplectomorphisms of $\mathbb{R}^{2n}$ contains transformations of the form (1.2.3) and, more generally, of the form

$$(x, y) \mapsto \left(Ax, \left(A^T \right)^{-1} y \right) \tag{1.3.1}$$

where A is any non-singular $(n \times n)$-matrix. It also contains the unitary group $\mathrm{U}(n)$. Translations are also symplectic, of course.

2. Products of area preserving embeddings. Every area and orientation preserving embedding of a domain in $\mathbb{R}^2$ into another domain in $\mathbb{R}^2$ is symplectic, and by Proposition 1 there are plenty of such embeddings. An example are the "inverse symplectic polar coordinates"

$$(x, y) \mapsto \left(\sqrt{2x} \cos y, \sqrt{2x} \sin y \right) \tag{1.3.2}$$

embedding $]0, a/2\pi[\times]0, 2\pi[$ into $B^2(a)$, which we met before. As we shall see in Section 3.1, symplectic embeddings of domains in $\mathbb{R}^2$ can be described in an almost explicit way. Taking products, we obtain almost explicit symplectic embeddings of domains in $\mathbb{R}^{2n}$.

3. Lifts. For convenience we write $(u, v, x, y) = (x_1, y_1, x_2, y_2)$. Using definition (1.1.3) we find that the Hamiltonian vector field of the Hamiltonian function $(u, v, x, y) \mapsto -x$ is $(0, 0, 0, 1)$, so that the time-1-map of the Hamiltonian flow is the translation $(u, v, x, y) \mapsto (u, v, x, y + 1)$. Choose a smooth function $f \colon \mathbb{R} \to [0, 1]$ such that

$$f(s) = 0 \text{ if } s \le 0 \quad \text{and} \quad f(s) = 1 \text{ if } s \ge 1.$$

The Hamiltonian vector field of the Hamiltonian function $(u, v, x, y) \mapsto -f(u)x$ is $\big(0, f'(u)x, 0, f(u)\big)$, so that its time-1-map

$$(u, v, x, y) \mapsto \big(u, v + f'(u)x, x, y + f(u)\big) \tag{1.3.3}$$

fixes the half space $\{u \le 0\}$ and lifts the space $\{u \ge 1\}$ by 1 in the y-direction. Choosing f such that

$$f(s) = 0 \text{ if } s \le 0 \text{ or } s \ge 3 \quad \text{and} \quad f(s) = 1 \text{ if } s \in [1, 2]$$

and looking at the time-1-map generated by the Hamiltonian function

$$(u, v, x, y) \mapsto -f(u)f(v)x \tag{1.3.4}$$

we find "true" lifts (called elevators in the US, I guess). $\diamond$

At first glance, these elementary symplectic embeddings look useless for Problem EB. Indeed, none of them embeds the ellipsoid $E(a_1, \ldots, a_n)$ into a ball $B^{2n}(A)$ with $A < a_n$. However, these elementary symplectic embeddings will serve as building blocks for all our embedding constructions: Each of the symplectic embedding constructions described in the sequel will be a composition of elementary symplectic embeddings as above!

The first quantitative embedding result addressing Problem EB was proved by Traynor in [81] by means of a symplectic wrapping construction. Given $\lambda > 0$ the ellipsoid $\lambda E(a_1, a_2)$ symplectically embeds into the ball $\lambda B^4(A)$ if and only if $E(a_1, a_2)$ symplectically embeds into $B^4(A)$. We can thus assume without loss of generality that $a_1 = \pi$.

Traynor's Wrapping Theorem. *There exists a symplectic embedding*

$$E\big(\pi, k(k-1)\pi\big) \hookrightarrow B^4(k\pi + \epsilon)$$

for every integer $k \ge 2$ and every $\epsilon > 0$.

The symplectic wrapping construction invented by Traynor is a composition of linear symplectomorphisms and products of area preserving embeddings. It first uses a product of area preserving embeddings to view an ellipsoid as a Lagrangian product $\triangle \times \square$ of a simplex and a square in $\mathbb{R}^2_+(x) \times \mathbb{R}^2_+(y)$, and then uses a map of the form (1.3.1) to wrap this product around the torus $T^2(y) = \mathbb{R}^2(y)/2\pi\mathbb{Z}^2$ in $\mathbb{R}^2_+(x) \times T^2(y)$. The point is then that the product of the area preserving map (1.3.2) extends to a symplectic embedding of $\mathbb{R}^2_+(x) \times T^2(y)$ into $\mathbb{R}^2(x) \times \mathbb{R}^2(y)$. Details and an extension of the symplectic wrapping construction to higher dimensions are given in Section 6.1.

The contribution to Problem EB made by Traynor's Wrapping Theorem is encoded in the piecewise linear function w_{EB} on $[\pi, \infty[$ drawn in Figure 1.1 below, in which we again assume $a_1 = \pi$ and write $a = a_2$. We in particular see that $w_{EB}(a) < a$ only for $a > 3\pi$, so that Traynor's Wrapping Theorem does not tell us whether Theorem 1 is sharp. On the other hand, the obstructions to symplectic embeddings found in Section 1.2.2 confirm our hopes that some kind of folding can be used to show that Theorem 1 is sharp: Arguing heuristically, we consider the two (symplectic) areas $s_1(U)$ and $s_2(U)$ of the projections of a domain U in $\mathbb{R}^4$ to the coordinate planes $\mathbb{R}^2(x_1, y_1)$ and $\mathbb{R}^2(x_2, y_2)$. The obstructions to symplectic embeddings found in Section 1.2.2 were symplectic areas of surfaces different from these projections, but numerically they are equal to s_1 in both the Nonsqueezing Theorem and the Symplectic Hedgehog Theorem and equal to s_2 in Theorem 1. Consider now an ellipsoid $E = E(a_1, a_2)$. When we "fold E appropriately" to E', the smaller projection will double, $s_1(E') = 2a_1$, while the larger projection should decrease, $s_2(E') < a_2$. If $a_2 \le 2a_1$, then $s_1(E') = 2a_1 \ge a_2 = s_1(B^4(a_2))$, so that E' does not fit into a ball $B^4(A)$ with $A < a_2$, as predicted by Theorem 1. If $a_2 > 2a_1$, then $s_1(E') = 2a_1 < a_2$ and $s_2(E') < a_2$, however, so that we can hope that folding can be achieved in such a way that E' fits into a ball $B^4(A)$ with $A < a_2$. This can indeed be done in a symplectic way.

Theorem 2. *Assume $a_n > 2a_1$. Then there exists a symplectic embedding of the ellipsoid $E(a_1, \ldots, a_1, a_n)$ into the ball $B^{2n}(a_n - \delta)$ for every $\delta \in]0, \frac{a_n}{2} - a_1[$.*

The reader might ask why we look at "skinny" ellipsoids with $a_{n-1} = a_1$ in Theorem 2. The reason is that for "flat" ellipsoids an analogous embedding result does not hold in general. For instance, the third Ekeland–Hofer capacity c_3 implies that for $n \ge 3$ the "flat" $2n$-dimensional ellipsoid $E(a, 3a, \ldots, 3a)$ does not symplectically embed into the ball $B^{2n}(A)$ if $A < 3a$. The second Ekeland–Hofer capacity c_2 implies that the "mixed" ellipsoid $E(a, 2a, 3a)$ does not symplectically embed into the ball $B^6(A)$ if $A < 2a$, but we do not know the answer to

Question 1. *Does the ellipsoid $E(a, 2a, 3a)$ symplectically embed into $B^6(A)$ for some $A < 3a$?*

Symplectic folding was invented by Lalonde and McDuff in [48] in order to prove the General Nonsqueezing Theorem stated in Remark 9.3.7 as well as an inequality between Gromov width and displacement energy implying that the Hofer metric on the group of compactly supported Hamiltonian diffeomorphisms is always non-degenerate. In the same work [48] Lalonde and McDuff also observed that symplectic folding can be used to prove Theorem 2 in the case $n = 2$. A refinement of their symplectic folding construction in dimension 4 will prove Theorem 2 in all dimensions.

The symplectic folding construction is a composition of products of area preserving embeddings and a lift. Viewing an ellipsoid $E(a_1, a_2)$ as fibred over the larger disc $B^2(a_2)$, this construction first separates the smaller fibres from the larger ones by a suitable area preserving embedding of $B^2(a_2)$ into $\mathbb{R}^2$, then lifts the smaller fibres by the lift (1.3.3), and finally turns these lifted fibres over the larger fibres via another area preserving embedding. An idea of the construction can be obtained from Figure 3.12 on page 50 and from Figure 4.2 on page 53. Theorem 2 can be substantially improved by folding more than once. An idea of multiple symplectic folding is given by Figure 4.3 on page 54.

In describing the results for Problem EB, we now restrict ourselves to dimension 4 for the sake of clarity. As before we can assume $a_1 = \pi$ and write $a = a_2$. The optimal values A for the embedding problems $E(\pi, a) \hookrightarrow B^4(A)$ are encoded in the "characteristic function" χ_{EB} on $[\pi, \infty[$ defined by

$$\chi_{EB}(a) = \inf \left\{ A \mid E(\pi, a) \text{ symplectically embeds into } B^4(A) \right\}.$$

We illustrate the results with the help of Figure 1.1. In view of Theorem 1 we have $\chi_{EB}(a) = a$ for $a \in [\pi, 2\pi]$. For $a > 2\pi$, the second Ekeland–Hofer capacity c_2 still implies that $\chi_{EB}(a) \geq 2\pi$. This information is vacuous if $a \geq 4\pi$, since the volume condition $\mathrm{Vol}\big(E(\pi, a)\big) \leq \mathrm{Vol}\big(B^4(\chi_{EB}(a))\big)$ translates to $\chi_{EB}(a) \geq \sqrt{\pi a}$. The estimate $\chi_{EB}(a) \leq a/2 + \pi$ stated in Theorem 2 is obtained by folding once. It will turn out that for $a > 2\pi$ and for each $k \geq 1$, folding $k + 1$ times embeds $E(\pi, a)$ into a strictly smaller ball than folding k times. The function f_{EB} on $]2\pi, \infty[$ defined by

$$f_{EB}(a) = \inf \left\{ A \mid E(\pi, a) \text{ embeds into } B^4(A) \text{ by multiple symplectic folding} \right\}$$

is therefore obtained by folding "infinitely many times". The graph of the function f_{EB} is computed by a computer program. The function w_{EB} encoding Traynor's Wrapping Theorem is alternatingly larger and smaller than f_{EB}.

We are particularly interested in the behaviour of $\chi_{EB}(a)$ as $a \to 2\pi^+$ and as $a \to \infty$. We shall prove that

$$\limsup_{\epsilon \to 0^+} \frac{f_{EB}(2\pi + \epsilon) - 2\pi}{\epsilon} \leq \frac{3}{7},$$

and so the same estimate holds for χ_{EB}.

Question 2. *How does $\chi_{EB}(a)$ look like near $a = 2\pi$? In particular,*

$$\limsup_{\epsilon \to 0^+} \frac{\chi_{EB}(2\pi + \epsilon) - 2\pi}{\epsilon} < \frac{3}{7} ?$$

We have $f_{EB}(a) < w_{EB}(a)$ for all $a \in \,]2\pi, 5.1622\pi]$. The computer program for f_{EB} yields the particular values

$$f_{EB}(3\pi) \approx 2.3801\pi \quad \text{and} \quad f_{EB}(4\pi) \approx 2.6916\pi.$$

We do not expect that $\chi_{EB}(3\pi) = f_{EB}(3\pi)$ and $\chi_{EB}(4\pi) = f_{EB}(4\pi)$.

Question 3. *Is it true that* $\chi_{EB}(3\pi) = \chi_{EB}(4\pi) = 2\pi$ *?*

The difference $w_{EB}(a) - \sqrt{\pi a}$ between w_{EB} and the volume condition is bounded by $(3 - \sqrt{3})\pi$. We shall also prove that $f_{EB}(a) - \sqrt{\pi a}$ is bounded. It follows that $\chi_{EB}(a) - \sqrt{\pi a}$ is bounded. We in particular have

$$\lim_{a \to \infty} \frac{\text{Vol}\,(E(\pi, a))}{\text{Vol}\left(B^4\left(\chi_{EB}(a)\right)\right)} = 1. \tag{1.3.5}$$

This means that the embedding obstructions encountered for small a more and more disappear as $a \to \infty$.

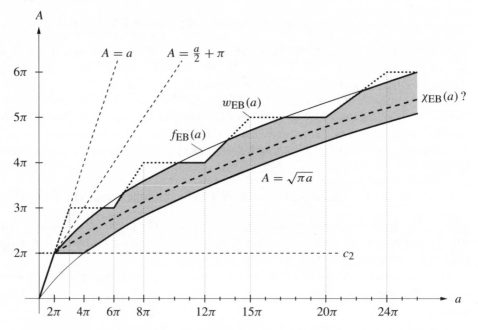

Figure 1.1. What is known about the embedding problem $E(\pi, a) \hookrightarrow B^4(A)$.

Denote by $D(a) = B^2(a)$ the open disc in $\mathbb{R}^2$ of area a centred at the origin. The open *symplectic polydisc* $P(a_1, \ldots, a_n)$ in $\mathbb{R}^{2n}$ is defined as

$$P(a_1, \ldots, a_n) = D(a_1) \times \cdots \times D(a_n).$$

The "n-cube" $P^{2n}(a, \ldots, a)$ will be denoted by $C^{2n}(a)$. Up to now, our model sets were ellipsoids, which we tried to symplectically embed into small balls. Starting from Sikorav's Nonsqueezing Theorem for the torus T^n and noticing that (the closure of) $C^{2n}(\pi)$ is the convex hull of T^n, we could equally well have taken polydiscs and cubes.

Problem PC. *What is the smallest cube $C^{2n}(A)$ into which $P(a_1, \ldots, a_n)$ symplectically embeds?*

For this problem, no interesting obstructions are known, however. The reason is that symplectic capacities only see the size $\min\{a_1, \ldots, a_n\}$ of the smallest disc of a polydisc and thus do not provide any obstruction for symplectic embeddings of polydiscs into cubes stronger than the volume condition. In particular, it is unknown whether the analogue of Theorem 1 for polydiscs holds true. On the other hand, both symplectic folding and symplectic wrapping can be used to construct interesting symplectic embeddings of polydiscs into cubes.

Somewhat more generally, we shall study symplectic embeddings of both ellipsoids and polydiscs into balls and cubes. While embedding an open set U into a minimal ball is related to minimizing its diameter (a 1-dimensional, metric quantity), embedding U into a minimal cube amounts to minimizing the areas of its projections to the symplectic coordinate planes (a 2-dimensional, "more symplectic" quantity). We refer to Appendix C for details on this.

1.3.2 Flexibility for skinny shapes.

Let us come back to Problem UV, which we reformulate as

Problem UV. *Consider an open set U in $\mathbb{R}^{2n}$ and a domain V in $\mathbb{R}^{2n}$. What is the largest λ such that λU symplectically embeds into V?*

As before, we shall eventually specialize U to an ellipsoid or a polydisc, but this time we take V to be an arbitrary domain in $\mathbb{R}^{2n}$ of finite volume. In fact, we shall look at symplectic embeddings into arbitrary connected symplectic manifolds of finite volume. A reader not familiar with smooth manifolds may skip the subsequent generalities on symplectic manifolds and take (M, ω) in Theorem 3 below to be a domain in $\mathbb{R}^{2n}$ of finite volume. As will become clear in Chapter 6 not much is lost thereby.

A differential 2-form ω on a smooth manifold M is called *symplectic* if ω is non-degenerate and closed. The pair (M, ω) is then called a *symplectic manifold*. The non-degeneracy of ω implies that M is even-dimensional, $\dim M = 2n$, and that $\Omega = \frac{1}{n!}\omega^n$ is a volume form on M, so that M is orientable. The non-degeneracy together with the closedness of ω imply that (M, ω) is locally isomorphic to $(\mathbb{R}^{2n}, \omega_0)$ with ω_0 as in (1.1.4):

Darboux's Theorem. *For every point $p \in M$ there exists a coordinate chart $\varphi_U : U \to \mathbb{R}^{2n}$ such that $\varphi_U(p) = 0$ and $\varphi_U^* \omega_0 = \omega$.*

Therefore, a symplectic manifold is a smooth $2n$-dimensional manifold admitting an atlas $\{(U, \varphi_U)\}$ such that all coordinate changes

$$\varphi_V \circ \varphi_U^{-1} : \varphi_U(U \cap V) \to \varphi_V(U \cap V)$$

are symplectic. Examples of symplectic manifolds are open subsets of $(\mathbb{R}^{2n}, \omega_0)$, the torus $\mathbb{R}^{2n}/\mathbb{Z}^{2n}$ endowed with the induced symplectic form, surfaces equipped with an area form, Kähler manifolds like complex projective space $\mathbb{CP}^n$ endowed with their Kähler form, and cotangent bundles with their canonical symplectic form. Many more examples are obtained by taking products and via the symplectic blow-up operation. We refer to the book [62] for more information on symplectic manifolds.

As before, we endow each open subset U of $\mathbb{R}^{2n}$ with the standard symplectic form ω_0. A smooth embedding $\varphi : U \hookrightarrow (M, \omega)$ is called *symplectic* if

$$\varphi^* \omega = \omega_0.$$

Problem UV generalizes to

Problem UM. *Consider an open set U in $\mathbb{R}^{2n}$ and a connected $2n$-dimensional symplectic manifold (M, ω). What is the largest number λ such that λU symplectically embeds into (M, ω)?*

A smooth embedding $\varphi : U \hookrightarrow (M, \omega)$ is called *volume preserving* if

$$\varphi^* \Omega = \Omega_0$$

where as before $\Omega_0 = \frac{1}{n!} \omega_0^n$ and $\Omega = \frac{1}{n!} \omega^n$. Of course, every symplectic embedding is volume preserving. A necessary condition for a symplectic embedding of U into (M, ω) is therefore

$$\mathrm{Vol}(U) \le \mathrm{Vol}(M, \omega)$$

where we set $\mathrm{Vol}(M, \omega) = \frac{1}{n!} \int_M \omega^n$. For volume preserving embeddings, this obvious condition is the only one in view of the following generalization of Proposition 1, a proof of which can again be found in Appendix B.

Proposition 2. *An open set U in $\mathbb{R}^{2n}$ embeds into (M, ω) by a volume preserving embedding if and only if $\mathrm{Vol}(U) \le \mathrm{Vol}(M, \omega)$.*

For symplectic embeddings of "round" shapes $U \subset \mathbb{R}^{2n}$ into (M, ω), however, there often are strong obstructions beyond the volume condition. We have already seen this in Section 1.2.2 in case that (M, ω) is a cylinder or a ball, and many more examples can be found in Chapters 4 and 9. To give one other example, we consider the product $(M, \omega) = (S^2 \times S^2, \sigma \oplus k\sigma)$, where σ is an area form on the 2-sphere S^2 of area $\int_{S^2} \sigma = \pi$ and where $k \geq 1$. Then the ball $B^4(a)$ symplectically embeds into (M, ω) if and only if $a \leq \pi$. For skinny shapes, though, the situation for symplectic embeddings is not too different from the one for volume preserving embeddings: We choose U to be a $2n$-dimensional ellipsoid $E(\pi, \ldots, \pi, a)$ or a $2n$-dimensional polydisc $P(\pi, \ldots, \pi, a)$, consider a connected $2n$-dimensional symplectic manifold (M, ω) of finite volume $\mathrm{Vol}(M, \omega) = \frac{1}{n!} \int_M \omega^n$, and define for each $a \geq \pi$ the numbers

$$p_a^E(M, \omega) = \sup_\lambda \frac{\mathrm{Vol}\big(\lambda E(\pi, \ldots, \pi, a)\big)}{\mathrm{Vol}(M, \omega)},$$

$$p_a^P(M, \omega) = \sup_\lambda \frac{\mathrm{Vol}\big(\lambda P(\pi, \ldots, \pi, a)\big)}{\mathrm{Vol}(M, \omega)},$$

where the supremum is taken over all those λ for which $\lambda E(\pi, \ldots, \pi, a)$ respectively $\lambda P(\pi, \ldots, \pi, a)$ symplectically embeds into (M, ω).

Theorem 3. *Assume that (M, ω) is a connected symplectic manifold of finite volume. Then*

$$\lim_{a \to \infty} p_a^E(M, \omega) = 1 \quad and \quad \lim_{a \to \infty} p_a^P(M, \omega) = 1.$$

This means that the obstructions encountered for symplectic embeddings of round shapes more and more disappear as we pass to skinny shapes. Notice that if (M, ω) is a 4-dimensional ball, the first statement in Theorem 3 is equivalent to the identity (1.3.5), which also followed from Traynor's Wrapping Theorem. Symplectic folding can be used to prove the full statement of Theorem 3. The second statement in Theorem 3 will be proved along the following lines. First, fill almost all of M with cubes. Using multiple symplectic folding, these cubes can then almost be filled with symplectically embedded thin polydiscs, see Figure 5.2 on page 85. Using the remaining space in M, these embeddings can finally be glued to a symplectic embedding of a very long and thin polydisc, see Figure 6.1 on page 108. The proof of the first statement is more involved and uses a non-elementary result of McDuff and Polterovich on filling a cube by balls.

1.3.3 A vanishing theorem. The Basic Problem and its variations asked for symplectic embeddings which were not required to have any additional properties. We now look at symplectic embeddings φ of the ball $B^{2n}(a)$ into the symplectic cylinder $Z^{2n}(\pi)$ whose image $\varphi(B^{2n}(a)) \subset Z^{2n}(\pi)$ has a specific property. By Gromov's Nonsqueezing Theorem there does not exist a symplectic embedding of $B^{2n}(a)$ into $Z^{2n}(\pi)$ if $a > \pi$. So fix $a \in]0, \pi]$. The simply connected hull $\hat{T}$ of a subset T of $\mathbb{R}^2$ is the union of its closure $\overline{T}$ and the bounded components of $\mathbb{R}^2 \setminus T$. We denote by μ the Lebesgue measure on $\mathbb{R}^2$, and we abbreviate $\hat{\mu}(T) = \mu(\hat{T})$. For the unit circle

S^1 in $\mathbb{R}^2$ we have $\mu(S^1) = 0 < \pi = \hat{\mu}(S^1)$. As is well-known, the Nonsqueezing Theorem is equivalent to each of the identities

$$a = \inf_{\varphi} \mu\left(p\{\varphi(B^{2n}(a))\}\right),$$

$$a = \inf_{\varphi} \hat{\mu}\left(p\{\varphi(B^{2n}(a))\}\right),$$

where φ varies over all symplectomorphisms of $\mathbb{R}^{2n}$ which embed $B^{2n}(a)$ into $Z^{2n}(\pi)$ and where $p\colon Z^{2n}(\pi) \to B^2(\pi)$ is the projection, see [22] and Appendix C.2. Following McDuff, [59], we consider *sections* of the image $\varphi(B^{2n}(a))$ instead of its projection, and define

$$\zeta(a) := \inf_{\varphi} \sup_{x} \mu\left(p\{\varphi(B^{2n}(a)) \cap D_x\}\right),$$

$$\hat{\zeta}(a) := \inf_{\varphi} \sup_{x} \hat{\mu}\left(p\{\varphi(B^{2n}(a)) \cap D_x\}\right),$$

where φ again varies over all symplectomorphisms of $\mathbb{R}^{2n}$ which embed $B^{2n}(a)$ into $Z^{2n}(\pi)$, and where $D_x \subset Z^{2n}(\pi)$ denotes the disc

$$D_x = B^2(\pi) \times \{x\}, \quad x \in \mathbb{R}^{2n-2}.$$

Clearly,

$$\zeta(a) \le \hat{\zeta}(a) \le a.$$

It is also well-known that the Nonsqueezing Theorem is equivalent to the identity

$$\hat{\zeta}(\pi) = \pi. \tag{1.3.6}$$

Indeed, the Nonsqueezing Theorem implies that for every symplectomorphism φ of $\mathbb{R}^{2n}$ which embeds $B^{2n}(\pi)$ into $Z^{2n}(\pi)$ there exists an x in $\mathbb{R}^{2n-2}$ such that $\varphi(B^{2n}(\pi)) \cap D_x$ contains the unit circle $S^1 \times \{x\}$. On her search for symplectic rigidity phenomena beyond the Nonsqueezing Theorem, McDuff therefore posed the following problem.

Problem ζ. *Find a non-trivial lower bound for the function $\zeta(a)$. In particular, is it true that $\zeta(a) \to \pi$ as $a \to \pi$?*

A further motivation for this problem comes from convex geometry and from the fact that on bounded convex subsets of $(\mathbb{R}^{2n}, \omega_0)$ normalized symplectic capacities agree up to a constant. It was known to Polterovich that $\zeta(a)/a \to 0$ as $a \to 0$. The following result answers the question in Problem ζ in the negative and completely solves Problem ζ.

Theorem 4. $\zeta(a) = 0$ *for all* $a \in \,]0, \pi]$ *and* $\hat{\zeta}(a) = 0$ *for all* $a \in \,]0, \pi[$.

The second assertion in Theorem 4 can be proved by symplectic folding. In order to prove the full theorem, we shall iterate the symplectic lifting construction briefly described in Section 1.3.1. An idea of the two proofs is given in Figure 8.1 on page 171 and in Figure 8.9 on page 182.

1.3.4 Symplectic packings. We finally look at the *symplectic packing problem*. Given a connected $2n$-dimensional symplectic manifold (M, ω) of finite volume and given a natural number k, this problem asks

Problem kBM. *What is the largest number a for which the disjoint union of k equal balls $B^{2n}(a)$ symplectically embeds into (M, ω)?*

Equivalently, one studies the *k'th symplectic packing number*

$$p_k(M, \omega) = \sup_a \frac{k \, \mathrm{Vol}\big(B^{2n}(a)\big)}{\mathrm{Vol}\,(M, \omega)}$$

where the supremum is taken over all those a for which $\coprod_{i=1}^{k} B^{2n}(a)$ symplectically embeds into (M, ω). Obstructions to full packings of a ball were already found by Gromov in [31], where he proved that $p_k(B^{2n}(\pi), \omega_0) \leq \frac{k}{2^n}$ for $2 \leq k \leq 2^n$. Later on, spectacular progress in the symplectic packing problem was made by McDuff and Polterovich in [61] and by Biran in [7], [8], who obtained symplectic packings via the symplectic blow-up operation. These works established many further packing obstructions for small values of k. For large k, however, it was shown in [61] that

$$\lim_{k \to \infty} p_k(M, \omega) = 1$$

for every connected symplectic manifold (M, ω) of finite volume, and it was shown in [7], [8] that for many symplectic 4-manifolds (M, ω) there exists a number k_0 such that $p_k(M, \omega) = 1$ for all $k \geq k_0$. This transition from rigidity for symplectic packings by few balls to flexibility for packings by many balls is reminiscent to the transition from rigidity to flexibility for symplectic embeddings of ellipsoids discussed in Sections 1.3.1 and 1.3.2.

Essentially all presently known packing numbers were obtained in [61], [7], [8]. The symplectic packings found in these works are not explicit, however. For some symplectic manifolds as balls and products of surfaces and for some values of k, *explicit* maximal symplectic packings were constructed by Karshon [41], Traynor [81], Kruglikov [45], and Maley, Mastrangeli and Traynor [54]. In the last chapter, we shall describe a very simple and explicit construction realizing the packing numbers $p_k(M, \omega)$ for those symplectic 4-manifolds (M, ω) and numbers k considered in [41], [81], [45], [54] as well as for ruled symplectic 4-manifolds and small values of k. For example, we shall see that maximal packings of the standard 4-ball by 5 or 6 balls and of the product $S^2 \times S^2$ of 2-spheres of equal area by 5 balls can be described by Figure 1.2.

These symplectic packings are simply obtained via products $\alpha_1 \times \alpha_2$ of suitable area preserving diffeomorphisms between a disc and a rectangle. Taking n-fold products we shall also construct a full packing of the standard $2n$-ball by l^n balls for each $l \in \mathbb{N}$ in a most simple way.

Figure 1.2. Maximal symplectic packings of the 4-ball by 5 or 6 balls and of $S^2 \times S^2$ by 5 balls.

Contrary to the symplectic embedding results described before, none of our symplectic packing results is new in view of the packings in [61], [7], [8]. We were just curious how maximal packings might look like, and we hope the reader will enjoy the pictures in Chapter 9, too. Moreover, in the range of k for which the explicit constructions in [41], [81], [45], [54] and our constructions fail to give maximal packings, they give a feeling that the balls in the packings from [61], [7], [8] must be "wild".

The book is organized as follows: In Chapter 2 we prove Theorem 1 and several other rigidity results for ellipsoids. In Chapter 3 we prove Theorem 2 by symplectic folding. In Chapter 4 we use multiple symplectic folding to obtain rather satisfactory results for symplectic embeddings of 4-dimensional ellipsoids and polydiscs into 4-dimensional balls and cubes. In Chapter 5 we look at higher dimensions. We will concentrate on embedding skinny ellipsoids into balls and skinny polydiscs into cubes. The results in this chapter form half of the proof of Theorem 3, which is completed in Chapter 6. In Chapter 6 we shall also notice that for certain symplectic manifolds our embedding methods can be used to improve Theorem 3. In Chapter 7 we recall the symplectic wrapping method invented by Traynor, and compare the results obtained by symplectic folding and wrapping. In Chapter 8 we review the motivations for Problem ζ and prove Theorem 4 and its generalizations by symplectic folding and by symplectic lifting. In Chapter 9 we give various motivations for the symplectic packing problem, collect the known symplectic packing numbers, and pack balls and ruled symplectic 4-manifolds by hand.

In Appendix A we give the well-known proof of the Extension after Restriction Principle and discuss an extension of this principle to unbounded domains. In Appendix B we prove Proposition 2. In Appendix C we clarify the relations between the invariants defined by Problem UB and Problem UC and other symplectic invariants. Appendix D provides computer programs necessary to compute the optimal embeddings of 4-dimensional ellipsoids into a 4-ball and a 4-cube which can be obtained by multiple symplectic folding. In Appendix E we describe some symplectic embedding problems not studied in this book; while some of them are almost solved, others are widely open.

Throughout this book we work in the C^∞-category, i.e., all manifolds and diffeomorphisms are assumed to be C^∞-smooth, and so are all symplectic forms and maps.

Chapter 2

Proof of Theorem 1

This chapter contains the rigidity results proved in this book. We start with giving a feeling for the difference between linear and non-linear symplectic embeddings. We then look at ellipsoids and use Ekeland–Hofer capacities to prove a generalization of Theorem 1. We finally notice that the polydisc analogues of our rigidity results for ellipsoids are either wrong or unknown.

We denote by $\mathcal{O}(n)$ the set of bounded domains in $\mathbb{R}^{2n}$ diffeomorphic to a ball. Each $U \in \mathcal{O}(n)$ is endowed with the standard symplectic structure $\omega_0 = \sum_{i=1}^{n} dx_i \wedge dy_i$. Orienting U by the volume form $\Omega_0 = \frac{1}{n!} \omega_0^n$ we write $|U| = \int_U \Omega_0$ for the usual volume of U. Let $\mathcal{D}(n)$ be the group of all symplectomorphisms of $\mathbb{R}^{2n}$ and let $\mathrm{Sp}(n; \mathbb{R})$ be its subgroup of linear symplectomorphisms of $\mathbb{R}^{2n}$. Define the following relations on $\mathcal{O}(n)$:

$$U \leq_1 V \iff \text{There exists a } \varphi \in \mathrm{Sp}(n; \mathbb{R}) \text{ with } \varphi(U) \subset V.$$
$$U \leq_2 V \iff \text{There exists a } \varphi \in \mathcal{D}(n) \text{ with } \varphi(U) \subset V.$$
$$U \leq_3 V \iff \text{There exists a symplectic embedding } \varphi \colon U \hookrightarrow V.$$

2.1 Comparison of the relations $\leq_i$

Clearly, $\leq_1 \Rightarrow \leq_2 \Rightarrow \leq_3$. It is, however, well known that the relations $\leq_l$ are different.

Proposition 2.1.1. *The relations $\leq_l$ are all different.*

Proof. In order to show that the relations $\leq_1$ and $\leq_2$ are different it suffices to find an area and orientation preserving diffeomorphism φ of $(\mathbb{R}^2, \omega_0)$ mapping the unit disc $D(\pi)$ to a set which is not convex. The existence of such a φ follows, for instance, from Lemma 3.1.5 below. For the products $U = D(\pi) \times \cdots \times D(\pi)$ and $V = \varphi(D(\pi)) \times \cdots \times \varphi(D(\pi))$ we then have $U \nleq_1 V$ and $U \leq_2 V$.

The construction of sets U and $V \in \mathcal{O}(n)$ with $U \leq_3 V$ but $U \nleq_2 V$ relies on the following simple observation. Suppose that $U \leq_2 V$ and in addition that $|U| = |V|$. If φ is a map realizing $U \leq_2 V$, no point of $\mathbb{R}^{2n} \setminus U$ can then be mapped to V, and we conclude that φ is a homeomorphism from the boundary ∂U of U to the boundary ∂V of V. Following Traynor, [81], we consider now the slit disc

$$SD(\pi) = D(\pi) \setminus \{(x, y) \mid x \geq 0, y = 0\},$$

and we set $U = C^{2n}(\pi) = D(\pi) \times \cdots \times D(\pi)$ and $V = SD(\pi) \times \cdots \times SD(\pi)$. By Proposition 1 of the introduction, $D(\pi) \leq_3 SD(\pi)$, see also Lemma 3.1.5 below. Therefore, $U \leq_3 V$, and clearly $|U| = |V|$. But ∂U and ∂V are not homeomorphic. $\square$

We wish to mention that for $n \geq 2$ more interesting examples showing that $\leq_2$ and $\leq_3$ are different were found by Eliashberg and Hofer, [23], and by Cieliebak, [15]. In order to describe their examples, we assume that ∂U is smooth. Recall that the characteristic line bundle $\mathcal{L}_U$ of ∂U is defined as

$$\mathcal{L}_U = \{(x, \xi) \in T\partial U \mid \omega_0(\xi, \eta) = 0 \text{ for all } \eta \in T_x \partial U\}. \tag{2.1.1}$$

The unparametrized integral curves of $\mathcal{L}_U$ are called *characteristics* and form the *characteristic foliation* of ∂U. Moreover, ∂U is said to be *of contact type* if on a neighbourhood of ∂U there exists a smooth vector field X which is transverse to ∂U and meets $\mathcal{L}_X \omega_0 = d\iota_X \omega_0 = \omega_0$. E.g., the radial vector field $X = \frac{1}{2} \frac{\partial}{\partial r}$ on $\mathbb{R}^{2n}$ meets $\mathcal{L}_X \omega_0 = \omega_0$, and so all starshaped domains in $\mathbb{R}^{2n}$ are of contact type. If $U \leq_2 V$, then the characteristic foliations of ∂U and ∂V are isomorphic, and ∂U is of contact type if and only if ∂V is of contact type. Theorem 1.1 in [23] and its proof show that there exist convex $U, V \in \mathcal{O}(n)$ with smooth boundaries such that U and V are symplectomorphic and C^∞-close to the ball $B^{2n}(\pi)$, but the characteristic foliation of ∂U contains an isolated closed characteristic while the one of ∂V does not. And Corollary A in [15] and its proof imply that given any $U \in \mathcal{O}(n)$, $n \geq 2$, with smooth boundary ∂U of contact type, there exists a symplectomorphic and C^0-close set $V \in \mathcal{O}(n)$ whose boundary is not of contact type. We in particular see that even for U being a ball, $U \leq_3 V$ does not imply $U \leq_2 V$.

2.2 Rigidity for ellipsoids

Proposition 2.1.1 shows that in order to detect some rigidity via the relations $\leq_l$ we must pass to a small subcategory of sets: Let $\mathcal{E}(n)$ be the collection of symplectic ellipsoids defined in Section 1.2.2,

$$\mathcal{E}(n) = \{E(a) = E(a_1, \ldots, a_n)\}, \quad a = (a_1, \ldots, a_n),$$

and write $\preccurlyeq_l$ for the restrictions of the relations $\leq_l$ to $\mathcal{E}(n)$. Notice again that

$$\preccurlyeq_1 \implies \preccurlyeq_2 \implies \preccurlyeq_3 . \tag{2.2.1}$$

The equivalence (2.2.5) below and the theorems in Section 1.3.1 combined with the Extension after Restriction Principle from Section 1.2.2 show that the relations $\preccurlyeq_1$ and $\preccurlyeq_2$ are different. The relations $\preccurlyeq_2$ and $\preccurlyeq_3$ are very similar: Since ellipsoids are starshaped, the Extension after Restriction Principle implies

$$E(a) \preccurlyeq_3 E(a') \implies E(\delta a) \preccurlyeq_2 E(a') \quad \text{for all } \delta \in \,]0, 1[. \tag{2.2.2}$$

It is, however, not known whether $\preccurlyeq_2$ and $\preccurlyeq_3$ are the same: While Theorem 2.2.4 proves this under an additional condition, the folding construction of Section 3.2 suggests that $\preccurlyeq_2$ and $\preccurlyeq_3$ are different in general. But let us first prove a general and common rigidity property of these relations:

Proposition 2.2.1. *The relations* $\preccurlyeq_l$ *are partial orderings on* $\mathcal{E}(n)$.

Proof. The relations are clearly reflexive and transitive, so we are left with identitivity, i.e.,

$$\left(E(a) \preccurlyeq_l E(a') \text{ and } E(a') \preccurlyeq_l E(a) \right) \implies E(a) = E(a').$$

Of course, the identitivity of $\preccurlyeq_3$ implies the one of $\preccurlyeq_2$ which, in turn, implies the one of $\preccurlyeq_1$. To prove the identitivity of $\preccurlyeq_3$ we use Ekeland–Hofer capacities introduced in [21].

Definition 2.2.2. An *extrinsic symplectic capacity on* $(\mathbb{R}^{2n}, \omega_0)$ is a map c associating with each subset S of $\mathbb{R}^{2n}$ a number $c(S) \in [0, \infty]$ in such a way that the following axioms are satisfied.

A1. Monotonicity: $c(S) \leq c(T)$ if there exists $\varphi \in \mathcal{D}(n)$ such that $\varphi(S) \subset T$.

A2. Conformality: $c(\lambda S) = \lambda^2 c(S)$ for all $\lambda \in \mathbb{R} \setminus \{0\}$.

A3. Nontriviality: $0 < c(B^{2n}(\pi))$ and $c(Z^{2n}(\pi)) < \infty$.

The Ekeland–Hofer capacities form a countable family $\{c_j\}$, $j \geq 1$, of extrinsic symplectic capacities on $\mathbb{R}^{2n}$. For a symplectic ellipsoid $E = E(a_1, \ldots, a_n)$ these invariants are given by the identity of sets

$$\{c_1(E) \leq c_2(E) \leq \ldots\} = \{ka_i \mid k = 1, 2, \ldots; \ i = 1, \ldots, n\}, \qquad (2.2.3)$$

see [21, Proposition 4]. Observe that for any $l = 1, 2, 3$ and $\lambda > 0$

$$E(a) \preccurlyeq_l E(a') \implies E(\lambda a) \preccurlyeq_l E(\lambda a'). \qquad (2.2.4)$$

This is seen by conjugating the given map φ with the dilatation by λ^{-1}. Recalling (2.2.2) we conclude that for any $\delta_1, \delta_2 \in]0, 1[$ the postulated relations

$$E(a) \preccurlyeq_3 E(a') \preccurlyeq_3 E(a)$$

imply

$$E(\delta_2\delta_1 a) \preccurlyeq_2 E(\delta_1 a') \preccurlyeq_2 E(a).$$

Now the monotonicity property (A1) of the capacities and the set of relations in (2.2.3) immediately imply that $a = a'$. This completes the proof of Proposition 2.2.1. $\square$

Remark 2.2.3. In the above proof we derived the identitivity of $\preccurlyeq_1$ and $\preccurlyeq_2$ from the one of $\preccurlyeq_3$. We find it instructive to give direct proofs.

It is well known from linear symplectic algebra [39, p. 40] that

$$E(a) \preccurlyeq_1 E(a') \iff a_i \le a'_i \quad \text{for all } i, \tag{2.2.5}$$

in particular $\preccurlyeq_1$ is identitive.

In order to give an elementary proof of the identitivity of $\preccurlyeq_2$ we look at the characteristic foliation on the boundary and at the actions $A(\gamma)$ of closed characteristics. To compute the characteristic foliation of $\partial E(a)$ recall that $E(a) = H^{-1}(1)$, where $H(z) = \sum_{i=1}^{n} \frac{\pi |z_i|^2}{a_i}$. Using definition (1.1.3) we find

$$X_H(z) = -2\pi J\left(\tfrac{1}{a_1}z_1, \ldots, \tfrac{1}{a_n}z_n\right)$$

where $J = \sqrt{-1} \in \mathbb{C} = \mathbb{R}^2(x, y)$ is the standard complex structure. The characteristic on $\partial E(a)$ through $z = z(0)$ can therefore by parametrized as $z(t) = (z_1(t), \ldots, z_n(t))$, where

$$z_i(t) = e^{-2\pi J t/a_i} z_i(0), \quad i = 1, \ldots, n.$$

If the n numbers $a_1, \ldots, a_n$ are linearly independent over $\mathbb{Z}$, then the only periodic orbits are $(0, \ldots, 0, z_i(t), 0, \ldots, 0)$ with

$$z_i(t) = e^{-2\pi J t/a_i} z_i(0) \quad \text{and} \quad \pi |z_i(0)|^2 = a_i.$$

In general, the traces of the closed characteristics on $\partial E(a)$ form the disjoint union

$$\partial E(a^1) \cup \cdots \cup \partial E(a^d)$$

where $a^1 \cup \cdots \cup a^d$ is the partition of $a = (a_1, \ldots, a_n)$ into maximal linearly dependent subsets. Recall that the action of a closed characteristic γ is defined as $A(\gamma) = \left| \int_{\gamma} \lambda \right|$, where $d\lambda = \omega_0$. Denoting by $a(\gamma)$ the smallest subset of a such that $\gamma \subset \partial E(a(\gamma))$, and choosing $\lambda = \sum_{i=1}^{n} x_i \, dy_i$, we readily compute that $A(\gamma)$ is the least common multiple of the elements in $a(\gamma)$,

$$A(\gamma) = \operatorname{lcm}(a(\gamma)). \tag{2.2.6}$$

Assume now that $E(a) \preccurlyeq_2 E(b)$ and $E(b) \preccurlyeq_2 E(a)$. Then there exists a symplectomorphism φ of $\mathbb{R}^{2n}$ such that $\varphi(E(a)) = E(b)$, and we see as in the proof of Proposition 2.1.1 that $\varphi(\partial E(a)) = \partial E(b)$. It follows easily from the definition (2.1.1) of $\mathcal{L}_{E(a)}$ and $\mathcal{L}_{E(b)}$ that φ maps the characteristic foliation on $\partial E(a)$ to the one of $\partial E(b)$. Moreover, the actions of closed characteristics are preserved. Indeed, if γ is a closed characteristic on $\partial E(a)$, we choose a smooth closed disc $D \subset \mathbb{R}^{2n}$ with boundary γ and find

$$\int_{\varphi(\gamma)} \lambda = \int_{\varphi(D)} \omega_0 = \int_D \varphi^* \omega_0 = \int_D \omega_0 = \int_{\gamma} \lambda,$$

so that $A\left(\varphi(\gamma)\right) = A(\gamma)$. Denoting the *simple action spectrum* of $E(a)$ by

$$\sigma\left(E(a)\right) = \{A\left(\gamma\right) \mid \gamma \text{ is a closed characteristic on } \partial E(a)\}$$

we in particular have $\sigma\left(E(a)\right) = \sigma\left(E(b)\right)$.

If $a_1, \ldots, a_n$ are linearly independent over $\mathbb{Z}$, then $\partial E(a)$ carries only the n closed characteristics $\partial E(a_i)$ with action a_i, so that

$$\{a_1, \ldots, a_n\} = \sigma\left(E(a)\right) = \sigma\left(E(b)\right) = \{b_1, \ldots, b_n\},$$

proving $E(a) = E(b)$. The proof of the general case is not much harder: Of course,

$$a_1 = \min\{\sigma\left(E(a)\right)\} = \min\{\sigma\left(E(b)\right)\} = b_1.$$

Arguing by induction we assume that $a_i = b_i$ for $i = 1, \ldots, k - 1$. Suppose that $a_k < b_k$. We then consider the subsets $\Gamma\left(a \mid a_k\right) \subset \partial E(a)$ and $\Gamma\left(b \mid a_k\right) \subset \partial E(b)$ formed by those closed characteristics whose action divides a_k. By the above discussion, $\varphi\left(\Gamma\left(a \mid a_k\right)\right) = \Gamma\left(b \mid a_k\right)$, so that $\Gamma\left(a \mid a_k\right)$ and $\Gamma\left(b \mid a_k\right)$ must be homeomorphic. On the other hand, let $\{a_{i_1}, \ldots, a_{i_l}\}$ be the set of those a_i in $\{a_1, \ldots, a_{k-1}\}$ which divide a_k. We then read off from (2.2.6) that $\Gamma\left(b \mid a_k\right) = \partial E(a_{i_1}, \ldots, a_{i_l})$. Similarly, $\Gamma\left(a \mid a_k\right) = \partial E(a_{i_1}, \ldots, a_{i_l}, a_k, \ldots, a_{k+m_k-1})$ where m_k is the multiplicity of a_k in a. In particular, $\dim \Gamma\left(b \mid a_k\right) < \dim \Gamma\left(a \mid a_k\right)$. This contradiction shows that $a_k \geq b_k$. Interchanging a and b we also find $a_k \leq b_k$, so that $a_k = b_k$. This completes the induction, and the identitivity of $\preccurlyeq_2$ is proved in an elementary way. $\diamond$

Recall that $\preccurlyeq_2$ does not imply $\preccurlyeq_1$ in general. However, a suitable pinching condition guarantees that "linear" and "non linear" coincide:

Theorem 2.2.4. *Let $\kappa \in \left]\frac{b}{2}, b\right[$. Then the following statements are equivalent:*

(i) $B^{2n}(\kappa) \preccurlyeq_1 E(a) \preccurlyeq_1 E(a') \preccurlyeq_1 B^{2n}(b)$,

(ii) $B^{2n}(\kappa) \preccurlyeq_2 E(a) \preccurlyeq_2 E(a') \preccurlyeq_2 B^{2n}(b)$,

(iii) $B^{2n}(\kappa) \preccurlyeq_3 E(a) \preccurlyeq_3 E(a') \preccurlyeq_3 B^{2n}(b)$.

We should mention that for $n = 2$, Theorem 2.2.4 was proved in [26]. Their proof uses a deep result of McDuff, [55], stating that the space of symplectic embeddings of a closed ball into a larger ball is connected, and then uses the isotopy invariance of symplectic homology. However, Ekeland–Hofer capacities provide an easy proof as we shall see. The crucial observation is that capacities have – in contrast to symplectic homology – the monotonicity property.

Proof of Theorem 2.2.4. In view of (2.2.1) it is enough to show the implication (iii) $\Rightarrow$ (i). We start with showing the implication (ii) $\Rightarrow$ (i). By assumption,

$$B^{2n}(\kappa) \preccurlyeq_2 E(a) \preccurlyeq_2 B^{2n}(b).$$

Hence, by the monotonicity of the first Ekeland–Hofer capacity c_1 we obtain

$$\kappa \le a_1 \le b, \tag{2.2.7}$$

and by the monotonicity of c_n

$$\kappa \le c_n(E(a)) \le b. \tag{2.2.8}$$

The estimates (2.2.7) and $\kappa > b/2$ imply $2a_1 > b$, whence the only Ekeland–Hofer capacities of $E(a)$ possibly smaller than b are $a_1, \ldots, a_n$. It follows therefore from (2.2.8) that $a_n = c_n(E(a))$, whence $c_i(E(a)) = a_i$ for $i = 1, \ldots, n$. Similarly we find $c_i(E(a')) = a_i'$ for $i = 1, \ldots, n$, and from $E(a) \preccurlyeq_2 E(a')$ we conclude $a_i \le a_i'$.

(iii) $\Rightarrow$ (i) now follows by a similar reasoning as in the proof of the identitivity of $\preccurlyeq_3$. Indeed, starting from

$$B^{2n}(\kappa) \preccurlyeq_3 E(a) \preccurlyeq_3 E(a') \preccurlyeq_3 B^{2n}(b),$$

the implication (2.2.2) shows that for any $\delta_1, \delta_2, \delta_3 \in \,]0, 1[$

$$B^{2n}(\delta_3\delta_2\delta_1\kappa) \preccurlyeq_2 E(\delta_2\delta_1 a) \preccurlyeq_2 E(\delta_1 a') \preccurlyeq_2 B^{2n}(b).$$

Choosing $\delta_1, \delta_2, \delta_3$ so large that $\delta_3\delta_2\delta_1\kappa > b/2$ we can apply the already proved implication to see

$$B^{2n}(\delta_3\delta_2\delta_1\kappa) \preccurlyeq_1 E(\delta_2\delta_1 a) \preccurlyeq_1 E(\delta_1 a) \preccurlyeq_1 B^{2n}(b),$$

and since $\delta_1, \delta_2, \delta_3$ can be chosen arbitrarily close to 1, the statement (i) follows in view of (2.2.5). This completes the proof of Theorem 2.2.4. $\square$

In Section 1.2.2 we gave a direct proof of Theorem 1. Here, we show how Theorem 1 follows from Theorem 2.2.4. In the notation of this section, Theorem 1 reads

Theorem 2.2.5. *Assume that* $E(a_1, \ldots, a_n) \preccurlyeq_3 B^{2n}(A)$ *for some* $A < a_n$. *Then* $a_n > 2a_1$.

Proof. Arguing by contradiction we assume $E(a_1, \ldots, a_n) \preccurlyeq_3 B^{2n}(A)$ for some $A < a_n$ and $a_n \le 2a_1$. A volume comparison shows $a_1 < A$. Hence, $a_1 \in \,]\frac{A}{2}, A[$. Therefore, $B^{2n}(a_1) \preccurlyeq_3 E(a_1, \ldots, a_n) \preccurlyeq_3 B^{2n}(A)$, Theorem 2.2.4 and the equivalence (2.2.5) imply that $a_n \le A$. This contradiction shows $a_n > 2a_1$, as claimed. $\square$

2.3 Rigidity for polydiscs ?

The rigidity results for symplectic embeddings of ellipsoids into ellipsoids found in the previous section were proved with the help of Ekeland–Hofer capacities. Recall that $P(a_1, \ldots, a_n)$ denotes the open symplectic polydisc. We may again assume $a_1 \leq a_2 \leq \cdots \leq a_n$. The Ekeland–Hofer capacities of a polydisc are given by

$$c_j(P(a_1, \ldots, a_n)) = ja_1, \quad j = 1, 2, \ldots, \tag{2.3.1}$$

[21, Proposition 5], and so they only see the smallest area a_1. Many of the polydisc analogues of the rigidity results for ellipsoids are therefore either wrong or much harder to prove. It is for instance not true anymore that $P(a_1, \ldots, a_n)$ embeds into $P(A_1, \ldots, A_n)$ by a linear symplectomorphism if and only if $a_i \leq A_i$ for all i, as the following example shows.

Lemma 2.3.1. *Assume $r > 1 + \sqrt{2}$. Then there exists $A < \pi r^2$ such that the polydisc $P^{2n}(\pi, \ldots, \pi, \pi r^2)$ embeds into the cube $C^{2n}(A) = P^{2n}(A, \ldots, A)$ by a linear symplectomorphism.*

Proof. It is enough to prove the lemma for $n = 2$. Consider the linear symplectomorphism given by

$$(z_1, z_2) \mapsto (z_1', z_2') = \frac{1}{\sqrt{2}}(z_1 + z_2, z_1 - z_2).$$

For $(z_1, z_2) \in P(\pi, \pi r^2)$ and $i = 1, 2$ we have

$$|z_i'|^2 \leq \frac{1}{2}\left(|z_1|^2 + |z_2|^2 + 2|z_1||z_2|\right) < \frac{1}{2} + \frac{r^2}{2} + r. \tag{2.3.2}$$

The right hand side of (2.3.2) is strictly smaller than r^2 provided that $r > 1 + \sqrt{2}$. $\square$

Moreover, it is not known whether the full analogue of Proposition 2.2.1 for polydiscs instead of ellipsoids holds true. Let $\mathcal{P}(n)$ be the collection of polydiscs

$$\mathcal{P}(n) = \{P(a_1, \ldots, a_n)\}$$

and write $\preceq_l$ for the restrictions of the relations $\leq_l$ to $\mathcal{P}(n), l = 1, 2, 3$. Again $\preceq_2$ and $\preceq_3$ are very similar, and again all the relations $\preceq_l$ are clearly reflexive and transitive. Furthermore, the smooth part of the boundary $\partial P(a_1, \ldots, a_n)$ is foliated by closed characteristics with actions $a_1, \ldots, a_n$, so that the identitivity of $\preceq_2$ and hence the one of $\preceq_1$ follows at once. The identitivity of $\preceq_2$ also follows from a result proved in [26] by using symplectic homology: *Symplectomorphic polydiscs are equal*. For $n = 2$, the identitivity of $\preceq_3$ follows from the monotonicity of any symplectic capacity, which show that the smaller discs are equal, and from the equality of the volumes, which then shows that also the larger discs are equal. For $n \geq 3$, however, we do not know whether the relation $\preceq_3$ is identitive. In particular, we have no answer to the following question.

Question 2.3.2. *Assume that there exist symplectic embeddings*

$$P(a_1, a_2, a_3) \hookrightarrow P(a_1, a_2', a_3') \quad and \quad P(a_1, a_2', a_3') \hookrightarrow P(a_1, a_2, a_3).$$

Is it then true that $a_2 = a_2'$ and $a_3 = a_3'$?

We also do not know whether the polydisc-analogue of Theorem 1 or of Theorem 2.2.4 holds true. The symplectic embedding results proved in the subsequent chapters will suggest, however, that the polydisc-analogue of Theorem 1 holds true, see Conjecture 7.2.4.

Chapter 3

Proof of Theorem 2

In this chapter we prove Theorem 2 by symplectic folding. After reducing Theorem 2 to a symplectic embedding problem in dimension 4, we construct essentially explicit symplectomorphisms between 2-dimensional simply connected domains. This construction is important for the symplectic folding construction, which is described in detail in Section 3.2. While symplectic folding will be the main tool until Chapter 8, the construction of explicit 2-dimensional symplectomorphisms will be basic also for the symplectic packing constructions given in Chapter 9. This chapter almost coincides with the paper [73].

3.1   Reformulation of Theorem 2

Recall from the introduction that the ellipsoid $E(a_1, \ldots, a_n)$ is defined by

$$E(a_1, \ldots, a_n) = \left\{ (z_1, \ldots, z_n) \in \mathbb{C}^n \mid \sum_{i=1}^n \frac{\pi |z_i|^2}{a_i} < 1 \right\}. \tag{3.1.1}$$

Theorem 2 in Section 1.3.1 clearly can be reformulated as follows.

Theorem 3.1.1. *Assume* $a > 2\pi$. *Then* $E^{2n}(\pi, \ldots, \pi, a)$ *symplectically embeds into* $B^{2n} \left(\frac{a}{2} + \pi + \epsilon \right)$ *for every* $\epsilon > 0$.

The symplectic folding construction of Lalonde and McDuff considers a 4-ellipsoid as a fibration of discs of varying size over a disc and applies the flexibility of volume preserving maps to both the base and the fibres. It is therefore purely four dimensional in nature. We will refine the method in such a way that it allows us to prove Theorem 3.1.1 for every $n \geq 2$.

We shall conclude Theorem 3.1.1 from the following proposition in dimension 4.

Proposition 3.1.2. *Assume* $a > 2\pi$. *Given* $\epsilon > 0$ *there exists a symplectic embedding*

$$\Phi \colon E(a, \pi) \hookrightarrow B^4 \left(\frac{a}{2} + \pi + \epsilon \right)$$

satisfying

$$\pi |\Phi(z_1, z_2)|^2 < \frac{a}{2} + \epsilon + \frac{\pi^2 |z_1|^2}{a} + \pi |z_2|^2 \quad \text{for all } (z_1, z_2) \in E(a, \pi).$$

We recall that $|\cdot|$ denotes the Euclidean norm. Postponing the proof, we first show that Proposition 3.1.2 implies Theorem 3.1.1.

Corollary 3.1.3. *Assume that Φ is as in Proposition* 3.1.2. *Then the composition of the permutation $E^{2n}(\pi, \ldots, \pi, a) \rightarrow E^{2n}(a, \pi, \ldots, \pi)$ with the restriction of $\Phi \times \mathrm{id}_{2n-4}$ to $E^{2n}(a, \pi, \ldots, \pi)$ embeds $E^{2n}(\pi, \ldots, \pi, a)$ into $B^{2n}\left(\frac{a}{2} + \pi + \epsilon\right)$.*

Proof. Let $z = (z_1, \ldots, z_n) \in E^{2n}(a, \pi, \ldots, \pi)$. By Proposition 3.1.2 and the definition (3.1.1) of the ellipsoid,

$$\pi \, |\Phi \times \mathrm{id}_{2n-4}(z)|^2 = \pi \left(|\Phi(z_1, z_2)|^2 + \sum_{i=3}^{n} |z_i|^2 \right)$$

$$< \frac{a}{2} + \epsilon + \frac{\pi^2 |z_1|^2}{a} + \pi \sum_{i=2}^{n} |z_i|^2$$

$$= \frac{a}{2} + \epsilon + \pi \left(\frac{\pi |z_1|^2}{a} + \sum_{i=2}^{n} \frac{\pi |z_i|^2}{\pi} \right)$$

$$< \frac{a}{2} + \epsilon + \pi,$$

as claimed.   □

It remains to prove Proposition 3.1.2. In order to do so, we start with some preparations.

The flexibility of 2-dimensional area preserving maps is crucial for the construction of the map Φ. We now make sure that we can describe such a map by prescribing it on an exhausting and nested family of embedded loops. Recall that $D(a)$ denotes the open disc of area a centred at the origin, and that $|U|$ denotes the area of a domain $U \subset \mathbb{R}^2$.

Definition 3.1.4. A family $\mathcal{L}$ of loops in a simply connected domain $U \subset \mathbb{R}^2$ is called *admissible* if there is a diffeomorphism $\beta \colon D(|U|) \setminus \{0\} \rightarrow U \setminus \{p\}$ for some point $p \in U$ such that

  (i) concentric circles are mapped to elements of $\mathcal{L}$,

  (ii) in a neighbourhood of the origin β is a translation.

Lemma 3.1.5. *Let U and V be bounded and simply connected domains in $\mathbb{R}^2$ of equal area and let $\mathcal{L}_U$ and $\mathcal{L}_V$ be admissible families of loops in U and V, respectively. Then there is a symplectomorphism between U and V mapping loops to loops.*

Remark 3.1.6. The regularity condition (ii) imposed on the families taken into consideration can be weakened. Some condition, however, is necessary. Indeed, if $\mathcal{L}_U$ is a family of concentric circles and $\mathcal{L}_V$ is a family of rectangles with smooth corners and width larger than a positive constant, then no bijection from U to V mapping loops to loops is continuous at the origin. $\diamond$

Proof of Lemma 3.1.5. Denote the concentric circle of radius r by $C(r)$. We may assume that $\mathcal{L}_U = \{C(r)\}$, $0 < r < R$. Let β be the diffeomorphism parametrizing $(V \setminus \{p\}, \mathcal{L}_V)$. After reparametrizing the r-variable by a diffeomorphism of $]0, R[$ which is the identity near 0 we may assume that β maps the loop $C(r)$ of radius r to the loop $L(r)$ in $\mathcal{L}_V$ which encloses the domain $V(r)$ of area πr^2. We denote the Jacobian of β at $re^{i\varphi}$ by $\beta'(re^{i\varphi})$. Since β is a translation near the origin and U is connected, $\det \beta'(re^{i\varphi}) > 0$. By our choice of β,

$$\pi r^2 = |V(r)| = \int_{D(\pi r^2)} \det \beta' = \int_0^r \rho \, d\rho \int_0^{2\pi} \det \beta'(\rho e^{i\varphi}) \, d\varphi.$$

Differentiating in r we obtain

$$2\pi = \int_0^{2\pi} \det \beta'(re^{i\varphi}) \, d\varphi. \tag{3.1.2}$$

Define the smooth function $h\colon]0, R[\times \mathbb{R} \to \mathbb{R}$ as the unique solution of the initial value problem

$$\left.\begin{aligned} \tfrac{d}{dt}h(r, t) &= 1/\det \beta'(re^{ih(r,t)}), & t &\in \mathbb{R} \\ h(r, t) &= 0, & t &= 0 \end{aligned}\right\} \tag{3.1.3}$$

depending on the parameter r. We claim that

$$h(r, t + 2\pi) = h(r, t) + 2\pi. \tag{3.1.4}$$

It then follows, since the function h is strictly increasing in the variable t, that for every r fixed the map $h(r, \cdot)\colon \mathbb{R} \to \mathbb{R}$ induces a diffeomorphism of the circle $\mathbb{R}/2\pi\mathbb{Z}$. In order to prove the claim (3.1.4) we denote by $t_0(r) > 0$ the unique solution of $h(r, t_0(r)) = 2\pi$. Substituting $\varphi = h(r, t)$ into formula (3.1.2) we obtain, using $\det \beta'(re^{ih(r,t)}) \cdot \tfrac{d}{dt}h(r, t) = 1$, that

$$2\pi = \int_0^{t_0(r)} dt = t_0(r).$$

Hence $h(r, 2\pi) = 2\pi$. Therefore, the two functions in t, $h(r, t+2\pi) - 2\pi$ and $h(r, t)$, solve the same initial value problem (3.1.3), and so the claim (3.1.4) follows. The desired diffeomorphism is now defined by

$$\alpha\colon U \setminus \{0\} \to V \setminus \{p\}, \quad re^{i\varphi} \mapsto \beta(re^{ih(r,\varphi)}).$$

It is area preserving. Indeed, representing α as the composition

$$re^{i\varphi} \mapsto (r, \varphi) \mapsto (r, h(r, \varphi)) \mapsto re^{ih(r,\varphi)} \mapsto \beta(re^{ih(r,\varphi)})$$

we obtain for the determinant of the Jacobian

$$\frac{1}{r} \cdot \frac{\partial h}{\partial \varphi}(r, \varphi) \cdot r \cdot \det \beta'(re^{ih(r,\varphi)}) = 1,$$

where we again have used (3.1.3). Finally, α is a translation in a punctured neighbour-hood of the origin and thus smoothly extends to the origin. This finishes the proof of Lemma 3.1.5.   □

Consider a bounded domain $U \subset \mathbb{C}$ and a continuous function $f : U \to \mathbb{R}_{>0}$. The set $\mathcal{F}(U, f)$ in $\mathbb{C}^2$ defined by

$$\mathcal{F}(U, f) = \left\{ (z_1, z_2) \in \mathbb{C}^2 \mid z_1 \in U, \, \pi |z_2|^2 < f(z_1) \right\}$$

is the trivial fibration over U having as fibre over z_1 the disc of capacity $f(z_1)$. Given two such fibrations $\mathcal{F}(U, f)$ and $\mathcal{F}(V, g)$, a symplectic embedding $\varphi : U \hookrightarrow V$ defines a symplectic embedding $\varphi \times \mathrm{id} : \mathcal{F}(U, f) \hookrightarrow \mathcal{F}(V, g)$ if and only if $f(z_1) \leq g(\varphi(z_1))$ for all $z_1 \in U$.

Examples 3.1.7. 1. The ellipsoid $E(a, b)$ can be represented as

$$E(a, b) = \mathcal{F}\left(D(a), f(z_1) = b\left(1 - \frac{\pi |z_1|^2}{a} \right) \right).$$

2. Define the open trapezoid $T(a, b)$ by $T(a, b) = \mathcal{F}(R(a), g)$, where

$$R(a) = \{ z_1 = (u, v) \mid 0 < u < a, \, 0 < v < 1 \}$$

is a rectangle and $g(z_1) = g(u) = b(1 - u/a)$. We set $T^4(a) = T(a, a)$. The example is inspired by [49, p. 54]. It will be very useful to think of $T(a, b)$ as depicted in Figure 3.1.   ◇

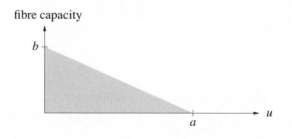

Figure 3.1. The trapezoid $T(a, b)$.

In order to reformulate Proposition 3.1.2 we shall prove the following lemma which later on allows us to work with more convenient "shapes".

Lemma 3.1.8. *Assume* $\epsilon > 0$. *Then*

(i) $E(a, b)$ *symplectically embeds into* $T(a + \epsilon, b + \epsilon)$,

(ii) $T^4(a)$ *symplectically embeds into* $B^4(a + \epsilon)$.

Proof. (i) Set $\epsilon' = a\epsilon^2/(ab + a\epsilon + b\epsilon)$. We are going to use Lemma 3.1.5 to construct an area preserving diffeomorphism $\alpha \colon D(a) \to R(a)$ such that for the first coordinate in the image $R(a)$,

$$u(\alpha(z_1)) \le \pi |z_1|^2 + \epsilon' \quad \text{for all } z_1 \in D(a), \tag{3.1.5}$$

see Figures 3.2 and 3.3.

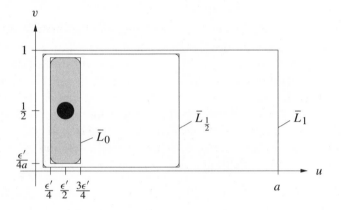

Figure 3.2. Constructing the embedding α.

In an "optimal world" we would choose the loops $\hat{L}_u$, $0 < u < a$, in the image $R(a)$ as the boundaries of the rectangles with corners $(0, 0)$, $(0, 1)$, $(u, 0)$, $(u, 1)$. If the family $\hat{\mathcal{L}} = \{\hat{L}_u\}$ induced a map $\hat{\alpha}$, we would then have $u(\hat{\alpha}(z_1)) \le \pi |z_1|^2$ for all $(z_1, z_2) \in R(a)$. The non admissible family $\hat{\mathcal{L}}$ can be perturbed to an admissible family $\mathcal{L}$ in such a way that the induced map α satisfies the estimate (3.1.5). Indeed, choose the translation disc appearing in the proof of Lemma 3.1.5 as the disc of radius $\epsilon'/8$ centred at $(u_0, v_0) = \left(\frac{\epsilon'}{2}, \frac{1}{2}\right)$. For $r < \epsilon'/8$ the loops $L(r)$ are therefore the circles centred at (u_0, v_0). In the following, all rectangles considered have edges parallel to the coordinate axes. We may thus describe a rectangle by specifying its lower left and upper right corner. Let $\bar{L}_0$ be the boundary of the rectangle with corners $\left(\frac{\epsilon'}{4}, \frac{\epsilon'}{4a}\right)$ and $\left(\frac{3\epsilon'}{4}, 1 - \frac{\epsilon'}{4a}\right)$, and let $\bar{L}_1$ be the boundary of $R(a)$. We define a family of loops $\bar{L}_s$ by linearly interpolating between $\bar{L}_0$ and $\bar{L}_1$, i.e., $\bar{L}_s$ is the boundary of the rectangle with corners

$$\left((1 - s)\frac{\epsilon'}{4}, (1 - s)\frac{\epsilon'}{4a}\right) \quad \text{and} \quad \left(u_s, 1 - \frac{\epsilon'}{4a} + \frac{\epsilon'}{4a}s\right), \quad s \in [0, 1],$$

where $u_s = \frac{3\epsilon'}{4} + s\left(a - \frac{3\epsilon'}{4}\right)$. Since $u_s < a$, the area enclosed by $\bar{L}_s$ is estimated from below by

$$\left(u_s - \frac{\epsilon'}{4}\right)\left(1 - 2\frac{\epsilon'}{4a}\right) > u_s - \frac{3\epsilon'}{4}. \tag{3.1.6}$$

Let $\{L_s\}$, $s \in [0, 1[$, be the smooth family of smooth loops obtained from $\{\bar{L}_s\}$ by smoothing the corners as indicated in Figure 3.2. By choosing the smooth corners of L_s more and more rectangular as $s \to 1$, we can arrange that the set $\bigsqcup_{0 < s < 1} L_s$ is the domain bounded by L_0 and $\bar{L}_1$. Moreover, by choosing all smooth corners rectangular enough, we can arrange that the area enclosed by L_s and $\bar{L}_s$ is less than $\epsilon'/4$. In view of (3.1.6), the area enclosed by L_s is then at least $u_s - \epsilon'$. Complete the families $\{L(r)\}$ and $\{L_s\}$ to an admissible family $\mathcal{L}$ of loops in $R(a)$ and let $\alpha \colon D(a) \to R(a)$ be the map defined by $\mathcal{L}$. Fix $(z_1, z_2) \in D(a)$. If $\alpha(z_1)$ lies on a loop in $\mathcal{L} \setminus \{L_s\}_{0 < s < 1}$, then $u(\alpha(z_1)) < \frac{3\epsilon'}{4} \leq \pi |z_1|^2 + \epsilon'$, and so the required estimate (3.1.5) is satisfied. If $\alpha(z_1) \in L_s$ for some $s \in]0, 1[$, then the area enclosed by L_s is $\pi |z_1|^2$, and so $\pi |z_1|^2 + \epsilon' > u_s \geq u(\alpha(z_1))$, whence (3.1.5) is again satisfied. This completes the construction of a symplectomorphism $\alpha \colon D(a) \to R(a)$ satisfying (3.1.5). In the sequel, we will illustrate a map like α by a picture like in Figure 3.3.

To continue the proof of (i) we shall show that $(\alpha(z_1), z_2) \in T(a + \epsilon, b + \epsilon)$ for every $(z_1, z_2) \in E(a, b)$, so that the symplectic map $\alpha \times \mathrm{id}$ embeds $E(a, b)$ into $T(a+\epsilon, b+\epsilon)$. Take $(z_1, z_2) \in E(a, b)$. Then, using the definition (3.1.1) of $E(a, b)$, the estimate (3.1.5) and the definition of ϵ' we find

$$\begin{aligned}
\pi |z_2|^2 &< b\left(1 - \frac{\pi |z_1|^2}{a}\right) \leq b\left(1 - \frac{u(\alpha(z_1))}{a} + \frac{\epsilon'}{a}\right) \\
&< b\left(1 - \frac{u(\alpha(z_1))}{a + \epsilon}\right) + b\frac{\epsilon'}{a} \\
&= b\left(1 - \frac{u(\alpha(z_1))}{a + \epsilon}\right) + \epsilon - \frac{\epsilon}{a + \epsilon}(a + \epsilon') \\
&\leq b\left(1 - \frac{u(\alpha(z_1))}{a + \epsilon}\right) + \epsilon - \frac{\epsilon}{a + \epsilon}u(\alpha(z_1)) \\
&= (b + \epsilon)\left(1 - \frac{u(\alpha(z_1))}{a + \epsilon}\right).
\end{aligned}$$

It follows that

$$(\alpha(z_1), z_2) \in T(a + \epsilon, b + \epsilon) = \mathcal{F}\left(R(a + \epsilon), (b + \epsilon)\left(1 - \frac{u}{a + \epsilon}\right)\right)$$

as claimed.

In order to prove (ii) we shall construct an area preserving diffeomorphism ω from a rectangular neighbourhood of $R(a)$ having smooth corners and area $a + \epsilon$ to $D(a + \epsilon)$ such that

$$\pi |\omega(z_1)|^2 \leq u + \epsilon \quad \text{for all } z_1 = (u, v) \in R(a). \tag{3.1.7}$$

Such a map ω can again be obtained with the help of Lemma 3.1.5. In an "optimal world" we would choose the loops $\hat{L}_u$ in the domain $R(a)$ as before. This time, we perturb this non admissible family to an admissible family $\mathcal{L}$ of loops as illustrated in Figure 3.3. If the smooth corners of all those loops in $\mathcal{L}$ which enclose an area greater than $\epsilon/2$ lie outside $R(a)$ and if the upper, left and lower edges of all these loops are close enough, then the induced map ω will satisfy (3.1.7).

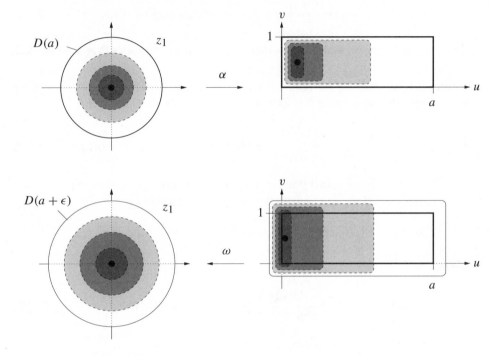

Figure 3.3. The first and the last base deformation.

Restricting ω to $R(a)$ we obtain a symplectic embedding $\omega \times \mathrm{id} \colon T^4(a) \hookrightarrow \mathbb{R}^4$. For $(z_1, z_2) \in T^4(a)$ we have $\pi |z_2|^2 < a(1 - u/a)$, where $z_1 = (u, v) \in R(a)$. In view of (3.1.7) we conclude that

$$
\pi \left(|\omega(z_1)|^2 + |z_2|^2 \right) < u + \epsilon + a \left(1 - \frac{u}{a} \right)
$$
$$
= u + \epsilon + a - u
$$
$$
= a + \epsilon,
$$

and so $(\omega \times \mathrm{id})(z_1, z_2) \in B^4(a + \epsilon)$ for all $(z_1, z_2) \in T^4(a)$. $\qquad \square$

Lemma 3.1.8 allows us to reformulate Proposition 3.1.2 as follows.

Proposition 3.1.9. *Assume $a > 2\pi$. Given $\epsilon > 0$, there exists a symplectic embedding*

$$\Psi \colon T(a, \pi) \hookrightarrow T^4 \left(\frac{a}{2} + \pi + \epsilon \right), \quad (z_1, z_2) \mapsto (z_1', z_2'),$$

$z_1 = (u, v)$ *and* $z_1' = (u', v')$, *satisfying*

$$u' + \pi |z_2'|^2 < \frac{a}{2} + \epsilon + \frac{\pi u}{a} + \pi |z_2|^2 \quad \text{for all } (u, v, z_2) \in T(a, \pi). \qquad (3.1.8)$$

Postponing the proof, we first show that Proposition 3.1.9 implies Proposition 3.1.2.

Corollary 3.1.10. *Assume the statement of Proposition 3.1.9 holds true. Then there exists a symplectic embedding* $\Phi \colon E(a, \pi) \hookrightarrow B^4 \left(\frac{a}{2} + \pi + \epsilon \right)$ *satisfying*

$$\pi |\Phi(z_1, z_2)|^2 < \frac{a}{2} + \epsilon + \frac{\pi^2 |z_1|^2}{a} + \pi |z_2|^2 \quad \text{for all } (z_1, z_2) \in E(a, \pi). \qquad (3.1.9)$$

Proof. Let $\epsilon' > 0$ be so small that $ca + \epsilon' > 2\pi$, where $c = 1 - \epsilon'/\pi$. As in the proof of Lemma 3.1.8 we can construct a symplectic embedding

$$\alpha \times \mathrm{id} \colon E(ca, c\pi) \hookrightarrow T(ca + \epsilon', c\pi + \epsilon') = T(ca + \epsilon', \pi)$$

satisfying the estimate

$$u(\alpha(z_1)) \le \pi |z_1|^2 + \frac{a(\epsilon')^2}{ca\pi + a\epsilon' + \pi \epsilon'} \quad \text{for all } z_1 \in D(ca) \qquad (3.1.10)$$

and another symplectic embedding

$$\omega \times \mathrm{id} \colon T^4 \left(\frac{ca}{2} + \pi + \epsilon' \right) \hookrightarrow B^4 \left(\frac{ca}{2} + \pi + 2\epsilon' \right)$$

satisfying

$$\pi |\omega(z_1)|^2 \le u + \epsilon' \quad \text{for all } z_1 = (u, v) \in R \left(\frac{ca}{2} + \pi + \epsilon' \right). \qquad (3.1.11)$$

Since $ca + \epsilon' > 2\pi$, Proposition 3.1.9 applied to $ca + \epsilon'$ replacing a and $\epsilon'/2$ replacing ϵ guarantees a symplectic embedding

$$\Psi \colon T(ca + \epsilon', \pi) \hookrightarrow T^4 \left(\frac{ca}{2} + \pi + \epsilon' \right),$$

$(z_1, z_2) \mapsto (\Psi_1(z_1, z_2), \Psi_2(z_1, z_2))$, satisfying

$$u\left(\Psi_1(\alpha(z_1), z_2)\right) + \pi |\Psi_2\left(\alpha(z_1), z_2\right)|^2 < \frac{ca}{2} + \epsilon' + \frac{\pi u(\alpha(z_1))}{ca + \epsilon'} + \pi |z_2|^2 \qquad (3.1.12)$$

for all $(u(\alpha(z_1)), v, z_2) \in T(ca + \epsilon', \pi)$. Set $\hat{\Phi} = (\omega \times \mathrm{id}) \circ \Psi \circ (\alpha \times \mathrm{id})$. Then $\hat{\Phi}$ symplectically embeds $E(ca, c\pi)$ into $B^4 \left(\frac{ca}{2} + \pi + 2\epsilon' \right)$. Moreover, if $(z_1, z_2) \in E(ca, c\pi)$, then

$$
\begin{aligned}
\pi \left| \hat{\Phi}(z_1, z_2) \right|^2 &= \pi \left| \omega \left(\Psi_1(\alpha(z_1), z_2) \right) \right|^2 + \pi \left| \Psi_2(\alpha(z_1), z_2) \right|^2 \\
&\overset{(3.1.11)}{\leq} u(\Psi_1(\alpha(z_1), z_2)) + \epsilon' + \pi \left| \Psi_2(\alpha(z_1), z_2) \right|^2 \\
&\overset{(3.1.12)}{<} \frac{ca}{2} + 2\epsilon' + \frac{\pi u(\alpha(z_1))}{ca + \epsilon'} + \pi |z_2|^2 \\
&\overset{(3.1.10)}{\leq} \frac{ca}{2} + 2\epsilon' + \frac{\pi^2 |z_1|^2}{ca + \epsilon'} + \frac{\pi}{ca + \epsilon'} \frac{a(\epsilon')^2}{ca\pi + a\epsilon' + \pi\epsilon'} + \pi |z_2|^2 \\
&< \frac{ca}{2} + 3\epsilon' + \frac{\pi^2 |z_1|^2}{ca} + \pi |z_2|^2
\end{aligned}
$$

where in the last step we again used $ca + \epsilon' > 2\pi$. Now choose $\epsilon' > 0$ so small that $\frac{\pi + 3\epsilon'}{c} < \pi + \epsilon$. We denote the dilatation by $\sqrt{c}$ in $\mathbb{R}^4$ also by $\sqrt{c}$, and define $\Phi \colon E(a, \pi) \to \mathbb{R}^4$ by $\Phi = \left(\sqrt{c} \right)^{-1} \circ \hat{\Phi} \circ \sqrt{c}$. Then Φ symplectically embeds $E(a, \pi)$ into $B^4 \left(\frac{a}{2} + \frac{\pi + 2\epsilon'}{c} \right) \subset B^4 \left(\frac{a}{2} + \pi + \epsilon \right)$, and since $\pi |z_1|^2 < a$ for all $(z_1, z_2) \in E(a, \pi)$ and by the choice of ϵ',

$$
\begin{aligned}
\pi |\Phi(z_1, z_2)|^2 &= \frac{\pi}{c} \left| \hat{\Phi} \left(\sqrt{c}\, z_1, \sqrt{c}\, z_2 \right) \right|^2 \\
&< \frac{1}{c} \left(\frac{ca}{2} + 3\epsilon' + \frac{\pi^2 |z_1|^2}{a} + \pi c |z_2|^2 \right) \\
&= \frac{a}{2} + \frac{3\epsilon'}{c} + \frac{1}{c} \frac{\pi^2 |z_1|^2}{a} + \pi |z_2|^2 \\
&< \frac{a}{2} + \epsilon + \frac{\pi^2 |z_1|^2}{a} + \pi |z_2|^2
\end{aligned}
$$

for all $(z_1, z_2) \in E(a, \pi)$. This proves the required estimate (3.1.9), and so the proof of Corollary 3.1.10 is complete. $\qquad \square$

It remains to prove Proposition 3.1.9. This is done in the following two sections.

3.2 The folding construction

The idea in the construction of an embedding Ψ as in Proposition 3.1.9 is to separate the small fibres from the large ones and then to fold the two parts on top of each other. As in the previous section we denote the coordinates in the base and the fibre by $z_1 = (u, v)$ and $z_2 = (x, y)$, respectively.

Step 1. Following [49, Lemma 2.1] we first separate the "low" regions over $R(a)$ from the "high" ones. We may do this using Lemma 3.1.5. We prefer, however, to give an explicit construction.

Let $\delta > 0$ be small. Set $\mathcal{F} = \mathcal{F}(U, f)$, where U and f are described in Figure 3.4, and write

$$P_1 = U \cap \left\{ u \le \frac{a}{2} + \delta \right\},$$

$$P_2 = U \cap \left\{ u \ge \frac{a + \pi}{2} + 11\delta \right\},$$

$$L = U \setminus (P_1 \cup P_2).$$

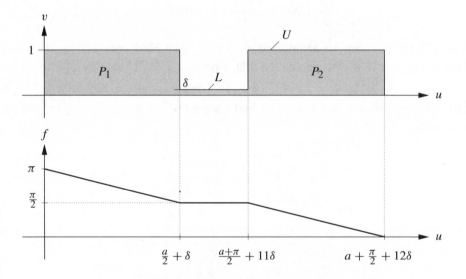

Figure 3.4. Separating the low fibres from the large fibres.

Hence, U is the disjoint union

$$U = P_1 \bigsqcup L \bigsqcup P_2.$$

Choose a smooth function $h \colon [0, a + \delta] \to \,]0, 1]$ as in Figure 3.5, i.e.

(i) $h(w) = 1$ for $w \in \left[0, \frac{a}{2}\right]$,

(ii) $h'(w) < 0$ for $w \in \,\left]\frac{a}{2}, \frac{a}{2} + \delta^2\right[$,

(iii) $h\left(\frac{a}{2} + \delta^2\right) = \delta$,

(iv) $h(w) = h(a - w)$ for all $w \in [0, a + \delta]$.

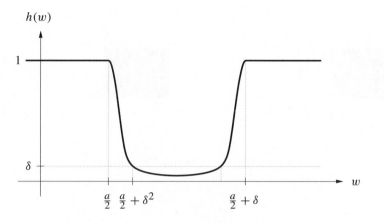

Figure 3.5. The function h.

By (ii), (iii) and (iv) we have that

$$\int_{\frac{a}{2}}^{\frac{a}{2}+\delta^2} \frac{1}{h(w)}\, dw < \delta \quad \text{and} \quad \int_{\frac{a}{2}+\delta-\delta^2}^{\frac{a}{2}+\delta} \frac{1}{h(w)}\, dw < \delta. \qquad (3.2.1)$$

We may thus further require that

(v) $h(w) < \delta$ for $w \in \left]\frac{a}{2}+\delta^2, \frac{a}{2}+\delta-\delta^2\right[$,

(vi) $\int_{\frac{a}{2}}^{\frac{a}{2}+\delta} \frac{1}{h(w)}\, dw = \frac{\pi}{2} + 12\delta$.

Consider the map

$$\beta\colon R(a) \to \mathbb{R}^2, \quad (u,v) \mapsto \left(\int_0^u \frac{1}{h(w)}\, dw,\, h(u)v\right).$$

Clearly, β is a symplectic embedding. We see from (i), (iv) and (vi) that

$$\beta\big|_{\{u<\frac{a}{2}\}} = \mathrm{id} \quad \text{and} \quad \beta\big|_{\{u>\frac{a}{2}+\delta\}} = \mathrm{id} + \left(\frac{\pi}{2}+11\delta, 0\right). \qquad (3.2.2)$$

These identities and the estimates (3.2.1) and (v) imply that β embeds $R(a)$ into U (cf. Figure 3.6, where the black region in $R(a)$ is mapped to the black region in U, and so on). Finally, by construction, $\beta \times \mathrm{id}$ symplectically embeds $T(a,\pi)$ into $\mathcal{F}$.

Step 2. We next map the fibres into a convenient shape. Using Lemma 3.1.5 in a similar way as it was used in the proof of Lemma 3.1.8 we find a symplectomorphism σ mapping $D(\pi)$ to the rectangle R_e and $D\left(\frac{\pi}{2}\right)$ to the rectangle with smooth corners R_i as specified in Figure 3.7. We require that for $z_2 \in D\left(\frac{\pi}{2}\right)$

$$\pi|z_2|^2 + 2\delta > y(\sigma(z_2)) - \left(-\frac{\pi}{2} - 2\delta\right),$$

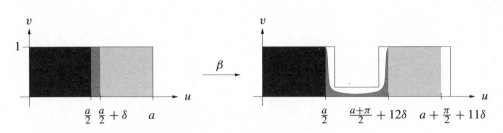

Figure 3.6. The embedding $\beta\colon R(a) \hookrightarrow U$.

i.e.

$$y(\sigma(z_2)) < \pi |z_2|^2 - \frac{\pi}{2} \quad \text{for } z_2 \in D\left(\frac{\pi}{2}\right). \tag{3.2.3}$$

Write for the resulting bundle $(\mathrm{id} \times \sigma)\mathcal{F}$ of rectangles with smooth corners

$$(\mathrm{id} \times \sigma)\mathcal{F} = \mathscr{S} = \mathscr{S}(P_1) \coprod \mathscr{S}(L) \coprod \mathscr{S}(P_2).$$

In order to fold $\mathscr{S}(P_2)$ over $\mathscr{S}(P_1)$ we first move $\mathscr{S}(P_2)$ along the y-axis and then turn it in the z_1-direction over $\mathscr{S}(P_1)$.

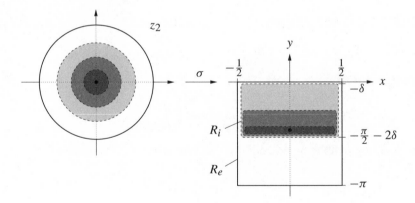

Figure 3.7. Preparing the fibres.

Step 3. In order to move $\mathscr{S}(P_2)$ along the y-axis we follow again [48, p. 355]. Let $c\colon \mathbb{R} \to [0,\, 1 - 2\delta]$ be a smooth cut off function as in Figure 3.8:

$$c(t) = \begin{cases} 0, & t \leq \frac{a}{2} + 2\delta \text{ and } t \geq \frac{a+\pi}{2} + 10\delta, \\ 1 - 2\delta, & \frac{a}{2} + 3\delta \leq t \leq \frac{a+\pi}{2} + 9\delta, \end{cases}$$

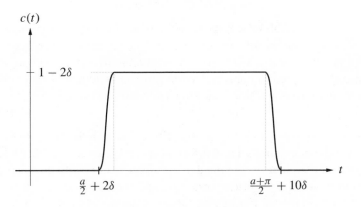

Figure 3.8. The cut off c.

Set $I(t) = \int_0^t c(s)\, ds$ and define the diffeomorphism $\varphi \colon \mathbb{R}^4 \to \mathbb{R}^4$ by

$$\varphi(u, x, v, y) = \left(u, x, v + c(u)\left(x + \frac{1}{2} \right), y + I(u) \right). \qquad (3.2.4)$$

We then find for the derivative

$$d\varphi(u, x, v, y) = \begin{bmatrix} \mathbb{I}_2 & 0 \\ A & \mathbb{I}_2 \end{bmatrix} \quad \text{with } A = \begin{bmatrix} * & c(u) \\ c(u) & 0 \end{bmatrix},$$

whence φ is a symplectomorphism in view of the criterion in [39, p. 5]. Moreover, defining the number I_∞ by $I_\infty = I\left(\frac{a+\pi}{2} + 10\delta \right)$, we find

$$\varphi\big|_{\{u \le \frac{a}{2} + 2\delta\}} = \mathrm{id} \quad \text{and} \quad \varphi\big|_{\{u \ge \frac{a+\pi}{2} + 10\delta\}} = \mathrm{id} + (0, 0, 0, I_\infty), \qquad (3.2.5)$$

and assuming that $\delta < \frac{1}{15}$ we compute with the help of Figure 3.8 that

$$\frac{\pi}{2} + 2\delta < I_\infty < \frac{\pi}{2} + 5\delta. \qquad (3.2.6)$$

The first inequality in (3.2.6) implies

$$\varphi(P_2 \times R_i) \cap (\mathbb{R}^2 \times R_e) = \emptyset. \qquad (3.2.7)$$

Remark 3.2.1. The lifting map φ is the crucial map of the folding construction. Indeed, φ is the only map in the construction which does not split as a product of 2-dimensional maps. It is a cut off in the u-direction of a translation separating R_i from R_e. As we have seen in Section 1.3.1, the lifting map φ is the time-1-map of the Hamiltonian function

$$H(u, v, x, y) = -I(u)\left(x + \tfrac{1}{2} \right).$$

While the set L has area about $\frac{\pi}{2}\delta$, the projection of $\varphi\left(\mathscr{S}(L)\right)$ to the (u, v)-plane almost fills the room between P_1 and P_2 and has area larger than $\frac{\pi}{2}$, see Figure 3.9. This must be so: As will become clear in the next step, a lifting map leading to a smaller projection could be used to construct symplectic embeddings of balls into cylinders which contradict Gromov's Nonsqueezing Theorem. $\diamond$

Step 4. We now turn $\varphi\left(\mathscr{S}(P_2)\right)$ over $\mathscr{S}(P_1)$ by folding in the base. From the definition (3.2.4) of the map φ and Figure 3.4 and Figure 3.7 we read off that the projection of $\varphi(\mathscr{S})$ to the (u, v)-plane is contained in the union $\mathcal{U}$ of U with the open set bounded by the graph of $u \mapsto \delta + c(u)$, the u-axis and the two lines $\{u = a/2 + \delta\}$ and $\{u = (a+\pi)/2 + 11\delta\}$, cf. Figure 3.9. Observe that $\delta + c(u) \leq 1 - \delta$. Define a local symplectic embedding γ of $\mathcal{U}$ into $\{(u, v) \mid 0 < u < (a+\pi)/2 + 11\delta, \, 0 < v < 1\}$

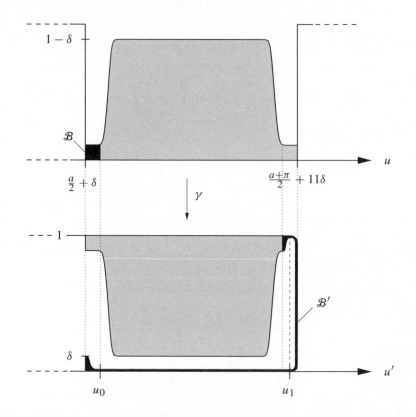

Figure 3.9. Folding in the base.

as follows: On $P_1 = \mathcal{U} \cap \{u \leq a/2 + \delta\}$ the map γ is the identity, and on $\mathcal{U} \cap \{u \geq a/2 + 2\delta\}$ it is the orientation preserving isometry which maps the right edge of $P_2 = \mathcal{U} \cap \{u \geq (a+\pi)/2 + 11\delta\}$ to the left edge of P_1. In particular, we

have for $z_1 = (u, v) \in P_2$,

$$u'(\gamma(z_1)) = a + \frac{\pi}{2} + 12\delta - u. \tag{3.2.8}$$

On the remaining black square $\mathcal{B} = \mathcal{U} \cap \{a/2 + \delta < u < a/2 + 2\delta\}$ the map γ looks as shown in Figure 3.9. We then have for $(u, v) \in \mathcal{U} \setminus (P_1 \cup P_2)$,

$$u'(\gamma(u, v)) - \left(\frac{a}{2} + \delta\right) < \frac{\pi}{2} + 10\delta - \left(u - \left(\frac{a}{2} + \delta\right)\right) + \delta,$$

i.e.

$$u'(\gamma(u, v)) < -u + \frac{\pi}{2} + a + 13\delta. \tag{3.2.9}$$

By (3.2.7) the map $\gamma \times \mathrm{id}$ is one-to-one on $\varphi(\mathcal{S})$.

The existence of an area and orientation preserving embedding as proposed in Figure 3.9 can be proved as follows: Set $u_0 = a/2 + 2\delta$ and $u_1 = (a + \pi)/2 + 21\delta/2$. Moreover, set $l = \pi/2 + 1 + 39\,\delta/4$ and choose $\lambda_3 > 0$ so small that $\lambda_3 l \leq \delta^2/3$. Similar to Figure 3.5 we choose a smooth function $h \colon \left[\frac{a}{2} + \delta, \frac{a}{2} + 2\delta\right] \to \,]0, 1]$ such that

(i) $h(u) = 1$ for u near $\frac{a}{2} + \delta$ and u near $\frac{a}{2} + 2\delta$,

(ii) $h(u) = \frac{\lambda_3}{\delta}$ for $u \in \left[\frac{a}{2} + \frac{3\delta}{2}, \frac{a}{2} + \frac{3\delta}{2} + \frac{\lambda_3 l}{\delta}\right]$,

(iii) $\int_{\frac{a}{2}+\delta}^{\frac{a}{2}+\frac{3\delta}{2}} \frac{1}{h(w)}\, dw = \delta$ and $\int_{\frac{a}{2}+\frac{3\delta}{2}+\frac{\lambda_3 l}{\delta}}^{\frac{a}{2}+2\delta} \frac{1}{h(w)}\, dw = \frac{\delta}{2}$.

The embedding $\gamma_\delta \colon \mathcal{B} \to \left[\frac{a}{2} + \delta, u_0 + l + \frac{\delta}{2}\right] \times [0, \delta]$ defined by

$$(u, v) \mapsto \left(\frac{a}{2} + \delta + \int_{\frac{a}{2}+\delta}^{u} \frac{1}{h(w)}\, dw, \, h(u)v\right)$$

and illustrated in Figure 3.10 is symplectic.

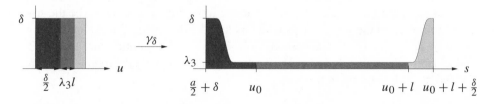

Figure 3.10. The map γ_δ.

We now map the image of γ_δ to a domain $\mathcal{B}'$ in the (u', v')-plane as painted in Figure 3.9: By the choice of l we may require that the part of the "outer" boundary of

$\mathcal{B}'$ between $(u_0, 0)$ and $(u_1, 1)$, which contains $(u_1, 0)$, is smooth, has length l, and is parametrized by $\zeta(s)$, where the parameter $s \in I := [u_0, u_0 + l]$ is arc length and

$$\zeta(s) = (s, 0) \qquad\qquad\qquad \text{on } [u_0, u_1],$$

$$\zeta(s) = (u_1 + u_0 + l - s, 1) \quad \text{on } \left[u_0 + l - \frac{\delta}{4}, u_0 + l \right]. \qquad (3.2.10)$$

Denote the inward pointing unit normal vector field along ζ by ν. We choose $\lambda_1 > 0$ so small that

$$\eta \colon I \times [0, \lambda_1] \to \mathbb{R}^2, \quad (s, t) \mapsto \zeta(s) + t\, \nu(s)$$

is an embedding. In order to make the map area preserving, we consider the initial value problem

$$\left. \begin{array}{l} \frac{\partial f}{\partial t}(s, t) = 1 / \det d\eta(s, f(s, t)) \\[2mm] f(s, 0) = 0 \end{array} \right\} \qquad (3.2.11)$$

in which $s \in I$ is a parameter. The existence and uniqueness theorem for ordinary differential equations with parameters yields a smooth solution f on $I \times [0, \lambda_2]$ for some $\lambda_2 > 0$. Then $f(s, t) < \lambda_1$ for all $(s, t) \in I \times [0, \lambda_2]$. This and the second equation in (3.2.11) imply that the composition

$$\gamma_\zeta \colon (s, t) \mapsto (s, f(s, t)) \overset{\eta}{\mapsto} (u', v')$$

is a diffeomorphism of $I \times [0, \lambda_2]$ onto half of a tubular neighbourhood of ζ. Moreover, by the first equation in (3.2.11),

$$\det \gamma_\zeta(s, t) = \frac{\partial f}{\partial t}(s, t) \det d\eta(s, f(s, t)) = 1,$$

i.e., γ_ζ is area preserving. In view of the identities (3.2.10) for ζ, the map γ_ζ is the identity in $\mathbb{R}^2$ for s near u_0 and $t \in [0, \lambda_2]$, and γ_ζ is an isometry for s near $u_0 + l$ and $t \in [0, \lambda_2]$.

We now choose the parameter $\lambda_3 > 0$ in the construction of γ_δ smaller than λ_2. Restrict γ_ζ to the gray region $I \times {]}0, \lambda_3[$ in the image of γ_δ, and let $\overline{\gamma}_\zeta$ be the smooth extension of γ_ζ to the image of γ_δ which is the identity on $\{u \le u_0\}$ and an isometry on $\{u \ge u_0 + l\}$. By (i), the composition $\overline{\gamma}_\zeta \circ \gamma_\delta$ is the identity near $u = a/2 + \delta$ and an isometry near $u = a/2 + 2\delta$. It thus smoothly fits with the map $\gamma|_{U \setminus \mathcal{B}}$ already defined at the beginning of this step.

Step 5. We finally adjust the fibres. In view of the constructions in Step 2 and Step 3, the projection of the image $\varphi(\mathcal{S})$ to the z_2-plane is contained in a tower shaped domain $\mathcal{T}$ (cf. Figure 3.11), and by the second inequality in (3.2.6) we have $\mathcal{T} \subset \{(x, y) \mid y < \frac{\pi}{2} + 4\delta\}$. Using once more our Lemma 3.1.5 we construct a symplectomorphism τ from a neighbourhood of $\mathcal{T}$ to a disc such that the preimages

of the concentric circles in the image are as in Figure 3.11. We require that for
$z_2 = (x, y)$,

$$\pi |\tau(z_2)|^2 < y + \frac{\pi}{2} + 3\delta \qquad\qquad \text{for } y \geq -\frac{\pi}{2} - 2\delta, \qquad (3.2.12)$$

$$\pi |\tau(z_2)|^2 < \pi |\sigma^{-1}(z_2)|^2 + \frac{\pi}{2} + 8\delta \quad \text{for } z_2 \in R_e, \qquad (3.2.13)$$

where $\sigma : D(\pi) \to R_e$ is the diffeomorphism constructed in Step 2.

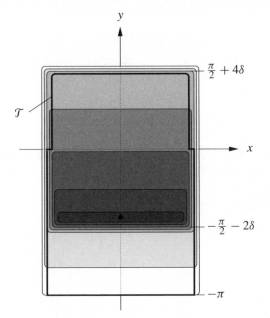

Figure 3.11. Mapping the tower to a disc.

Step 1 to Step 5 are the ingredients of our folding construction. The folding map
$\Psi : T(a, \pi) \hookrightarrow \mathbb{R}^4$ is defined as the composition of maps

$$\Psi = (\text{id} \times \tau) \circ (\gamma \times \text{id}) \circ \varphi \circ (\text{id} \times \sigma) \circ (\beta \times \text{id}) = (\gamma \times \tau) \circ \varphi \circ (\beta \times \sigma). \quad (3.2.14)$$

3.3   End of the proof

Recall that it remains to prove Proposition 3.1.9. So let $\epsilon > 0$ be as in Proposition 3.1.9
and set $\delta = \min\left\{\frac{1}{15}, \frac{\epsilon}{15}\right\}$. We define the desired map Ψ as in (3.2.14). It remains to
verify that Ψ meets the required estimate (3.1.8). So let $z = (z_1, z_2) = (u, v, x, y) \in$
$T(a, \pi)$ and write $\Psi(z) = (u', v', z_2')$. By the choice of δ it suffices to show that for
all $(u, v, z_2) \in T(a, \pi)$

$$u' - \frac{\pi u}{a} + \pi |z_2'|^2 - \pi |z_2|^2 < \frac{a}{2} + 15\delta. \qquad (3.3.1)$$

We distinguish three cases according to the locus of the image $\beta(z_1)$ in the set $U = P_1 \coprod L \coprod P_2$ (see Figure 3.4 and Figure 3.6). We denote the u-coordinate of $\beta(z_1) = \beta(u, v)$ by $u''(\beta(u, v))$.

Case 1. $\beta(z_1) \in P_1$. The first identity in (3.2.5) implies $\varphi|_{\mathcal{S}(P_1)} = \mathrm{id}$, and Step 4 implies $\gamma|_{\mathcal{S}(P_1)} = \mathrm{id}$. Therefore, $u' = u''(\beta(u, v))$. Moreover, $u''(\beta(u, v)) < u + \delta$. Indeed, the definition of the map β illustrated in Figure 3.6 shows that if $u''(\beta(u, v)) \leq \frac{a}{2}$, then $u''(\beta(u, v)) = u$, and if $u''(\beta(u, v)) \in]\frac{a}{2}, \frac{a}{2} + \delta]$, then $u > \frac{a}{2}$. Summarizing, we have

$$u' < u + \delta.$$

Using again $\varphi|_{\mathcal{S}(P_1)} = \mathrm{id}$ we find $\sigma(z_2) \in R_e$ and $z_2' = \tau(\sigma(z_2))$. Hence, the estimate (3.2.13) for the map τ yields

$$\pi|z_2'|^2 = \pi|\tau(\sigma(z_2))|^2 < \pi|\sigma^{-1}(\sigma(z_2))|^2 + \frac{\pi}{2} + 8\delta = \pi|z_2|^2 + \frac{\pi}{2} + 8\delta.$$

Finally, we have $u \leq \frac{a}{2} + \delta$. Indeed, if $u > \frac{a}{2} + \delta$, then the second identity in (3.2.2) implies $\beta(u, v) \in P_2$. Altogether we can estimate

$$u' - \frac{\pi u}{a} + \pi|z_2'|^2 - \pi|z_2|^2 < u\left(1 - \frac{\pi}{a}\right) + \delta + \frac{\pi}{2} + 8\delta$$

$$< \frac{a}{2}\left(1 - \frac{\pi}{a}\right) + \frac{\pi}{2} + 10\delta$$

$$= \frac{a}{2} + 10\delta.$$

Case 2. $\beta(z_1) \in P_2$. We have $\varphi|_{\mathcal{S}(P_2)} = \mathrm{id} + (0, 0, 0, I_\infty)$ by the second identity in (3.2.5), and so, in view of the identity (3.2.8), $u' = u'(\gamma(\beta(z_1))) = a + \frac{\pi}{2} + 12\delta - u''(\beta(u, v))$. Moreover, $u''(\beta(u, v)) > u + \frac{\pi}{2} + 10\delta$. Indeed, the definition of β shows that if $u''(\beta(u, v)) \geq \frac{a+\pi}{2} + 12\delta$, then $u''(\beta(u, v)) = u + \frac{\pi}{2} + 11\delta$, and if $u''(\beta(u, v)) \in [\frac{a+\pi}{2} + 11\delta, \frac{a+\pi}{2} + 12\delta[$, then $u < \frac{a}{2} + \delta$. Summarizing, we have

$$u' < a - u + 2\delta.$$

Step 2 shows $\sigma(z_2) \in R_i$, and so $y\big(\sigma(z_2) + (0, I_\infty)\big) \geq -\frac{\pi}{2} - 2\delta$. Hence, the estimates (3.2.12), (3.2.3) and (3.2.6) imply

$$\pi|z_2'|^2 = \pi\left|\tau\big(\sigma(z_2) + (0, I_\infty)\big)\right|^2$$

$$< y(\sigma(z_2)) + I_\infty + \frac{\pi}{2} + 3\delta$$

$$< \left(\pi|z_2|^2 - \frac{\pi}{2}\right) + \left(\frac{\pi}{2} + 5\delta\right) + \frac{\pi}{2} + 3\delta$$

$$= \pi|z_2|^2 + \frac{\pi}{2} + 8\delta.$$

Finally, we have $u \geq \frac{a}{2}$. Indeed, if $u < \frac{a}{2}$, then the first identity in (3.2.2) implies $\beta(u, v) \in P_1$. Altogether we can estimate

$$u' - \frac{\pi u}{a} + \pi |z_2'|^2 - \pi |z_2|^2 < a - u\left(1 + \frac{\pi}{a}\right) + 2\delta + \frac{\pi}{2} + 8\delta$$

$$\leq a - \frac{a}{2}\left(1 + \frac{\pi}{a}\right) + \frac{\pi}{2} + 10\delta$$

$$= \frac{a}{2} + 10\delta.$$

Case 3. $\beta(z_1) \in L$. Using the definition of φ, the estimate (3.2.9) implies

$$u' < -u''(\beta(u, v)) + \frac{\pi}{2} + a + 13\delta.$$

Since $\pi |z_2|^2 < \frac{\pi}{2}$, we have $\sigma(z_2) \in R_i$, cf. Figure 3.7. In particular, $y\big(\sigma(z_2) + (0, I(u''(\beta(u, v))))\big) \geq -\frac{\pi}{2} - 2\delta$. Hence, the estimates (3.2.12) and (3.2.3) and the estimate $I(t) < (1 - 2\delta)(t - (\frac{a}{2} + 2\delta))$ read off from Figure 3.8 yield

$$\pi |z_2'|^2 = \pi \left|\tau\big(x(\sigma(z_2)), y(\sigma(z_2)) + I(u''(\beta(u, v)))\big)\right|^2$$

$$< y(\sigma(z_2)) + I(u''(\beta(u, v))) + \frac{\pi}{2} + 3\delta$$

$$< \left(\pi |z_2|^2 - \frac{\pi}{2}\right) + (1 - 2\delta)\left(u''(\beta(u, v)) - \frac{a}{2} - 2\delta\right) + \frac{\pi}{2} + 3\delta$$

$$= \pi |z_2|^2 + u''(\beta(u, v)) - \frac{a}{2} - 2\delta - 2\delta u''(\beta(u, v)) + \delta a + 4\delta^2 + 3\delta.$$

Finally, we have $u''(\beta(u, v)) > \frac{a}{2} + \delta$ by the definition of L, and $u \geq \frac{a}{2}$ by the first identity in (3.2.2). Altogether we can estimate

$$u' - \frac{\pi u}{a} + \pi |z_2'|^2 - \pi |z_2|^2 < -u''(\beta(u, v)) + \frac{\pi}{2} + a + 13\delta - \frac{\pi}{a}\frac{a}{2}$$

$$+ u''(\beta(u, v)) - \frac{a}{2} - 2\delta - 2\delta\left(\frac{a}{2} + \delta\right)$$

$$+ \delta a + 4\delta^2 + 3\delta$$

$$= \frac{a}{2} + 14\delta + 2\delta^2$$

$$< \frac{a}{2} + 15\delta,$$

where in the last step we have used that $2\delta^2 < \delta$ which follows from $\delta < \frac{1}{15}$.

We have verified that the estimate (3.3.1) holds for all $(u, v, z_2) \in T(a, \pi)$, and the proof of Proposition 3.1.9 is complete.   □

Recall that by Corollary 3.1.10, Proposition 3.1.9 implies Proposition 3.1.2, and so, in view of Corollary 3.1.3, the proof of Theorem 3.1.1 is complete.

Remarks 3.3.1. 1. As the verifications done in this section showed, the specific choice of the maps $\beta, \sigma, \varphi, \gamma$ and τ constructed in the previous section is crucial for obtaining the required estimate (3.1.8).

2. We recall that the symplectic embedding $\Phi: E(a, \pi) \hookrightarrow B^4(\frac{a}{2} + \pi + \epsilon)$ in our construction is the composition

$$\Phi = c^{-1} \circ (\omega \times \mathrm{id}) \circ \Psi \circ (\alpha \times \mathrm{id}) \circ c$$
$$= c^{-1} \circ (\omega \times \mathrm{id}) \circ (\mathrm{id} \times \tau) \circ (\gamma \times \mathrm{id}) \circ \varphi \circ (\mathrm{id} \times \sigma) \circ (\beta \times \mathrm{id}) \circ (\alpha \times \mathrm{id}) \circ c,$$

where c is the dilatation by a number close to 1.

3. In view of the Extension after Restriction Principle stated in Section 1.2.2 the symplectic embedding Φ (restricted to a slightly smaller ellipsoid) is induced by a Hamiltonian diffeomorphism of $\mathbb{R}^{2n}$. Since each step in the folding construction can be achieved by a symplectic isotopy in a quite canonical way, not only the final map Φ but the whole construction is induced by a Hamiltonian flow (see Appendix A). This remark also applies to the multiple symplectic folding embeddings of ellipsoids and polydiscs into domains in $\mathbb{R}^{2n}$ constructed in the three subsequent chapters.

4. The folding map $\Psi: T(a, \pi) \hookrightarrow T^4(A)$ can be visualized as in Figure 3.12, in which the pictures are to be understood in the same sense as the picture in Figure 3.1: The horizontal direction is the u-direction and refers to the base, while the vertical direction indicates the locus of the fibres. In the first two pictures and in the last one, the fibres are (contained in) discs, and in the other three pictures they are (contained in) rectangles.

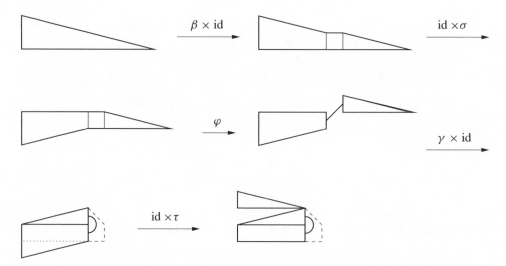

Figure 3.12. Folding an ellipsoid into a ball.

As illustrated in Figure 3.13, the map Ψ essentially restricts to the identity on the black rectangle and maps the triangle $\{u > \frac{a}{2}\}$ to the light triangle and the triangle $\{\pi |z_2|^2 > \frac{\pi}{2}\}$ to the dark triangle. Notice that the wording "folding" is not completely appropriate for what is going on: The triangle $\{u > \frac{a}{2}\}$ is not really "turned over", but rather first "lifted" by φ and then "turned around" by $\gamma \times \mathrm{id}$. We did not find a better wording, though. $\diamondsuit$

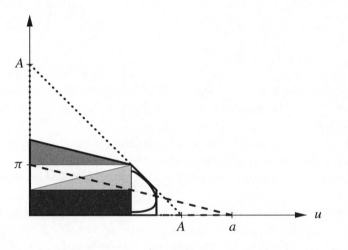

Figure 3.13. How the parts of the ellipsoid are mapped.

Chapter 4

Multiple symplectic folding in four dimensions

In four dimensions we shall exploit the great flexibility of symplectic maps which only depend on the fibre coordinates to provide rather satisfactory embedding results for simple shapes.

We first discuss a modification of the symplectic folding construction described in Section 3.2, then explain multiple folding, and finally calculate the optimal embeddings of ellipsoids and polydiscs into balls and cubes which can be obtained by these methods.

In order not to disturb the exposition unnecessarily with the arbitrarily small δ-terms (arising from "rounding off corners" and so on) we shall skip them in the sequel. Since all the sets under consideration will be bounded and all constructions will involve only finitely many steps, we will not lose control of the δ-terms.

4.1 Modification of the folding construction

The map σ in Step 2 of the folding construction given in Section 3.2 was dictated by the estimate (3.1.8) used to obtain the $2n$-dimensional result, $n > 2$. As a consequence, the map φ of Step 3 had to disjoin the z_2-projection of $\mathcal{S}(P_2)$ from the one of $\mathcal{S}(P_1)$, and we ended up with the unused white sandwiched triangle in Figure 3.13. In order to use this room for $n = 2$, in which case the estimate (3.1.8) is not necessary, we modify the folding construction as follows: Replace the map σ of Step 2 by the map σ given by Figure 4.1. If we define φ as in (3.2.4), the z_2-projection of the image of φ will almost coincide with the image of σ. Choose now γ as in Step 4 and define the final map τ on a neighbourhood of the image of σ such that it restricts to σ^{-1} on the image of σ. If all the δ's were chosen appropriately, the composite map Ψ defined as in (3.2.14) will be one-to-one, and the image of Ψ will be contained in $T^4(a/2 + \pi + \epsilon)$ for some small ϵ. The map Ψ can be visualized as in Figure 4.2. Essentially, Ψ restricts to the identity on the floor F_1 and maps the white triangle with vertices $\left(\frac{a}{2}, 0\right)$, $(a, 0)$, $\left(\frac{a}{2}, \frac{\pi}{2}\right)$ to the floor F_2.

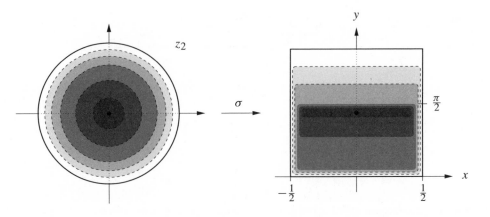

Figure 4.1. The modified map σ.

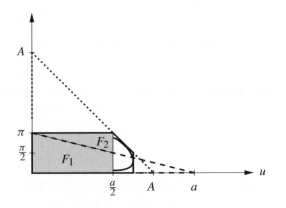

Figure 4.2. Folding in four dimensions.

4.2 Multiple folding

Neither Theorem 2 nor Traynor's Wrapping Theorem tells us whether the ellipsoid $E(\pi, 4\pi)$ symplectically embeds into the ball $B^4(A)$ for some $A \leq 3\pi$ (cf. Figure 7.1, p. 159). Multiple symplectic folding, which is explained in this section, will provide a symplectic embedding of $E(\pi, 4\pi)$ into $B^4(2.692\,\pi)$. To understand the general construction it is enough to look at a 3-fold. Up to the final fibre adjusting map τ, the folding map Ψ is then the composition of maps explained in Figure 4.3, in which the pictures are to be understood as in Figure 3.12: The horizontal direction refers to the base and the vertical direction indicates the fibres. Here are the details: Fix $a > \pi$ and view the ellipsoid $E(a, \pi)$ as the trapezoid $T(a, \pi)$, which fibres over the rectangle $R(a) = \{(u, v) \mid 0 < u < a, \ 0 < v < 1\}$. Pick $u_1, \ldots, u_4 \in \mathbb{R}_{>0}$ satisfying $\sum_{i=1}^{4} u_i = a$; the u_i will be specified in 4.3.1 for embedding $E(a, \pi)$ into a ball and

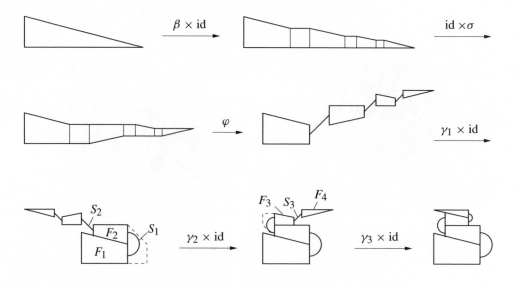

Figure 4.3. Multiple folding.

in 4.4.1 for embedding $E(a, \pi)$ into a cube. Then define the heights l_i, $i = 1, 2, 3$, by

$$l_i = \pi - \frac{\pi}{a} \sum_{j=1}^{i} u_j, \quad i = 1, 2, 3.$$

Step 1 (Separating smaller fibres from larger ones). Let U and f be as in Figure 4.4. Proceeding as in Step 1 of the folding construction in Section 3.2 we find a symplectic embedding $\beta \colon R(a) \hookrightarrow U$ such that $(\beta \times \mathrm{id})(T(a, \pi)) \subset \mathcal{F}(U, f)$.

Step 2 (Preparing the fibres). The map σ is explained in Figure 4.5. More precisely, σ maps the central black disc to the black disc D, and up to some neglected δ-terms we have

$$y(\sigma(z_2)) = \begin{cases} l_1 - l_2 + l_3 - \pi |z_2|^2 & \text{for most } z_2 \in D(l_3) \setminus D, \\ l_1 - l_2 + \pi |z_2|^2 & \text{for most } z_2 \in D(l_2) \setminus D(l_3), \\ l_1 - \pi |z_2|^2 & \text{for most } z_2 \in D(l_1) \setminus D(l_2), \\ \pi |z_2|^2 & \text{for most } z_2 \in D(\pi) \setminus D(l_1). \end{cases}$$

In general, when folding n times, σ maps the circles in the $n + 1$ annuli around a small central disc alternately to rectangular loops with essentially constant maximal but decreasing minimal y-coordinate and to rectangular loops with essentially constant minimal but increasing maximal y-coordinate.

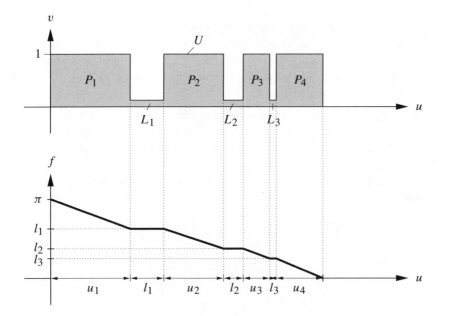

Figure 4.4. $\mathcal{F}(U, f)$.

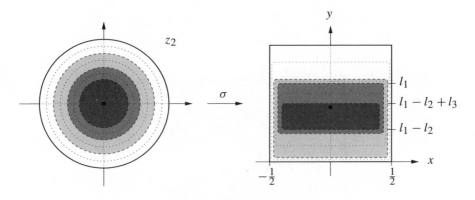

Figure 4.5. The map σ.

Step 3 (Lifting the fibres). Choose cut off functions c_i over L_i, $i = 1, 2, 3$, and abbreviate $c(t) = \sum_{i=1}^{3} c_i(t)$ and $I(t) = \int_0^t c(s)\, ds$. The symplectic embedding $\varphi : (\beta \times \sigma)(T(a, \pi)) \hookrightarrow \mathbb{R}^4$ is defined as in (3.2.4) by

$$\varphi(u, x, v, y) = \left(u, x, v + c(u) \left(x + \frac{1}{2} \right), y + I(u) \right).$$

Step 4 (Folding). Step 4 in Section 3.2 now requires three steps.

1. The folding map γ_1 is essentially the map γ of Section 3.2: On the part of the base denoted by P_1 it is the identity, for $u_1 \leq u \leq u_1 + l_1$ it looks like the map in Figure 3.9, and for $u > u_1 + l_1$ it is an isometry. Observe that by construction, the slope of the stairs S_2 is 1, while the slope of the upper edge of the floor F_1 is $\pi/a < 1$. The sets S_2 and F_1 are thus disjoint.

2. The map $\gamma_2 \times$ id is not really a global product map, but restricts to a product on certain pieces of its domain: It is the identity on $F_1 \coprod S_1 \coprod F_2$, and it is the product $\gamma_2 \times$ id on the remaining domain, where γ_2 is explained in Figure 4.6: It is the identity on the gray part of its domain, maps the black square to the black part of its image, and is an isometry on $\{u \leq 0\}$. The map γ_2 is constructed the same way as the map γ in Section 3.2.

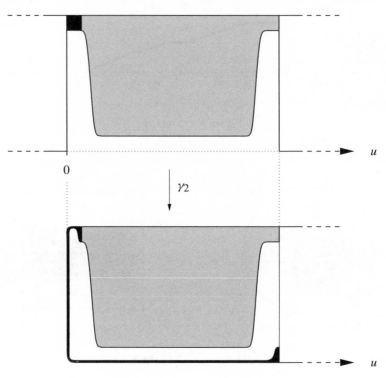

Figure 4.6. Folding on the left.

3. The map $\gamma_3 \times$ id, which turns F_4 over F_3, is analogous to the map $\gamma_1 \times$ id: It is an isometry on F_4, looks like the map given by Figure 3.9 on S_3, and restricts to the identity everywhere else.

Step 5 (Adjusting the fibres). The z_2-projection of the image of φ is a tower shaped domain $\mathcal{T}$. The final map τ is a symplectomorphism from a small neighbourhood of

$\mathcal{T}$ to a disc. It is enough to choose τ in such a way that up to some neglected δ-term we have for any $z_2 = (x, y)$, $z_2' = (x', y') \in \mathcal{T}$

$$y < y' \implies |\tau(z_2)|^2 < |\tau(z_2')|^2.$$

This finishes the 3-fold folding construction. The n-fold folding construction is analogous. Here, we have $u_1, \ldots, u_{n+1}$, heights

$$l_i = \pi - \frac{\pi}{a} \sum_{j=1}^{i} u_j, \quad i = 1, \ldots, n, \tag{4.2.1}$$

floors $F_1, \ldots, F_{n+1}$, and stairs $S_1, \ldots, S_n$. The thin stairs require some locus in space specified in the following lemma, which is proved in Steps 4.1 and 4.2 above.

Folding Lemma 4.2.1. *Let S_i be the stairs connecting the floors F_i and F_{i+1} of minimal respectively maximal height l_i, $i = 1, \ldots, n$.*

(i) *If i is odd, then the stairs S_i are contained in a trapezoid with horizontal lower edge of length l_i, left edge of length $2l_i$, and right edge of length l_i, cf. Figure 4.7 (i); moreover, no smaller trapezoid contains S_i.*

(ii) *If i is even, then the stairs S_i are contained in a trapezoid with horizontal upper edge of length l_i, left edge of length l_i, and right edge of length $2l_i$, cf. Figure 4.7 (ii); moreover, no smaller trapezoid contains S_i.*

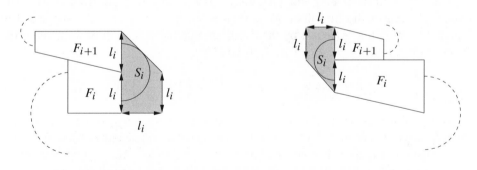

(i) Folding on the right. (ii) Folding on the left.

Figure 4.7. The locus of the stairs.

4.3 Embeddings into balls

In this section we use multiple symplectic folding to construct good embeddings of four dimensional ellipsoids and polydiscs into four dimensional balls.

4.3.1   Embedding ellipsoids into balls.   Recall from Theorem 1 that if $a \leq 2\pi$ then the ellipsoid $E(\pi, a)$, which is symplectomorphic to $E(a, \pi)$, does not symplectically embed into $B^4(A)$ if $A < a$. We therefore fix $a > 2\pi$. We again think of $E(a, \pi)$ as $T(a, \pi)$ and of $B^4(A)$ as $T^4(A) = T(A, A)$. In order to find the smallest trapezoid $T^4(A)$ into which $T(a, \pi)$ embeds via multiple symplectic folding we have to choose the u_i's, which appeared in Section 4.2, optimally. Our strategy to do so is as follows. We shall describe a procedure which associates with each $u_1 \in \,]0, a[$ the number

$$A(a, u_1) = 2\pi + \left(1 - \frac{2\pi}{a}\right) u_1, \tag{4.3.1}$$

and either a finite sequence $u_2, u_3, \ldots, u_{N+1}$ and the attribute *admissible*, or an empty or finite sequence $u_2, u_3, \ldots, u_N$ and the attribute *non-admissible*. In both cases the number $N = N(u_1)$ will depend only on u_1, and the procedure will describe a symplectic embedding of $T(a, \pi)$ into $T^4(A(a, u_1))$ by (N-fold) symplectic folding if and only if u_1 is admissible. In view of $a > 2\pi$ and equation (4.3.1) we have to look for the smallest admissible u_1. As we shall see, u_1 is non-admissible if $u_1 \leq a\pi/(a+\pi)$, and u_1 is admissible if $u_1 > a/2$. Moreover, we will show that if u_1 is admissible, then u_1' is admissible for any $u_1' > u_1$. It follows that there is a unique $u_0 = u_0(a)$ such that u_1 is admissible if $u_1 > u_0$ and u_1 is non-admissible if $u_1 < u_0$. Therefore, our procedure yields a symplectic embedding of $T(a, \pi)$ into $T^4(A(a, u_0) + \epsilon)$ for any $\epsilon > 0$. Finally, we will explain why our procedure is optimal, i.e., multiple symplectic folding does not yield an embedding of $T(a, \pi)$ into $T^4(A)$ if $A \leq A(a, u_0)$.

We start with describing our procedure. Fix $u_1 \in \,]0, a[$ and fold at u_1. The minimal height of the first floor F_1 is then $l_1 = \pi - (\pi/a)u_1$. As suggested in Figure 4.8, we define $A(a, u_1)$ by the condition that the second floor F_2 touches the "upper right boundary" of $T^4(A(a, u_1))$, i.e.,

$$A(a, u_1) = u_1 + 2l_1 = 2\pi + \left(1 - \frac{2\pi}{a}\right) u_1.$$

We now successively try to choose u_i, $i \geq 2$, in such a way that u_i is maximal and the stairs S_i are contained in $T^4(A(a, u_1))$. Define the remaining length r_1 by $r_1 = a - u_1$. The image of $T(a, \pi)$ after folding at u_1 is contained in $T^4(A(a, u_1))$ if and only if $r_1 < u_1$, i.e., $u_1 > a/2$. If $r_1 < u_1$, we set $u_2 = r_1$, and u_1 is admissible. Indeed, the data u_1, u_2 then describe a symplectic embedding of $T(a, \pi)$ into $T^4(A(a, u_1))$ obtained by folding once. Since $u_1 > a/2$, these embeddings are not better than the one constructed in Section 3.2. If $r_1 \geq u_1$, we are forced to fold a second time. Assume that we fold at $u_1 - u_2 > 0$, i.e., the second floor F_2 has length u_2. Then the height of F_2 at $u_1 - u_2$ is $l_2 = l_1 - (\pi/a)u_2$. If $l_1 \geq u_1$, then $l_2 \geq u_1 - (\pi/a)u_2 > u_1 - u_2$, and so the Folding Lemma 4.2.1 (ii) shows that the stairs S_2 are not contained in $\{u > 0\}$. A necessary condition that S_2 is contained in $T^4(A(a, u_1))$ is therefore $l_1 < u_1$. In view of $l_1 = \pi - (\pi/a)u_1$ this condition is equivalent to the condition on u_1

$$u_1 > \frac{a\pi}{a + \pi}. \tag{4.3.2}$$

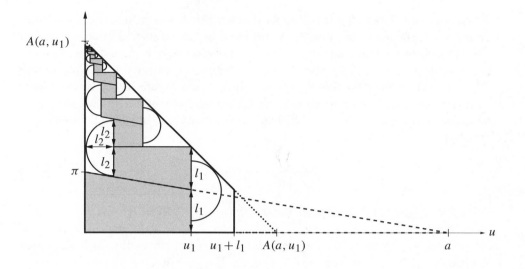

Figure 4.8. A 12-fold.

If this condition is not met, then u_1 is non-admissible. If $l_1 < u_1$, then the Folding Lemma 4.2.1 (ii) shows that for u_2 small enough we have $S_2 \subset \{u > 0\}$, and that the maximal such u_2 is characterized by the equation $l_2 + u_2 = u_1$, which by (4.2.1) translates into the formula

$$u_2 = \frac{a + \pi}{a - \pi} u_1 - \frac{a\pi}{a - \pi}. \tag{4.3.3}$$

We define u_2 by (4.3.3). Then $S_2 \subset \{u > 0\}$, but it is still possible that S_2 is not contained in $T^4(A(a, u_1))$, in which case u_1 is non-admissible. Assume that $S_2 \subset T^4(A(a, u_1))$. We denote the length $a - u_1 - u_2 = r_1 - u_2$ of the new remainder by r_2. If $r_2 < u_2$, we set $u_3 = r_2$, and u_1 is admissible. Indeed, the data u_1, u_2, u_3 then describe a symplectic embedding of $T(a, \pi)$ into $T^4(A(a, u_1))$ obtained by folding twice. If $r_2 \geq u_2$, we are forced to fold a third time. Assume that we fold at $l_2 + u_3$, i.e., the third floor F_3 has length u_3, and its height at $l_2 + u_3$ is $l_3 = l_2 - (\pi/a)u_3$. If $u_2 \leq 2\,l_2$, then $u_2 - u_3 \leq 2\,l_2 - u_3 < 2\,l_2 - (2\pi/a)u_3 = 2\,l_3$, and so the Folding Lemma 4.2.1 (i) shows that the stairs S_3 are not contained in $T^4(A(a, u_1))$. A necessary condition that S_3 is contained in $T^4(A(a, u_1))$ is therefore $u_2 > 2\,l_2$, and if this condition is not met, u_1 is non-admissible. If $u_2 > 2\,l_2$, the Folding Lemma 4.2.1 (i) shows that $S_3 \subset T^4(A(a, u_1))$ whenever $u_2 - u_3 > 2\,l_2 - (2\pi/a)u_3 = 2\,l_3$. Since $a > 2\pi$, the maximal such u_3 is characterized by the equation $u_2 - u_3 = 2\,l_2 - (2\pi/a)u_3$, i.e.,

$$u_3 = \frac{a}{a - 2\pi}(u_2 - 2\,l_2). \tag{4.3.4}$$

We define u_3 by (4.3.4).

Assume now that we folded already i times, where i is even, that $u_2, \ldots, u_i$ have been chosen maximal, and that u_1 is still possibly admissible. Denote the length $a - \sum_{j=1}^{i} u_j$ of the remainder by r_i. If $r_i < u_i$, we set $u_{i+1} = r_i$, and u_1 is admissible. Indeed, $u_1, u_2, \ldots, u_{i+1}$ then describe a symplectic embedding of $T(a, \pi)$ into $T^4(A(a, u_1))$ obtained by folding i times. If $r_i \geq u_i$, we are forced to fold again. As in the case $i = 2$, the Folding Lemma 4.2.1 (i) shows that if $u_i \leq 2 l_i$, then u_1 is non-admissible, and that if $u_i > 2 l_i$, then the maximal u_{i+1} for which the stairs S_{i+1} are contained in $T^4(A(a, u_1))$ is

$$u_{i+1} = \frac{a}{a - 2\pi}(u_i - 2 l_i). \tag{4.3.5}$$

We define u_{i+1} by (4.3.5). Denote the length $a - \sum_{j=1}^{i+1} u_j = r_i - u_{i+1}$ of the new remainder by r_{i+1}. If $r_{i+1} < u_{i+1} + l_i$, we set $u_{i+2} = r_{i+1}$, and u_1 is admissible. Indeed, $u_1, u_2, \ldots, u_{i+2}$ then describe a symplectic embedding of $T(a, \pi)$ into $T^4(A(a, u_1))$ obtained by folding $i + 1$ times. If $r_{i+1} \geq u_{i+1} + l_i$, we are forced to fold again. The Folding Lemma 4.2.1 (ii) shows that for u_{i+2} small enough we have $S_{i+2} \subset \{u > 0\}$, and that the maximal such u_{i+2} is characterized by the equation $u_{i+2} + l_{i+2} = u_{i+1} + l_i$, which by (4.2.1) translates into the formula

$$u_{i+2} = \frac{a + \pi}{a - \pi} u_{i+1}. \tag{4.3.6}$$

We define u_{i+2} by (4.3.6). Then $S_{i+2} \subset \{u > 0\}$, but possibly S_{i+2} is not contained in $T^4(A(a, u_1))$, in which case u_1 in non-admissible. If $S_{i+2} \subset T^4(A(a, u_1))$, we go on as before.

Proceeding this way we try to decide for each $u_1 \in {]}0, a{[}$ whether it is admissible or not. The value u_1 is admissible if and only if our procedure leads to an embedding of $T(a, \pi)$ into $T^4(A(a, u_1))$ after a finite number $N(u_1)$ of folds, and u_1 is non-admissible if and only if the procedure leads to an embedding obstruction after finitely many folds or if the procedure does not terminate. As we have already seen, u_1 is non-admissible if $u_1 \leq a\pi/(a + \pi)$, and u_1 is admissible if $u_1 > a/2$. Also recall that we have to look for the smallest admissible u_1. The following lemma implies that there exists a unique $u_0 = u_0(a)$ in the interval

$$I(a) := \left[\frac{a\pi}{a + \pi}, \frac{a}{2} \right] \tag{4.3.7}$$

such that u_1 is admissible if $u_1 > u_0$ and u_1 is non-admissible if $u_1 < u_0$.

Lemma 4.3.1. *Assume that $u_1, u_1' \in {]}0, a{[}$ and that $u_1 < u_1'$. If u_1 is admissible, then u_1' is admissible, and $N(u_1') \leq N(u_1)$.*

Proof. We abbreviate $N = N(u_1)$ and $N' = N(u_1')$. Assume that the embedding procedure associated with u_1 is described by $u_1, \ldots, u_{N+1}$. We shall prove the lemma by going through our procedure and checking that the folding conditions for u_1' are met whenever they are met for u_1.

If u_1' is admissible and $N' = 1$, we are done. Otherwise, $u_1 < u_1' < a/2$, and so $N \geq 2$. Then $u_1' > u_1 > a\pi/(a + \pi)$. Observe that the next condition $S_2 \subset T^4(A(a, u_1))$ is equivalent to $l_2 < u_2$. Equation (4.3.3) shows $u_2' > u_2$, and now equation (4.2.1) shows $l_2' < l_2$. Therefore, $l_2' < l_2 < u_2 < u_2'$, and so $S_2' \subset T^4(A(a, u_1'))$. If u_1' is admissible and $N' = 2$, we are done. Otherwise, $u_2 < u_2' < r_2' < r_2$, and so $N \geq 3$. In view of (4.2.1) we then have $u_2' > u_2 > 2l_2 > 2l_2'$, and equation (4.3.4) shows $u_3' > u_3$. If u_1' is admissible and $N' = 3$, we are done. Otherwise, $r_3' > u_3' + l_2'$, i.e., $a - \pi > (1 - \pi/a)(u_1' + u_2') + 2u_3'$, whence $a - \pi > (1 - \pi/a)(u_1 + u_2) + 2u_3$, i.e., $r_3 > u_3 + l_2$, and so $N > 3$. We proceed in this way using equations (4.3.5) and (4.3.6) and observing that the condition $S_i \subset T^4(A(a, u_1))$ for i even is equivalent to $l_i < u_i$. The lemma then follows. $\square$

Define the function $f_{\mathrm{EB}}(a)$ on $]2\pi, \infty[$ by

$$f_{\mathrm{EB}}(a) = 2\pi + \left(1 - \frac{2\pi}{a}\right)u_0(a). \qquad (4.3.8)$$

Lemma 4.3.1 shows that the ellipsoid $E(\pi, a)$ symplectically embeds into the ball $B^4\left(f_{\mathrm{EB}}(a) + \epsilon\right)$ for any $\epsilon > 0$.

We next explain why our procedure is optimal in the sense that we cannot embed $E(\pi, a)$ into a ball smaller than $B^4\left(f_{\mathrm{EB}}(a)\right)$ by multiple symplectic folding. Indeed, observe that our procedure can equivalently be described as follows: To $A \in]2\pi, a[$ associate the number $u_1 = u_1(a, A)$ defined by $A = 2\pi + (1 - 2\pi/a)u_1$, and then choose u_i, $i \geq 2$, maximal. In other words, for each A we successively choose $u_1, u_2, \ldots$ maximal with respect to the condition of staying inside $T^4(A)$. The only possible way of improving our procedure is therefore to choose some of the u_i smaller. So let $A < f_{\mathrm{EB}}(a)$, let $u_1, u_2, \ldots, u_N$ be the sequence associated with A by our procedure, and try to circumvent the embedding obstruction arising after folding N times by choosing $u_i' \leq u_i$. Then, however, the proof of Lemma 4.3.1 shows that $l_i' \geq l_i$ and $r_i' \geq r_i$, and so the modified procedure leads to an embedding obstruction after $N' \leq N$ folds.

Lemma 4.3.2. *We have $N(u_1) \to \infty$ as $u_1 \searrow u_0$. The value u_0 is non-admissible, and the procedure associated with u_0 does not terminate.*

Proof. Arguing by contradiction, assume that the first statement of the lemma is wrong. By Lemma 4.3.1, there exists N such that $N(u_1) = N$ for all $u_1 > u_0$ with $u_1 - u_0$ small enough. In the sequel we assume that $u_1 > u_0$ and that $u_1 - u_0$ is so small that $N(u_1) = N$. By the proof of Lemma 4.3.1, the functions $u_i(u_1)$, $i = 2, \ldots, N - 1$ are decreasing as $u_1 \searrow u_0$, and the functions $l_i(u_1)$, $i = 1, \ldots, N$, and $u_N(u_1)$ are increasing as $u_1 \searrow u_0$ and bounded. We set

$$u_i^0 = \lim_{u_1 \searrow u_0} u_i(u_1), \quad i = 2, \ldots, N, \quad \text{and} \quad l_i^0 = \lim_{u_1 \searrow u_0} l_i(u_1), \quad i = 1, \ldots, N.$$

Assume first that during the first N folds there is no embedding obstruction at u_0.

Case 1. $N = 1$. If $u_0 > u_2^0$, then $T(a, \pi)$ embeds into $T^4(A(a, u_0))$ by folding once, and if $u_0 = u_2^0$, then the Folding Lemma 4.2.1 (i) shows that $T(a, \pi)$ embeds into $T^4(A(a, u_0))$ by folding twice.

Case 2. $N \geq 3$ and N odd. If $u_N^0 + l_{N-1}^0 > u_{N+1}^0$, then $T(a, \pi)$ embeds into $T^4(A(a, u_0))$ by folding N times, and if $u_N^0 + l_{N-1}^0 = u_{N+1}^0$, then the Folding Lemma 4.2.1 (i) shows that $T(a, \pi)$ embeds into $T^4(A(a, u_0))$ by folding $N + 1$ times.

Case 3. N even. If $u_N^0 > u_{N+1}^0$, then $T(a, \pi)$ embeds into $T^4(A(a, u_0))$ by folding N times, and if $u_N^0 = u_{N+1}^0$, then the Folding Lemma 4.2.1 (ii) and $a > 2\pi$ imply that $T(a, \pi)$ embeds into $T^4(A(a, u_0))$ by folding $N + 1$ times.

Therefore, u_0 is admissible with $N(u_0) = N$ or $N(u_0) = N + 1$. Since all conditions in our procedure are open conditions, it follows that if $u_1 < u_0$ and $u_0 - u_1$ is small enough, then u_1 is admissible with $N(u_1) = N$ or $N(u_1) = N + 1$. This contradicts the definition of u_0. So assume that there is an embedding obstruction at u_0 appearing at the latest at the N'th fold. We shall conclude the proof of Lemma 4.3.2 by showing that an embedding obstruction at u_0 appearing at the latest at the N'th fold implies an embedding obstruction at all u_1 near u_0.

Case I. $u_0 \leq a\pi/(a + \pi)$. Then $u_0 = a\pi/(a + \pi)$, i.e., $l_1(u_0) = l_1^0 = u_0$. Therefore, $u_2^0 = 0$ and so $N \geq 2$ and $l_2^0 = l_1^0 = u_0$.

If $N = 2$, we find $u_3(u_1) > u_2(u_1)$ for u_1 near u_0, a contradiction.

If $N > 2$, we find $u_2(u_1) < 2l_2(u_1)$ for u_1 near u_0, i.e., there is an embedding obstruction for u_1 near u_0; this is another contradiction.

Case II. $S_i(u_0) \not\subset T^4(A(a, u_0))$ for some even i, i.e., $l_i^0 \geq u_i^0$ for some even i.

If $i = N$, we find $u_{N+1}(u_1) > u_N(u_1)$ for u_1 near u_0, a contradiction.

If $i < N$, we find $u_i(u_1) < 2l_i(u_1)$ for u_1 near u_0, i.e., there is an embedding obstruction for u_1 near u_0; this is another contradiction.

Case III. $u_i(u_0) \leq 2l_i(u_0)$ for some even i. Then $u_i^0 = 2l_i^0$.

If $i = N$, we find in view of $a > 2\pi$ that $u_{N+1}(u_1) > u_N(u_1)$ for u_1 near u_0, a contradiction.

If $i = N - 1$, then $u_{N+1}^0 \leq l_{N-1}^0$, and so we find $u_{N+1}(u_1) < u_{N-1}(u_1)$ for u_1 near u_0, contradicting $N > i$.

If $i < N - 1$, we find $u_{i+1}^0 = u_{i+2}^0 = 0$ and $l_i^0 = l_{i+1}^0 = l_{i+2}^0$, whence $l_i^0 + l_{i+1}^0 + l_{i+2}^0 > u_i^0$. This contradicts $S_{i+2}(u_1) \subset T^4(A(a, u_1))$ for u_1 near u_0.

We conclude that $N(u_1) \to \infty$ as $u_1 \searrow u_0$. The value u_0 is therefore non-admissible, and as we have seen in Case I, Case II and Case III above, the procedure associated with u_0 cannot lead to an embedding obstruction. Lemma 4.3.2 is thus proved. $\qquad\square$

The computer program in Appendix D.1 computing $u_0(a)$ is based on the following lemma, which implies that the value u_0 is the only value for which our procedure does not terminate.

Lemma 4.3.3. *Assume that $u_1 < u_0$. Then the procedure associated with u_1 leads to an embedding obstruction after finitely many folds.*

Proof. By Lemma 4.3.2 our procedure associated with u_0 generates the infinite sequences $u_i(u_0)$ and $l_i(u_0)$, $i = 1, 2, \ldots$, of lengths and heights of the floors $F_i(u_0)$. Set $h(u_0) := \sum_{i=1}^{\infty} l_i(u_0)$. Then $h(u_0) \leq A(a, u_0)$. We claim that

$$h(u_0) = A(a, u_0). \tag{4.3.9}$$

In other words, the image of $T(a, \pi)$ in $T^4(A(a, u_0))$ obtained by folding first at u_0 and then infinitely many times touches the upper vertex of $T^4(A(a, u_0))$, i.e., the sequence $F_i(u_0)$ of floors creeps into the upper corner of $T^4(A(a, u_0))$, cf. Figure 4.8. In order to prove the identity (4.3.9), we argue by contradiction and assume $h(u_0) < A(a, u_0)$. Then $w := A(a, u_0) - h(u_0) > 0$. The Folding Lemma 4.2.1 (ii) shows that $l_{2i}(u_0) + u_{2i}(u_0) > w$ for all $i = 1, 2, \ldots$. Since $h(u_0) < \infty$, there exists $j \in \mathbb{N}$ such that $l_{2i}(u_0) < w/2$ for $i \geq j$. Then $u_{2i}(u_0) > w - l_{2i}(u_0) > w/2$ for $i \geq j$, and so $a = \sum_{i=1}^{\infty} u_i(u_0) > \sum_{i=1}^{\infty} u_{2i}(u_0) = \infty$. This contradiction proves the identity (4.3.9).

Assume now that Lemma 4.3.3 is wrong for some $u_1 < u_0$. Since u_1 is non-admissible, the procedure associated with u_1 does not terminate and generates the infinite sequence $l_i(u_1)$, $i = 1, 2, \ldots$. The proof of Lemma 4.3.1 shows that $l_i(u_1) > l_i(u_0)$, $i = 1, 2, \ldots$. Therefore,

$$h(u_1) := \sum_{i=1}^{\infty} l_i(u_1) > \sum_{i=1}^{\infty} l_i(u_0) = h(u_0) = A(a, u_0) > A(a, u_1).$$

The contradiction $h(u_1) > A(a, u_1)$ shows that Lemma 4.3.3 holds true. $\square$

While the computer program in Appendix D.1 computes for each $a > 2\pi$ and each $\epsilon > 0$ the value of $f_{\mathrm{EB}}(a)$ up to accuracy ϵ, the following lemma gives some qualitative insight into the function $f_{\mathrm{EB}}(a)$.

Lemma 4.3.4. *The function f_{EB} on $]2\pi, \infty[$ is strictly increasing and hence almost everywhere differentiable. Moreover, f_{EB} is Lipschitz continuous with Lipschitz constant at most 1.*

Proof. Assume $2\pi < a < a'$. In view of the procedures associated with the pairs $(a, u_0(a'))$ and $(a', u_0(a'))$, the inequalities

$$l_1(a, u_0(a')) = \pi - \frac{\pi}{a} u_0(a') < \pi - \frac{\pi}{a'} u_0(a') = l_1(a', u_0(a'))$$

and $a - u_0(a') < a' - u_0(a')$ imply that $u_0(a) \leq u_0(a')$. Therefore,

$$f_{EB}(a) = 2\pi + \left(1 - \frac{2\pi}{a}\right) u_0(a) < 2\pi + \left(1 - \frac{2\pi}{a'}\right) u_0(a') = f_{EB}(a').$$

Since $a < a'$ were arbitrary, we conclude that the function f_{EB} is strictly increasing, and so, as every increasing real function, almost everywhere differentiable.

Assume again $2\pi < a < a'$. We set $\delta = a' - a$ and $u_1 = u_0(a) + \delta$. Since $u_0(a) < a$, we have

$$\frac{u_0(a)}{a} < \frac{u_1}{a'}. \tag{4.3.10}$$

It follows that

$$l_1(a', u_1) = \pi - \frac{\pi}{a'} u_1 < \pi - \frac{\pi}{a} u_0(a) = l_1(a, u_0(a)).$$

In view of the procedures associated with the pairs $(a, u_0(a))$ and (a', u_1), this inequality and the equality $a' - u_1 = a - u_0(a)$ imply that $u_1 \geq u_0(a')$. Using definition (4.3.8) and inequality (4.3.10), we may therefore estimate

$$\begin{aligned}
f_{EB}(a') - f_{EB}(a) &= \left(1 - \frac{2\pi}{a'}\right) u_0(a') - \left(1 - \frac{2\pi}{a}\right) u_0(a) \\
&\leq \left(1 - \frac{2\pi}{a'}\right) u_1 - \left(1 - \frac{2\pi}{a}\right) u_0(a) \\
&= u_1 - u_0(a) - 2\pi \left(\frac{u_1}{a'} - \frac{u_0(a)}{a}\right) \\
&< u_1 - u_0(a) \\
&= a' - a.
\end{aligned}$$

Since $a < a'$ were arbitrary, we conclude that the function f_{EB} is Lipschitz continuous with Lipschitz constant at most 1. $\square$

We do not know whether the function f_{EB} is differentiable on any open interval. We next investigate the behaviour of the function $f_{EB}(a)$ as $a \to 2\pi^+$.

Proposition 4.3.5. *We have*

$$\limsup_{\epsilon \to 0^+} \frac{f_{EB}(2\pi + \epsilon) - 2\pi}{\epsilon} \leq \frac{3}{7}.$$

Proof. Fix $a > 2\pi$. By the proof of Lemma 4.3.1 and by Lemma 4.3.2 there exists a unique value $u_{1,2} = u_{1,2}(a)$ such that $N(u_{1,2}) = 2$ and $u_3(u_{1,2}) = u_2(u_{1,2})$. Then $u_{1,2} + 2 u_2(u_{1,2}) = a$. In view of equation (4.3.3) we find $u_{1,2} = \frac{a^2 + a\pi}{3a + \pi}$. Therefore,

$$A_2(a) := A(a, u_{1,2}) = 2\pi + (a - 2\pi)\frac{a + \pi}{3a + \pi},$$

and so $\frac{d}{da}A_2(2\pi) = \frac{3}{7}$. Since $f_{EB}(a) \le A_2(a)$ for all $a > 2\pi$, the proposition follows. $\square$

Remark 4.3.6. The function $A_2(a)$ constructed in the above proof describes the optimal embedding of $E(\pi, a)$ into a ball obtainable by folding exactly twice. More generally, we can compute the function $A_N(a)$ describing the optimal embedding obtainable by folding exactly N times as follows. For each $a > 2\pi$ and each $N = 1, 2, \ldots$ there exists a unique value $u_{1,N} = u_{1,N}(a)$ such that $N(u_{1,N}) = N$ and such that $u_{N+1}(u_{1,N}) = u_N(u_{1,N}) + l_{N-1}(u_{1,N})$ if N is odd and $u_{N+1}(u_{1,N}) = u_N(u_{1,N})$ if N is even. If N is odd, we replace $l_{N-1}(u_{1,N})$ by the expression given in formula (4.2.1). Plugging the expression for $u_{N+1}(u_{1,N})$ into the equation

$$u_{1,N} + u_2(u_{1,N}) + \cdots + u_{N+1}(u_{1,N}) = a$$

and then alternately using equations (4.3.5) and (4.3.6), we compute $u_{1,N}$ as a rational function of a. We finally find

$$A_N(a) = A(a, u_{1,N}) = 2\pi + \left(1 - \frac{2\pi}{a}\right) u_{1,N}(a).$$

For instance,

$$A_1(a) = 2\pi + (a - 2\pi)\frac{1}{2}, \quad A_2(a) = 2\pi + (a - 2\pi)\frac{a+\pi}{3a+\pi}$$

and

$$A_3(a) = 2\pi + (a - 2\pi)\frac{(a+\pi)(a+2\pi)}{4(a^2 + a\pi + \pi^2)}.$$

By Lemma 4.3.1, $u_{1,N+1}(a) < u_{1,N}(a)$ for every N and every $a > 2\pi$, and arguing as in the proof of Lemma 4.3.4, we see that the function $u_{1,N}(a)$ is increasing for every N. The family $\{A_N\}$, $N = 1, 2, \ldots$, is therefore a strictly decreasing family of strictly increasing smooth rational functions on $]2\pi, \infty[$ converging to $f_{EB}(a)$. In view of Dini's Theorem, the convergence is uniform on bounded sets.

One might try to improve the estimate given in Proposition 4.3.5 by showing $\frac{d}{da}A_N(2\pi) < \frac{3}{7}$ for some N. However, $\frac{d}{da}A_N(2\pi) = \frac{3}{7}$ for all $N \ge 2$. Indeed, for all $a > 2\pi$ and $N > 2$ we have

$$u_2(u_{1,2}(a)) > u_2(u_{1,N}(a)) > 2\,l_2(u_{1,N}(a)) > 2\,l_2(u_{1,2}(a)) = \frac{2\pi}{a}u_2(u_{1,2}(a)),$$

and so $\lim_{a\to 2\pi^+} u_2(u_{1,N}(a)) = \lim_{a\to 2\pi^+} u_2(u_{1,2}(a))$. In view of formula (4.3.3), we conclude that

$$\lim_{a\to 2\pi^+} u_{1,N}(a) = \lim_{a\to 2\pi^+} u_{1,2}(a).$$

Next, the identity

$$A_N(a) = 2\pi + \left(1 - \frac{2\pi}{a}\right) u_{1,N}(a) \quad \text{for all } a > 2\pi$$

implies that

$$\frac{d}{da} A_N(a) = \frac{2\pi}{a^2} u_{1,N}(a) + \left(1 - \frac{2\pi}{a}\right) \frac{d}{da} u_{1,N}(a) \quad \text{for all } a > 2\pi.$$

The formal expression for $u_{1,N}(a)$ defines a rational function on $\mathbb{R}$. Since 2π is not a singularity of $u_{1,N}(a)$, the rational function $\frac{d}{da} u_{1,N}(a)$ is bounded near $a = 2\pi$. Taking the limit $a \to 2\pi^+$ we therefore find

$$\frac{d}{da} A_N(2\pi) = \frac{1}{2\pi} \lim_{a \to 2\pi^+} u_{1,N}(a) = \frac{1}{2\pi} \lim_{a \to 2\pi^+} u_{1,2}(a) = \frac{d}{da} A_2(2\pi) = \frac{3}{7},$$

as claimed. $\diamond$

We do not know how to analyze the asymptotic behaviour of the function $f_{EB}(a)$ as $a \to \infty$ directly. We shall prove the following proposition at the end of the subsequent section by comparing the optimal multiple folding embedding of $E(\pi, a)$ into a ball with the optimal multiple folding embedding of the polydisc $P(\pi, \frac{a}{2} + \pi)$ into a ball.

Proposition 4.3.7. *We have*

$$f_{EB}(a) - \sqrt{\pi a} \leq 2\pi \quad \textit{for all } a > 2\pi \,.$$

The function $f_{EB}(a)$ is further discussed and compared with the result yielded by symplectic wrapping in 7.2.1.1.

4.3.2 Embedding polydiscs into balls. As we have seen in the proof of Lemma 3.1.8, the disc $D(a)$ is symplectomorphic to the rectangle $R(a)$. The polydisc $P(\pi, a) = D(\pi) \times D(a)$ is therefore symplectomorphic to $R(a) \times D(\pi)$. Since the fibre $D(\pi)$ over each point $(u, v) \in R(a)$ is the same, the optimal embedding into a ball obtainable by multiple symplectic folding is easier to compute for a polydisc than for an ellipsoid. In contrast to our optimal embeddings of an ellipsoid into a ball, which were obtained by folding "more and more often", the optimal embedding of a polydisc into a ball obtainable by multiple folding will turn out to be described by a picture as in Figure 4.9. Our embedding result stated in Proposition 4.3.9 below is readily read off from such a picture. We aim, however, to show that this embedding result is the best one obtainable by multiple folding. We therefore proceed in a systematic way.

We again think of the ball $B^4(A)$ as the trapezoid $T^4(A)$. Fix $a \geq \pi$. Folding $R(a) \times D(\pi)$ first at $u_1 \in {]}0, a{[}$ determines $T^4(A(a, u_1))$ by the condition that the second floor F_2 touches the "upper right boundary" of $T^4(A(a, u_1))$. Then $A(a, u_1) =$

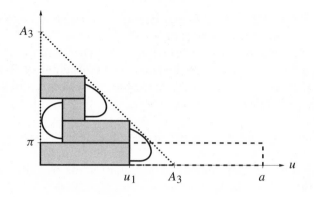

Figure 4.9. The optimal embedding $P(\pi, a) \hookrightarrow B^4(A)$ for $a = 10\pi$.

$u_1 + 2\pi$. We then successively choose u_i, $i \geq 2$, maximal with respect to the condition of staying inside $T^4(A(a, u_1))$.

The Folding Lemma 4.2.1 (ii) shows that a condition for folding a second time, if necessary, is $u_1 > \pi$, and that then the stairs S_2 are contained in $T^4(A(a, u_1))$ if and only if $u_1 > 2\pi$. The only condition on u_1 for folding a second time is therefore $u_1 > 2\pi$. The Folding Lemma 4.2.1 (i) shows that folding a third time, if necessary, is then possible whenever $u_2 > 2\pi$, i.e., $u_1 > 3\pi$. For $N \geq 2$, the only condition on u_1 for folding an N'th time, if necessary, is $u_1 > N\pi$.

If our procedure leads to an embedding obstruction after N folds, then choosing $u_i' \leq u_i$ leads to an embedding obstruction after $N' \leq N$ folds. It is therefore enough to compare embeddings obtained from our procedure.

We say that u_1 is *admissible* if the procedure associated with u_1 leads to an embedding of $R(a) \times D(\pi)$ into $T^4(A(a, u_1))$. We then write $N(u_1)$ for the number of folds needed. If $u_1 < u_1'$ and u_1 is admissible, then u_1' is admissible and $N(u_1') \leq N(u_1)$.

Lemma 4.3.8. *Assume that u_1 is admissible and that $N(u_1)$ is even or that $N(u_1)$ is odd and the last floor $F_{N(u_1)+1}$ does not touch $\{u = 0\}$. Then there exists an admissible u_1' such that $u_1' < u_1$ and such that $N(u_1')$ is odd and $F_{N(u_1')+1}$ touches $\{u = 0\}$.*

Proof. Set $N = N(u_1)$. Observe that on those admissible u_1's for which $N(u_1) = N$, the functions $u_2(u_1), \ldots, u_N(u_1)$ are continuous and increasing in u_1, and $u_{N+1}(u_1)$ is continuous and decreasing in u_1.

Assume first that N is odd and F_{N+1} does not touch $\{u = 0\}$, i.e., $u_2 < u_1$ if $N = 1$ and $u_{N+1} < u_N + \pi$ if $N \geq 3$. If $N = 1$, shrinking u_1 leads to an admissible u_1' such that $u_2' = u_1'$. If $N \geq 3$, shrinking u_1 either leads to an admissible u_1' such that $u_{N+1}' = u_N + \pi$, or to $u_1' = N\pi$. In the second case, however, folding at u_1' would already lead to an embedding after $N - 1$ folds, i.e., $u_1' < u_1$ would be admissible with $N(u_1') < N(u_1)$, a contradiction.

Assume next that N is even. After shrinking u_1, if necessary, we may assume that u_{N+1} touches the "upper right boundary" of $T^4(A(a, u_1))$, i.e., $u_{N+1} + \pi = u_N$. We have $u_{N+1} > \pi$, since otherwise $N(u_1) \leq N - 1$. Therefore $u_N > 2\pi$, and so we can fold another time at $u = u_N - \pi$ and obtain an embedding of $R(a) \times D(\pi)$ into $T^4(A(a, u_1))$ with $u_{N+2} < u_{N+1} + \pi$. As we have seen above, shrinking u_1 leads to an admissible u_1' such that $u_{N+2}' = u_{N+1} + \pi$. $\qquad\square$

Assume that $a \in [\pi, 2\pi]$ and that u_1 is admissible. Then $N(u_1) = 1$. By Lemma 4.3.8, we may assume that $u_1 = a/2$. Since $A(a, a/2) = a/2 + 2\pi \geq a + \pi$, we see that for $a \in [\pi, 2\pi]$ multiple symplectic folding does not provide a better embedding result than the inclusion $P(\pi, a) \hookrightarrow B^4(\pi + a)$.

So assume $a > 2\pi$. By Lemma 4.3.8 it suffices to analyze those embeddings for which the number of folds is $N = 2k - 1$ and F_{N+1} touches $\{u = 0\}$. The optimal embedding obtainable by folding once is therefore described by $A_1(a) = a/2 + 2\pi$. If $N \geq 3$, we read off from Figure 4.9 that

$$
\begin{aligned}
a &= u_1 + u_2 + \cdots + u_{N+1} \\
&= \pi + 2(u_1 - \pi) + 2(u_1 - 3\pi) + \cdots + 2(u_1 - N\pi) + \pi \\
&= 2\pi + 2ku_1 - 2k^2\pi
\end{aligned}
$$

provided that $u_1 > N\pi$. Solving for u_1 and using the formula $A(a, u_1) = u_1 + 2\pi$, we find that the optimal embedding of $R(a) \times D(\pi)$ into a ball obtainable by folding $N = 2k - 1$ times is described by

$$
A_k(a) = \frac{a - 2\pi}{2k} + (k + 2)\pi \quad \text{provided that } a > 2(k^2 - k + 1)\pi.
$$

Observe that this formula also holds true for $N = 1$. Define the function $f_{\mathrm{PB}}(a)$ on $]2\pi, \infty[$ by

$$
f_{\mathrm{PB}}(a) = \min \left\{ A_k(a) \mid k = 1, 2, \ldots; \, a > 2(k^2 - k + 1)\pi \right\},
$$

cf. Figure 4.10. We in particular have proved

Proposition 4.3.9. *Assume $a > 2\pi$. Then the polydisc $P(\pi, a)$ symplectically embeds into the ball $B^4\left(f_{\mathrm{PB}}(a) + \epsilon\right)$ for every $\epsilon > 0$, where*

$$
f_{\mathrm{PB}}(a) = \frac{a - 2\pi}{2k} + (k + 2)\pi
$$

for the unique integer k for which $2(k^2 - k + 1) < a/\pi \leq 2(k^2 + k + 1)$.

Remark 4.3.10. Let $d_{\mathrm{PB}}(a) = f_{\mathrm{PB}}(a) - \sqrt{2\pi a}$ be the difference between f_{PB} and the volume condition. The function d_{PB} attains its local maxima at $a_k = 2(k^2 - k + 1)\pi$, where $d_{\mathrm{PB}}(a_k) = (2k + 1)\pi - 2\pi\sqrt{k^2 - k + 1}$. This is an increasing sequence converging to 2π. $\qquad\diamond$

Extend the above function $f_{PB}(a)$ to a function on $[\pi, \infty[$ by setting $f_{PB}(a) = a + \pi$ for $a \in [\pi, 2\pi]$. The problem considered in this section was to understand the characteristic function χ_{PB} on $[\pi, \infty[$ defined by

$$\chi_{PB}(a) = \inf \left\{ A \mid P(\pi, a) \text{ symplectically embeds into } B^4(A) \right\}.$$

The following proposition summarizes what we know about this function.

Proposition 4.3.11. *The function* $\chi_{PB} \colon [\pi, \infty[\ \to \ \mathbb{R}$ *is bounded from below and above by*

$$\max \left(2\pi, \sqrt{2\pi a} \right) \leq \chi_{PB}(a) \leq f_{PB}(a),$$

see Figure 4.10. *It is monotone increasing and hence almost everywhere differentiable. Moreover,* χ_{PB} *is Lipschitz continuous with Lipschitz constant at most* 2; *more precisely,*

$$\chi_{PB}(a') - \chi_{PB}(a) \leq \frac{f_{PB}(a)}{a} (a' - a) \quad \text{for all } a' \geq a \geq \pi.$$

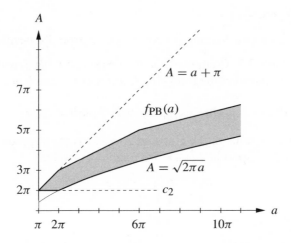

Figure 4.10. What is known about $\chi_{PB}(a)$.

Proof. In view of the monotonicity axiom for the second Ekeland–Hofer capacity, the identities (2.2.3) and (2.3.1) imply $\chi_{PB}(a) \geq 2\pi$ for all $a \geq \pi$. The volume condition $|P(\pi, a)| \leq |B^4(\chi_{PB}(a))|$ translates to $\chi_{PB}(a) \geq \sqrt{2\pi a}$. We conclude that $\max \left(2\pi, \sqrt{2\pi a} \right) \leq \chi_{PB}(a)$. For $a \leq 2\pi$, the estimate $\chi_{PB}(a) \leq f_{PB}(a)$ is provided by the inclusion $P(\pi, a) \hookrightarrow B^4(a + \pi)$, and for $a > 2\pi$ by Proposition 4.3.9.

  Assume $a < a'$. If φ symplectically embeds $P(\pi, a')$ into $B^4(A)$, then $\varphi|_{P(\pi,a)}$ symplectically embeds $P(\pi, a)$ into $B^4(A)$. Therefore χ_{PB} is increasing, and so, as every increasing real function, almost everywhere differentiable. Denote the dilatation by a'/a by λ. Assume that ψ symplectically embeds $P(\pi, a)$ into $B^4(A)$. Then the composition

$$P(\pi, a') \xrightarrow{\lambda^{-1}} P\left(\frac{a}{a'}\pi, a\right) \xrightarrow{\psi} B^4(A) \xrightarrow{\lambda} B^4\left(\frac{a'}{a}A\right)$$

is a symplectic embedding. Therefore $\chi_{\mathrm{PB}}(a') \leq \frac{a'}{a}\chi_{\mathrm{PB}}(a)$. We conclude that

$$\chi_{\mathrm{PB}}(a') - \chi_{\mathrm{PB}}(a) \leq \chi_{\mathrm{PB}}(a)\left(\frac{a'}{a} - 1\right) \leq \frac{f_{\mathrm{PB}}(a)}{a}(a' - a) \leq 2(a' - a)$$

as claimed. $\square$

For $a = 2\pi$, multiple symplectic folding does not provide a better upper bound of $\chi_{\mathrm{PB}}(a)$ than the inclusion $P(\pi, a) \hookrightarrow B^4(\pi + a)$, and Ekeland–Hofer capacities do not provide a better lower bound of $\chi_{\mathrm{PB}}(a)$ than the volume condition $|P(\pi, a)| \leq |B^4(\chi_{\mathrm{PB}}(a))|$. We therefore would like to know the answer to

Question 4.3.12. $\chi_{\mathrm{PB}}(2\pi) < 3\pi$?

We end this section by deriving Proposition 4.3.7 from Proposition 4.3.9.

Proof of Proposition 4.3.7. Computer calculations suggest that $f_{\mathrm{EB}}(a) < f_{\mathrm{PB}}\left(\frac{a}{2}\right)$ for all $a > 2\pi$. For our purpose, the following result will be sufficient.

Lemma 4.3.13. *We have* $f_{\mathrm{EB}}(a) < f_{\mathrm{PB}}\left(\frac{a}{2} + \pi\right)$ *for all* $a > 2\pi$.

Proof. As in 4.3.1 we think of the ellipsoid $E(\pi, a)$ as the trapezoid $T(a, \pi)$ and of the ball $B^4(A)$ as the trapezoid $T^4(A)$. We fix $a > 2\pi$ and let

$$\mathcal{F} = \coprod_{i=1}^{N+1} F_i \cup \coprod_{i=1}^{N} S_i$$

be the image of the "optimal" embedding of $P\left(\pi, \frac{a}{2} + \pi\right)$ into $T^4\left(f_{\mathrm{PB}}\left(\frac{a}{2} + \pi\right)\right)$. We recall from Lemma 4.3.8 that the number of folds N is odd and that for $N = 3$ the set $\mathcal{F}$ looks as in Figure 4.9. We define $A \in \,]2\pi, \infty[$ as the unique real number for which

$$f_{\mathrm{EB}}(A) = f_{\mathrm{PB}}\left(\frac{a}{2} + \pi\right)$$

and we let

$$\mathcal{F}' = \coprod_{i=1}^{\infty} F_i' \cup \coprod_{i=1}^{\infty} S_i'$$

be the image of the "optimal" embedding of the trapezoid $T(A, \pi)$ into $T^4\left(f_{\mathrm{EB}}(A)\right)$, cf. Figure 4.8.

In the sequel we shall compare the volume of $\mathcal{F}$ with the volume of $\mathcal{F}'$. Since the embeddings of $P(\pi, \frac{a}{2} + \pi)$ and $T(A, \pi)$ are both "optimal", the volumes of the stairs $\coprod S_i$ and $\coprod S_i'$ "vanish". We shall therefore neglect the stairs of both sets.

Recall from 4.3.1 that l_i denotes the minimal height of the floor F_i' and that the width and the height of the i'th "triangle" T_i' in $T^4(f_{EB}(A)) \setminus \mathcal{F}'$ is $2l_{2i-1}$. Also recall that

$$\pi > l_1 > l_2 > \cdots . \tag{4.3.11}$$

This and the Folding Lemma 4.2.1 imply that $F_i \subset \mathcal{F}'$ for each odd $i \geq 3$, and that $\mathcal{F} \setminus \mathcal{F}'$ is the disjoint union of the thin "triangle" $Q_1 = F_1 \setminus (F_1' \cup F_2')$, the "rectangles" $Q_i \subset F_i$, i even, each contained in a different triangle $T_{j(i)}'$, and the set Q_0 lying in the left end of the floor F_{N+1}, see Figure 4.11.

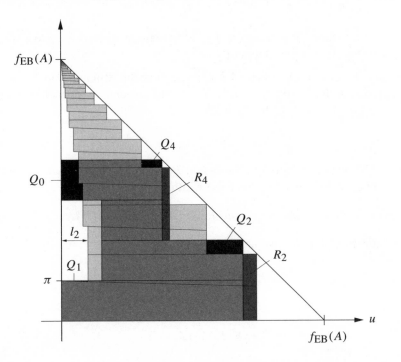

Figure 4.11. The sets $Q_i \subset \mathcal{F} \setminus \mathcal{F}'$ and the sets $R_{j(i)} \subset \mathcal{F}' \setminus \mathcal{F}$.

Using the definition (4.2.1) of l_1 and the estimate (4.3.2) we find

$$l_2 < l_1 = \pi - \frac{\pi}{A} u_1 < \pi - \frac{\pi^2}{A + \pi}$$

and so

$$|Q_0| + |Q_1| \leq l_2 \pi + \frac{1}{2} l_2 \frac{\pi}{A} l_2 \leq \pi^2. \tag{4.3.12}$$

We shall prove that

$$|Q_2| + |Q_4| + \cdots + |Q_{N+1}| < |\mathcal{F}' \setminus \mathcal{F}|. \tag{4.3.13}$$

The estimates (4.3.12) and (4.3.13) yield

$$\begin{aligned}
|\mathcal{F}| &= |\mathcal{F} \setminus \mathcal{F}'| + |\mathcal{F} \cap \mathcal{F}'| \\
&< \pi^2 + |\mathcal{F}' \setminus \mathcal{F}| + |\mathcal{F}' \cap \mathcal{F}| \\
&= \pi^2 + |\mathcal{F}'|.
\end{aligned}$$

Therefore,

$$\pi^2 + \frac{\pi a}{2} = \left| P\left(\pi, \frac{a}{2} + \pi\right) \right| = |\mathcal{F}| < \pi^2 + |\mathcal{F}'| = \pi^2 + |T(A, \pi)| = \pi^2 + \frac{\pi A}{2},$$

i.e., $a < A$. Since the function f_{EB} is monotone increasing, we conclude that $f_{\mathrm{EB}}(a) \leq f_{\mathrm{PB}}\left(\frac{a}{2} + \pi\right)$ as claimed.

In order to prove the estimate (4.3.13) we denote the "triangles" in $T^4(f_{\mathrm{EB}}(A)) \setminus \mathcal{F}$ of height and width 2π by T_i, $i = 1, 2, \ldots$, and associate with each rectangle Q_i in $T'_{j(i)}$, $i = 2, 4, \ldots, N+1$, the rectangle

$$R_i \subset \mathcal{F}' \cap T_{i/2} \subset \mathcal{F}' \setminus \mathcal{F}$$

whose width $w(R_i)$ is equal to the height $h(Q_i)$ of Q_i and whose height $h(R_i)$ is $2\pi - h(Q_i)$, cf. Figure 4.11. Since the width $w(Q_i)$ of Q_i is

$$2l_{2j(i)-1} - w(R_i) = 2l_{2j(i)-1} - h(Q_i)$$

we find together with the inequalities (4.3.11) that

$$\begin{aligned}
|Q_i| &= w(Q_i) h(Q_i) \\
&= \left(2l_{2j(i)-1} - h(Q_i)\right) h(Q_i) \\
&< (2\pi - h(Q_i)) h(Q_i) \\
&= h(R_i) w(R_i) \\
&= |R_i|.
\end{aligned}$$

The estimate (4.3.13) thus follows, and so the proof of Lemma 4.3.13 is complete. $\square$

Proposition 4.3.7 follows from Lemma 4.3.13 and Proposition 4.3.9. Indeed, in view of Proposition 4.3.9 the function

$$d(a) := f_{\mathrm{PB}}\left(\frac{a}{2} + \pi\right) - \sqrt{\pi a}$$

on $]2\pi, \infty[$ has its local maxima at

$$a_k = 2(2(k^2 - k + 1) - 1)\pi, \quad k = 1, 2, \ldots,$$

where

$$d(a_k) = (2k + 1)\pi - \sqrt{\pi a_k}.$$

The sequence $d(a_k)$ is monotone increasing to 2π. Together with Lemma 4.3.13 we conclude that

$$f_{\mathrm{EB}}(a) - \sqrt{\pi a} \le f_{\mathrm{PB}}\left(\frac{a}{2} + \pi\right) - \sqrt{\pi a} \le 2\pi$$

and so the proof of Proposition 4.3.7 is complete. $\qquad\square$

4.4 Embeddings into cubes

In this section we use multiple symplectic folding to construct symplectic embeddings of four dimensional ellipsoids and polydiscs into four dimensional cubes. While in Section 2.3 the lack of convenient invariants made it impossible to get good rigidity results for embeddings into cubes, multiple symplectic folding provides us with rather satisfactory flexibility results.

4.4.1 Embedding ellipsoids into cubes. Fix $a > \pi$. We think of the ellipsoid $E(\pi, a)$ as $T(a, \pi)$ and of the cube $C^4(A)$ as $R(A) \times D(A)$. In order to find the smallest A for which $T(a, \pi)$ embeds into $R(A) \times D(A)$ via multiple symplectic folding, we proceed as follows. We fix $u_1 \in]0, a[$ and fold at u_1. By the Folding Lemma 4.2.1 (i),

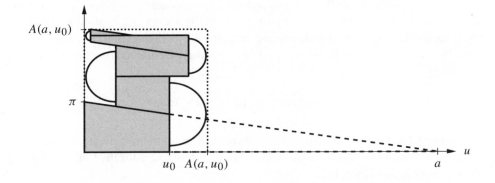

Figure 4.12. The optimal embedding $E(\pi, a) \hookrightarrow C^4(A)$ for $a = 7\pi$.

the stairs S_1 are contained in $\{u < w(a, u_1)\}$, where

$$w(a, u_1) = u_1 + l_1 = \pi + \left(1 - \frac{\pi}{a}\right) u_1. \qquad (4.4.1)$$

We then choose u_i, $i \ge 2$, maximal with respect to the condition of staying inside $\{0 < u < w(a, u_1)\}$. If the remainder $r_1 = a - u_1$ is smaller than u_1, we obtain an embedding of $T(a, \pi)$ into $\{0 < u < w(a, u_1)\}$ by folding once. If $r_1 \ge u_1$, we are forced to fold a second time. The same discussion as in 4.3.1 shows that the only condition for doing so is $l_1 < u_1$, i.e.,

$$u_1 > \frac{a\pi}{a + \pi}. \tag{4.4.2}$$

If this condition is met, we define u_2 by formula (4.3.3). The sequence $l_i, i = 1, 2, \ldots,$ of the widths of the stairs S_i is decreasing. Hence, there are no further conditions at the subsequent folds, and if there is an i'th fold, then $u_i > u_{i-1}$ for all $i \geq 3$. Under the condition (4.4.2), our procedure therefore embeds $T(a, \pi)$ into $\{0 < u < w(a, u_1)\}$ by folding finitely many times. We denote the number of folds needed by $N(u_1)$. Recall that $l_i = l_i(a, u_1)$ is the width of the stairs S_i as well as the minimal height of the floor F_i as well as the maximal height of $F_{i+1}, i = 1, \ldots, N(u_1)$. We set $l_i(a, u_1) = 0$ if $i > N(u_1)$. For $a > \pi$ fixed and $u_1 \in \left]\frac{a\pi}{a+\pi}, a\right[$, the functions $l_i(a, u_1)$ are decreasing continuous functions of u_1. Therefore, the height of the image

$$h(a, u_1) = \sum_{i=1}^{N(u_1)} l_i(a, u_1)$$

is a decreasing continuous function of u_1 for $u_1 \in \left]\frac{a\pi}{a+\pi}, a\right[$. For $u_1 \geq a/2$ we have $h(a, u_1) = \pi$, and for $u_1 \searrow a\pi/(a + \pi)$ we have $N(u_1) \to \infty$ and $u_i \to 0$, $l_i(a, u_1) \to l_1(a, u_1)$ for all $i \geq 2$, and so $h(a, u_1) \to \infty$ as $u_1 \searrow a\pi/(a+\pi)$. On the other hand, $w(a, u_1) = \pi + (1 - \pi/a)u_1$ is a strictly increasing continuous function of u_1 such that $w(a, 0) = \pi$ and $w(a, a) = a$. It follows that $w(a, u_1) = h(a, u_1)$ for exactly one $u_1 \in \left]\frac{a\pi}{a+\pi}, a\right[$, which we call $u_0 = u_0(a)$. Define the function $f_{\mathrm{EC}}(a)$ on $]\pi, \infty[$ by

$$f_{\mathrm{EC}}(a) = \pi + \left(1 - \frac{\pi}{a}\right)u_0(a),$$

cf. Figure 4.13. We have shown that the ellipsoid $E(\pi, a)$ symplectically embeds into the cube $C^4(f_{\mathrm{EC}}(a) + \epsilon)$ for any $\epsilon > 0$. A computer program for the function $f_{\mathrm{EC}}(a)$ is presented in Appendix D.2.

Our procedure is optimal in the sense that we cannot embed $E(\pi, a)$ into a cube smaller than $C^4(f_{\mathrm{EC}}(a))$ by multiple symplectic folding. This follows from an argument similar to the one given in 4.3.1. Indeed, our procedure can equivalently be described as follows: For each $A \in]\pi, a[$ we successively choose $u_1, u_2, \ldots$ maximal with respect to the condition of staying inside $\{0 < u < A\}$. Then $f_{\mathrm{EC}}(a)$ is the smallest A for which we do not run into an embedding obstruction and for which the height of the image is smaller than A. The only way of improving our procedure is therefore to choose some of the u_i smaller. So let $A < f_{\mathrm{EC}}(a)$, and choose $u_i' \leq u_i$. We then either run into an embedding obstruction, or the height of the image of the modified embedding is larger than $h(a, u_1)$ and hence larger than A.

Remarks 4.4.1. 1. We are going to investigate the function $f_{EC}(a)$ in more detail.

We have $N(u_1) = 1$ if and only if $u_1 \in [a/2, a[$. Then $h(a, u_1) = \pi < \pi + (1 - \pi/a)u_1 = w(a, u_1)$, and so $w(a, a/2) = (a + \pi)/2$, and $h(a, u_1) < w(a, u_1)$ for all $u_1 \in [a/2, a[$. It follows that by folding once we can embed $E(\pi, a)$ into $C^4\left(\frac{a+\pi}{2} + \epsilon\right)$ for any $\epsilon > 0$, and that folding only once never yields an optimal embedding, i.e., $f_{EC}(a) < (a + \pi)/2$ for all $a > \pi$.

We have $N(u_1) = 2$ if and only if $u_1 < a/2$ and $l_2 + u_3 = l_2 + (a/\pi)l_2 \le w(a, u_1)$. Using the formulas (4.4.1), (4.2.1) and (4.3.3) for w, l_2 and u_2, we find that the second inequality is equivalent to the condition on u_1

$$\frac{a(a^2 + \pi^2)}{3a^2 + \pi^2} \le u_1. \tag{4.4.3}$$

If $N(u_1) = 2$, then $h(a, u_1) = 2l_1 + l_2$. Plugging the identity $u_2 = u_1 - l_2$ into the identity $a = u_1 + u_2 + u_3 = u_1 + u_2 + (a/\pi)l_2$ and solving for l_2 we find

$$h(a, u_1) = 2\pi - \frac{2\pi}{a}u_1 + \frac{\pi(a - 2u_1)}{a - \pi}.$$

The equation $h = w$ thus yields

$$u_0(a) = \frac{a\pi(2a - \pi)}{a^2 + 2a\pi - \pi^2} \tag{4.4.4}$$

provided that $u_0(a)$ meets condition (4.4.2), that $u_0(a) < a/2$, and that $u_0(a)$ meets condition (4.4.3). We compute that $u_0(a)$ meets condition (4.4.2) and that $u_0(a) < a/2$ whenever $a > \pi$, and that $u_0(a)$ meets condition (4.4.3) if and only if $\pi \le a \le 3\pi$. It follows that (4.4.4) holds for all $a \in]\pi, 3\pi]$.

In fact, the identity (4.4.4) also holds true for all those a for which the optimal embedding of $T(a, \pi)$ obtainable by multiple folding is a 3-fold for which the height is still $h = 2l_1 + l_2$, i.e., for which $u_4(u_0(a)) \le u_3(u_0(a))$. The largest a for which (4.4.4) holds true is characterized by the identity $u_4(u_0(a)) = u_3(u_0(a))$. Using the identity $u_3 = \frac{a+\pi}{a-\pi}u_2$ we compute that the equation $a = u_0(a) + u_2(u_0(a)) + 2u_3(u_0(a))$ translates into

$$a = \frac{a\pi(5a^2 - 2a\pi + \pi^2)}{(a - \pi)(a^2 + 2a\pi - \pi^2)},$$

i.e., $a = \left(2 + \sqrt{5}\right)\pi$. Therefore,

$$f_{EC}(a) = \frac{a\pi(3a - \pi)}{a^2 + 2a\pi - \pi^2} \quad \text{for } \pi < a \le \left(2 + \sqrt{5}\right)\pi.$$

In general, $f_{EC}(a)$ is a piecewise rational function. Its singularities are those a for which $u_{N(a)}(u_0(a)) = u_{N(a)+1}(u_0(a))$, $N(a) = 3, 5, 7, \ldots$, and those a for which the "endpoint" of $F_{N(a)+1}$ touches the "axis" $\{u = 0\}$; here, we set $N(a) = N(u_0(a))$.

Denoting the two sets of singularities by $a_3, a_5, a_7, \ldots$ and $a_4, a_6, a_8, \ldots$, we have that the singular set of f_{EC} is the strictly increasing, diverging sequence (a_k), $k \geq 3$.

2. Let $d_{EC}(a) = f_{EC}(a) - \sqrt{\pi a / 2}$ be the difference between f_{EC} and the volume condition. Set $a_2 = \pi$. Computer calculations suggest that the function d_{EC} attains exactly one local maximum M_k between a_{2k} and a_{2k+1}, that d_{EC} attains its local minima m_k at a_{2k+1}, and that d_{EC} strictly increases between a_{2k+1} and a_{2k+2}, $k \geq 1$. Moreover, they suggest that both (m_k) and (M_k) are strictly increasing and converging to $(2/3)\pi$. In particular, we seem to have $\lim_{a \to \infty} f_{EC}(a) - \sqrt{\pi a / 2} = (2/3)\pi$. $\diamond$

As in the case of the function f_{EB} studied in 4.3.1 we do not know how to analyze the asymptotic behaviour of the function $f_{EC}(a)$ as $a \to \infty$ directly. We shall prove the following proposition at the end of the subsequent section by comparing the optimal multiple folding embedding of $E(\pi, a)$ into a cube with the optimal multiple folding embedding of the polydisc $P(\pi, \frac{a}{2} + \pi)$ into a cube.

Proposition 4.4.2. *We have*

$$f_{EC}(a) - \sqrt{\frac{\pi a}{2}} \leq \frac{3}{2} \pi \quad \text{for all } a > 2\pi.$$

Extend the function $f_{EC}(a)$ to a function on $[\pi, \infty[$ by setting $f_{EC}(\pi) = \pi$. The problem considered in this section was to understand the characteristic function χ_{EC} on $[\pi, \infty[$ defined by

$$\chi_{EC}(a) = \inf \left\{ A \mid E(\pi, a) \text{ symplectically embeds into } C^4(A) \right\}.$$

The following proposition summarizes what we know about this function.

Proposition 4.4.3. *The function* $\chi_{EC} \colon [\pi, \infty[\to \mathbb{R}$ *is bounded from below and above by*

$$\max \left(\pi, \sqrt{\frac{\pi a}{2}} \right) \leq \chi_{EC}(a) \leq f_{EC}(a),$$

see Figure 4.13. *It is monotone increasing and hence almost everywhere differentiable. Moreover,* χ_{EC} *is Lipschitz continuous with Lipschitz constant at most* 1; *more precisely,*

$$\chi_{EC}(a') - \chi_{EC}(a) \leq \frac{f_{EC}(a)}{a} (a' - a) \quad \text{for all } a' \geq a \geq \pi.$$

Proof. In view of the monotonicity axiom for the first Ekeland–Hofer capacity, the identities (2.2.3) and (2.3.1) imply $\chi_{EC}(a) \geq \pi$ for all $a \geq \pi$, and the volume condition $|E(\pi, a)| \leq |C^4(\chi_{EC}(a))|$ translates to $\chi_{EC}(a) \geq \sqrt{\pi a / 2}$. The first claim thus follows. The remaining claims follow as in the proof of Proposition 4.3.11. $\square$

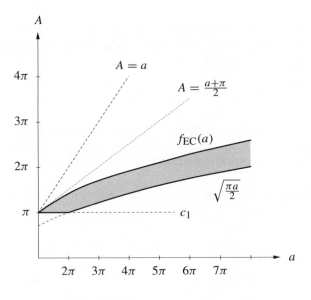

Figure 4.13. What is known about $\chi_{EC}(a)$.

4.4.2 Embedding polydiscs into cubes. Fix $a > \pi$. We think of the polydisc $P(\pi, a)$ as $R(a) \times D(\pi)$ and of the cube $C^4(A)$ as $R(A) \times D(A)$. In order to find the smallest A for which $R(a) \times D(\pi)$ embeds into $R(A) \times D(A)$ via multiple symplectic folding, we proceed as follows. We fix $u_1 \in {]}0, a[$ and fold at u_1. By the Folding Lemma 4.2.1 (i), the stairs S_1 are contained in $\{u < u_1 + \pi\}$. We then choose $u_i, i \geq 2$, maximal with respect to the condition of staying inside $\{0 < u < u_1 + \pi\}$.

The Folding Lemma 4.2.1 (ii) shows that the only condition for folding a second time, if necessary, is $u_1 > \pi$. For $N \geq 2$, the only condition on u_1 for folding an N'th time, if necessary, is $u_1 > \pi$.

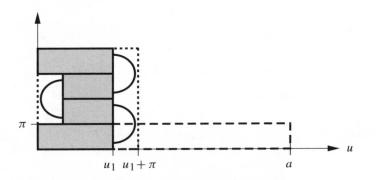

Figure 4.14. Folding $P(\pi, a)$ three times.

We say that u_1 is *admissible* if $u_1 \geq a/2$ or $u_1 > \pi$. It follows that if u_1 is admissible, then our procedure embeds $R(a) \times D(\pi)$ into $R(A(a, u_1)) \times D(A(a, u_1))$ by a finite number $N(u_1)$ of folds. Here,

$$A(a, u_1) = \max \{u_1 + \pi, (N(u_1) + 1)\pi\}. \tag{4.4.5}$$

Let $u_2, \ldots, u_{N(u_1)+1}$ be the lengths associated with some admissible u_1. Choosing some of the u_i, $i = 2, \ldots, N(u_1)$, smaller would lead to an embedding by folding at least $N(u_1)$ times. It is therefore enough to compare embeddings obtained from our procedure.

Assume that $a \in]\pi, 2\pi]$. Then $A(a, u_1) \geq 2\pi$ for every admissible u_1. It follows that for $a \in]\pi, 2\pi]$ multiple symplectic folding does not provide a better embedding result than the inclusion $P(\pi, a) \hookrightarrow C^4(a)$.

So assume $a > 2\pi$. Suppose that u_1 is admissible and that $N := N(u_1)$ is even. We claim that if the last floor F_{N+1} does not touch $\{u = u_1 + \pi\}$, then there exists an admissible u_1' such that $u_1' < u_1$, $N(u_1') = N$ and F_{N+1}' touches $\{u = u_1' + \pi\}$. Indeed, shrinking u_1 either leads to a u_1' as claimed or to $u_1' = \pi$. In the second case, however, we find $a \leq 2\pi$, a contradiction. We may therefore assume that F_{N+1} touches $\{u = u_1 + \pi\}$. A similar argument shows that we may also assume that $F_{N(u_1)+1}$ touches $\{u = 0\}$ if $N(u_1)$ is odd.

The optimal embedding obtainable by folding only once is therefore described by $A_1(a) = \max\{\frac{a}{2} + \pi, 2\pi\} = \frac{a}{2} + \pi$, and if the number of folds is $N \geq 2$, we read off from Figure 4.14 that

$$a = 2\pi + (N + 1)(u_1 - \pi)$$

provided that $u_1 > \pi$. Since $a > 2\pi$, we see that this condition is met. Solving for u_1 and using formula (4.4.5) we then find that the optimal embedding of $R(a) \times D(\pi)$ into a cube obtainable by folding N times is described by

$$A_N(a) = \max\left\{\frac{a + 2N\pi}{N+1}, (N+1)\pi\right\}.$$

Observe that this formula also holds true for $N = 1$. Define the function $f_{PC}(a)$ on $]2\pi, \infty[$ by

$$f_{PC}(a) = \min\{A_N(a) \mid N = 1, 2, \ldots\},$$

cf. Figure 7.2. We in particular have proved

Proposition 4.4.4. *Assume $a > 2\pi$. Then the polydisc $P(\pi, a)$ symplectically embeds into the cube $C^4(f_{PC}(a) + \epsilon)$ for every $\epsilon > 0$, where*

$$f_{PC}(a) = \begin{cases} (N+1)\pi & \text{if } (N-1)N + 2 < \frac{a}{\pi} \leq N^2 + 1, \\ \frac{a+2N\pi}{N+1} & \text{if } N^2 + 1 < \frac{a}{\pi} \leq N(N+1) + 2. \end{cases}$$

The function $f_{PC}(a)$ is compared with the result yielded by symplectic wrapping in 7.2.2.1.

We end this section by deriving Proposition 4.4.2 from Proposition 4.4.4.

Proof of Proposition 4.4.2. We proceed as in the proof of Proposition 4.3.7.

Lemma 4.4.5. *We have* $f_{EC}(a) < f_{PC}\left(\frac{a}{2} + \pi\right)$ *for all* $a > 2\pi$.

Proof. As in 4.4.1 we think of the ellipsoid $E(\pi, a)$ as the trapezoid $T(a, \pi)$. We fix $a > 2\pi$ and let

$$\mathcal{F} = \coprod_{i=1}^{N+1} F_i \cup \coprod_{i=1}^{N} S_i$$

be the image of the "optimal" embedding of $P\left(\pi, \frac{a}{2} + \pi\right)$ into $C^4\left(f_{PC}\left(\frac{a}{2} + \pi\right)\right)$. For $N = 3$ the set $\mathcal{F}$ looks as in Figure 4.14. We define $A \in \,]2\pi, \infty[$ as the unique real number for which

$$f_{EC}(A) = f_{PC}\left(\frac{a}{2} + \pi\right)$$

and we let

$$\mathcal{F}' = \coprod_{i=1}^{N'+1} F_i' \cup \coprod_{i=1}^{N'} S_i'$$

be the image of the "optimal" embedding of $T(A, \pi)$ into $C^4(f_{EC}(A))$, cf. Figure 4.12.

As in the proof of Lemma 4.3.13 we shall neglect the stairs of $\mathcal{F}$ and $\mathcal{F}'$. Recall from 4.4.1 that l_i denotes the minimal height of the floor F_i' and that

$$\pi > l_1 > l_2 > \cdots .$$

This and the Folding Lemma 4.2.1 imply that

$$\mathcal{F} \setminus \mathcal{F}' = F_1 \setminus \mathcal{F}' \cup F_{N+1} \setminus \mathcal{F}'.$$

The set $Q_1 := F_1 \setminus \mathcal{F}' = F_1 \setminus \left(F_1' \cup F_2'\right)$, which is analogous to the set Q_1 in Figure 4.11, has volume

$$|Q_1| = \frac{1}{2} l_2 \frac{\pi}{A} l_2.$$

We decompose the set $Q_0 := F_{N+1} \setminus \mathcal{F}'$ into the sets Q_0' and $Q_0'' = Q_0 \setminus Q_0'$, where

$$Q_0' := \begin{cases} \{(u, v, x, y) \in Q_0 \mid u < l_1\} & \text{if } N \text{ is odd,} \\ \{(u, v, x, y) \in Q_0 \mid u > f_{EC}(A) - l_1\} & \text{if } N \text{ is even.} \end{cases}$$

Using the definition (4.2.1) of l_1 and the estimate (4.4.2) we find

$$l_2 < l_1 = \pi - \frac{\pi}{A} u_1 < \pi - \frac{\pi^2}{A + \pi}$$

and so

$$|Q'_0| + |Q_1| \le l_1 \pi + \frac{1}{2} l_2 \frac{\pi}{A} l_2 \le \pi^2. \tag{4.4.6}$$

We shall prove that

$$|Q''_0| < |\mathcal{F}' \setminus \mathcal{F}|. \tag{4.4.7}$$

The estimates (4.4.6) and (4.4.7) and the same argument as in the proof of Lemma 4.3.13 yield $a < A$. Since the function f_{EC} is monotone increasing, we conclude that $f_{EC}(a) \le f_{PC}\left(\frac{a}{2} + \pi\right)$ as claimed.

We are left with proving the estimate (4.4.7). The length of the set Q''_0 is $f_{EC}(A) - \pi - l_1$. We assume first that N' is even. Recall that $l_{N'}$ is the height of the floor $F'_{N'+1}$. Since the length of $F'_{N'+1}$ is at most $f_{EC}(A) - l_{N'}$, the height of Q''_0 is at most $\frac{\pi}{A}(f_{EC}(A) - l_{N'})$. Therefore,

$$|Q''_0| \le (f_{EC}(A) - \pi - l_1) \frac{\pi}{A} (f_{EC}(A) - l_{N'}). \tag{4.4.8}$$

Let R be the union of maximal "rectangles" in $\mathcal{F}' \setminus \mathcal{F}$ based over

$$\{(u, v) \mid f_{EC}(A) - \pi \le u < f_{EC}(A) - l_1\}.$$

If N is odd, R has one component, whose height is at least $f_{EC}(A) - l_{N'}$. If N is even, R has one or two components, whose total height is at least $f_{EC}(A) - \pi - l_{N'}$. Together with $\pi - l_1 = \frac{\pi}{A}(f_{EC}(A) - l_1)$ we conclude that

$$|\mathcal{F}' \setminus \mathcal{F}| > |R| \ge \frac{\pi}{A} (f_{EC}(A) - l_1)(f_{EC}(A) - \pi - l_{N'}). \tag{4.4.9}$$

Since $l_1 > l_{N'}$, the right hand side in (4.4.9) is larger than the one in (4.4.8), and so the estimate (4.4.7) follows. Assume now that N' is odd. Then the above argument with $l_{N'}$ replaced by $l_{N'-1}$ goes through. The estimate (4.4.7) is thus proved, and so the proof of Lemma 4.4.5 is complete. $\qquad\square$

Proposition 4.4.2 follows from Lemma 4.4.5 and Proposition 4.4.4. Indeed, in view of Proposition 4.4.4 the function

$$d(a) := f_{PC}\left(\frac{a}{2} + \pi\right) - \sqrt{\frac{\pi a}{2}}$$

on $]2\pi, \infty[$ has its local maxima at

$$a_N = 2(N^2 - N + 1)\pi, \quad N = 1, 2, \ldots,$$

where

$$d(a_N) = \left(N + 1 - \sqrt{N^2 - N + 1}\right)\pi.$$

The sequence $d(a_N)$ is monotone increasing to $\frac{3}{2}\pi$. Together with Lemma 4.4.5 we conclude that

$$f_{EC}(a) - \sqrt{\frac{\pi a}{2}} < f_{PC}\left(\frac{a}{2} + \pi\right) - \sqrt{\frac{\pi a}{2}} \leq \frac{3}{2}\pi$$

and so the proof of Proposition 4.4.2 is complete. $\square$

Chapter 5

Symplectic folding in higher dimensions

Even though symplectic folding is a four dimensional process, we can use it to prove interesting symplectic embedding results in higher dimensions as well. The reason is that we can fold into $n - 1$ different symplectic directions of the $(2n - 2)$-dimensional fibre over the 2-dimensional symplectic base. We will concentrate on embedding skinny polydiscs into cubes and skinny ellipsoids into balls. The results of this chapter will be used in Chapter 6 to prove Theorem 3.

5.1 Four types of folding

In Chapter 4 we folded on the right and on the left into the y-direction. In the multiple folding procedures considered in this chapter we shall also fold into the $(-y)$-direction. Hence there will be four types of folding. This section reviews these four types. As usual, we shall neglect those terms in the constructions which can be chosen arbitrarily small.

We define $F \subset \mathbb{R}^4$ by

$$F := \{\, (u, v, x, y) \in \mathbb{R}^4 \mid u \in \mathbb{R},\ 0 < v < 1,\ 0 < x < 1,\ 0 < y < \pi \,\}.$$

Fix a "folding point" $u_1 \in \mathbb{R}$ and choose a (right-)cut off function $c_r \colon \mathbb{R} \to [0, 1]$ with support $[u_1, u_1 + \pi]$ and a (left-)cut off function $c_l \colon \mathbb{R} \to [0, 1]$ with support $[u_1 - \pi, u_1]$.

1. Folding on the right into the y-direction. We fold F on the right at $u = u_1$ into the y-direction by applying the symplectic map

$$\phi_{r+} := (\gamma_1 \times \mathrm{id}) \circ \varphi_1 \circ (\beta_1 \times \mathrm{id}).$$

Here, the maps β_1 and γ_1, which are constructed the same way as the maps β and γ in Step 1 and Step 4 of Section 3.2, are explained in the first column of Figure 5.1, and the lifting map φ_1 is defined by

$$\varphi_1(u, v, x, y) = \left(u, x, v + c_r(u)x, y + \int_{u_1}^{u} c_r(s)\, ds \right).$$

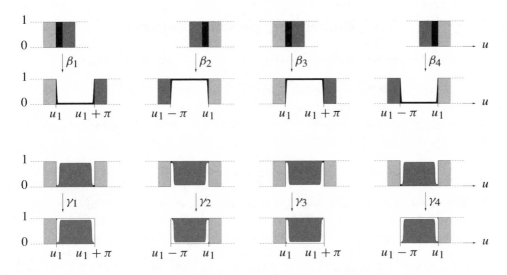

Figure 5.1. The maps β_i and γ_i, $i = 1, 2, 3, 4$.

2. Folding on the left into the y-direction. We fold F on the left at $u = u_1$ into the y-direction by applying the map

$$\phi_{l+} := (\gamma_2 \times \mathrm{id}) \circ \varphi_2 \circ (\beta_2 \times \mathrm{id}).$$

Here, the maps β_2 and γ_2 are explained in the second column of Figure 5.1, and

$$\varphi_2(u, v, x, y) = \left(u, x, v - c_l(u)x, y + \int_u^{u_1} c_l(s)\, ds \right).$$

3. Folding on the right into the $(-y)$-direction. We fold F on the right at $u = u_1$ into the $(-y)$-direction by applying the map

$$\phi_{r-} := (\gamma_3 \times \mathrm{id}) \circ \varphi_3 \circ (\beta_3 \times \mathrm{id}).$$

Here, the maps β_3 and γ_3 are explained in the third column of Figure 5.1, and

$$\varphi_3(u, v, x, y) = \left(u, x, v - c_r(u)x, y - \int_{u_1}^u c_r(s)\, ds \right).$$

4. Folding on the left into the $(-y)$-direction. We fold F on the left at $u = u_1$ into the $(-y)$-direction by applying the map

$$\phi_{l-} := (\gamma_4 \times \mathrm{id}) \circ \varphi_4 \circ (\beta_4 \times \mathrm{id}).$$

Here, the maps β_4 and γ_4 are explained in the fourth column of Figure 5.1, and

$$\varphi_4(u, v, x, y) = \left(u, x, v + c_l(u)x, y - \int_u^{u_1} c_l(s)\, ds \right).$$

5.2   Embedding polydiscs into cubes

In this section we shall study symplectic embeddings of skinny polydiscs

$$P^{2n}(\pi, \ldots, \pi, a) = D(\pi) \times \cdots \times D(\pi) \times D(a)$$

into cubes $C^{2n}(A)$ for $n \geq 2$. As before, we shall work with more convenient shapes. Define the rectangle $R(a, b)$ by

$$R(a, b) = \{(x, y) \mid 0 < x < a, \ 0 < y < b\}.$$

We denote the $2n$-dimensional set $R(a, 1) \times R(1, b) \times \cdots \times R(1, b)$ by

$$R^n(a, b) = R(a, 1) \times R(1, b) \times \cdots \times R(1, b).$$

If $b = a$, we abbreviate $R^n(a) = R^n(a, a)$. In view of Lemma 3.1.5, the disc $D(a)$ is symplectomorphic to $R(a, 1)$ and the disc $D(\pi)$ is symplectomorphic to $R(1, \pi)$. Therefore, the polydisc $P^{2n}(\pi, \ldots, \pi, a)$, which is symplectomorphic to $P^{2n}(a, \pi, \ldots, \pi)$, is symplectomorphic to $R^n(a, \pi)$. Similarly, the cube $C^{2n}(A)$ is symplectomorphic to $R^n(A)$. The symplectic coordinates will be denoted by

$$(u, v, x_2, y_2, \ldots, x_n, y_n) \equiv (z_1, z_2, \ldots, z_n) \in \mathbb{R}^{2n}$$

where we set again $(u, v) = (x_1, y_1)$. We abbreviate $x = (x_2, \ldots, x_n)$ and $y = (y_2, \ldots, y_n)$ as well as

$$1_y = (1, \ldots, 1) \in \mathbb{R}^{n-1}(y).$$

We shall associate with each triple $a > 2\pi$, $N \in \mathbb{N}$, $n \geq 2$ a symplectic embedding procedure

$$\phi_N^n(a) \colon R^n(a, \pi) \hookrightarrow \mathbb{R}^{2n}.$$

In the following description we again neglect the arbitrarily small δ-terms appearing in the actual construction.

If $n = 2$, we proceed as in 4.4.2, cf. Figure 4.14: For $a > 2\pi$ and $N \in \mathbb{N}$ we define $u_1 = u_{1,N}^2(a)$ by

$$a = 2\pi + (N + 1)(u_1 - \pi). \tag{5.2.1}$$

Then $u_1 > \pi$. We first fold $R^2(a, \pi)$ on the right into the y_2-direction by applying the map

$$(z_1, z_2) \mapsto \phi_{r+}(z_1, z_2)$$

at $u = u_1$. Here, ϕ_{r+} is the restriction to $R^2(a, \pi)$ of the map ϕ_{r+} introduced in 5.1.1. If $N = 1$, the embedding procedure $\phi_N^2(a)$ terminates at this point. Indeed, in view of definition (5.2.1), "a is used up" and the front face of the second floor of the image touches the hyperplane $\{u = 0\}$. If $N \geq 2$, the inequality $u_1 > \pi$ implies that we can then fold the second floor

$$\left\{ (u, v, x_2, y_2) \in \phi_{r+}(R^2(a, \pi)) \mid u < u_1, \; y_2 > \pi \right\}$$

of the image on the left into the y_2-direction by applying the map

$$(z_1, z_2) \mapsto \phi_{l+}(z_1, z_2)$$

to the second floor at $u = \pi$. Going on this way, we altogether fold N times into the y_2-direction by alternatingly folding the last floor of the image on the right at $u = u_1$ and on the left at $u = \pi$. At this point the embedding procedure $\phi_N^2(a)$ terminates. Indeed, in view of definition (5.2.1), "a is used up" and the front face of the last floor of the image touches the hyperplane $\{u = u_1 + \pi\}$ if N is even and the hyperplane $\{u = 0\}$ if N is odd.

If $n = 3$, the embedding procedure $\phi_N^3(a)$ can be visualized by Figure 5.2. For $a > 2\pi$ and $N \in \mathbb{N}$ we define $u_1 = u_{1,N}^3(a)$ by

$$a = 2\pi + (N + 1)^2 (u_1 - \pi). \tag{5.2.2}$$

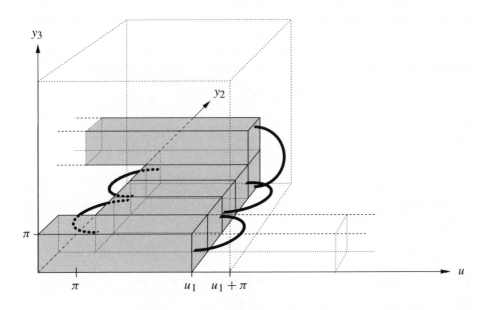

Figure 5.2. The first 5 folds of an embedding $\phi_N^3(a) \colon R^3(a, \pi) \hookrightarrow \mathbb{R}^{2n}$ with $N = 4$.

Then $u_1 > \pi$. Set $a' = 2\pi + (N + 1)(u_1 - \pi)$. The first N folds of the embedding procedure $\phi_N^3(a)$ yield a symplectic embedding of $R^3(a, \pi)$ into $\mathbb{R}^6$ whose restriction to $R^3(a', \pi)$ is

$$\phi_N^2(a') \times \mathrm{id} \colon R^2(a', \pi) \times R(1, \pi) \hookrightarrow \mathbb{R}^4 \times R(1, \pi).$$

We next fold once into the y_3-direction. If N is even, we do this by applying the map

$$(z_1, z_2, z_3) \mapsto (z_1', z_2', z_3') \quad \text{where } (z_1', z_3') = \phi_{r+}(z_1, z_3) \text{ and } z_2' = z_2$$

to the $N+1$'st floor of the image at $u = u_1$, and if N is odd, we do this by applying the map

$$(z_1, z_2, z_3) \mapsto (z_1', z_2', z_3') \quad \text{where } (z_1', z_3') = \phi_{l+}(z_1, z_3) \text{ and } z_2' = z_2$$

to the $N+1$'st floor of the image at $u = \pi$, see Figure 5.2. We then fold the part of the image on which $y_3 > \pi$ exactly N times into the $(-y_2)$-direction by using restrictions of the maps ϕ_{r-} and ϕ_{l-} and thereby fill a second z_1-z_2-layer. If $N = 1$, the embedding procedure $\phi_N^3(a)$ terminates at this point. Indeed, in view of definition (5.2.2), "a is used up" and the front face of the last floor of the image touches the hyperplane $\{u = 0\}$. If $N \geq 2$, we fold a second time into the y_3-direction, and fill a third z_1-z_2-layer. Going on this way, we altogether fold $(N+1)^2 - 1$ times, in which we fold N times into the y_3-direction, and thereby fill $N+1$ z_1-z_2-layers. At this point the embedding procedure $\phi_N^3(a)$ terminates. Indeed, in view of definition (5.2.2), "a is used up" and the front face of the last floor of the image touches the hyperplane $\{u = u_1 + \pi\}$ if N is even and the hyperplane $\{u = 0\}$ if N is odd.

In order to describe the embedding procedure $\phi_N^n(a)$ for $n \geq 4$, we proceed by induction and assume that we have described the symplectic embeddings

$$\phi_N^{n-1}(a') \colon R^{n-1}(a', \pi) \hookrightarrow \mathbb{R}^{2n-2}, \quad a' > 2\pi.$$

We define $u_1 = u_{1,N}^n(a)$ by

$$a = 2\pi + (N+1)^{n-1}(u_1 - \pi). \tag{5.2.3}$$

Then $u_1 > \pi$. Set $a' = 2\pi + (N+1)^{n-2}(u_1 - \pi)$. The first $(N+1)^{n-2} - 1$ folds of the embedding procedure $\phi_N^n(a)$ yield a symplectic embedding of $R^n(a, \pi)$ into $\mathbb{R}^{2n}$ whose restriction to $R^n(a', \pi)$ is

$$\phi_N^{n-1}(a') \times \mathrm{id} \colon R^{n-1}(a', \pi) \times R(1, \pi) \hookrightarrow \mathbb{R}^{2n-2} \times R(1, \pi).$$

We next fold once into the y_n-direction. If N is even, we do this by applying the map

$$(z_1, \ldots, z_n) \mapsto (z_1', \ldots, z_n'), \quad (z_1', z_n') = \phi_{r+}(z_1, z_n), \ z_i' = z_i, \ i = 2, \ldots, n-1,$$

to the last floor of the image at $u = u_1$, and if N is odd, we do this by applying the map

$$(z_1, \ldots, z_n) \mapsto (z_1', \ldots, z_n'), \quad (z_1', z_n') = \phi_{l+}(z_1, z_n), \ z_i' = z_i, \ i = 2, \ldots, n-1,$$

to the last floor of the image at $u = \pi$. We then fill a second z_1-$\cdots$-z_{n-1}-layer by folding the part of the image on which $y_n > \pi$ exactly $(N+1)^{n-2} - 1$ times. If $N = 1$, the embedding procedure $\phi_N^n(a)$ terminates at this point in view of definition (5.2.3). If $N \geq 2$, we fold a second time into the y_n-direction, and fill a third z_1-$\cdots$-z_{n-1}-layer. Going on this way, we altogether fold $(N+1)^{n-1} - 1$ times, in which we fold N times into the y_n-direction, and thereby fill $N + 1$ z_1-$\cdots$-z_{n-1}-layers. At this point the embedding procedure $\phi_N^n(a)$ terminates in view of definition (5.2.3).

The following proposition generalizes Proposition 4.4.4 to arbitrary dimension.

Proposition 5.2.1. *Assume $a > 2\pi$. Then the polydisc $P^{2n}(\pi, \ldots, \pi, a)$ symplectically embeds into the cube $C^{2n}(f_{PC}^{2n}(a) + \epsilon)$ for every $\epsilon > 0$, where*

$$f_{PC}^{2n}(a) = \begin{cases} (N+1)\pi, & (N-1)N^{n-1} < \frac{a}{\pi} - 2 \leq (N-1)(N+1)^{n-1}, \\ \frac{a-2\pi}{(N+1)^{n-1}} + 2\pi, & (N-1)(N+1)^{n-1} < \frac{a}{\pi} - 2 \leq N(N+1)^{n-1}. \end{cases}$$

Proof. Fix $a > 2\pi$ and $N \in \mathbb{N}$. We define u_1 by equation (5.2.3). The previously described embedding procedure yields a symplectic embedding

$$\phi_N^n(a) : R^n(a, \pi) \hookrightarrow R^n(A_N(a) + \epsilon)$$

where

$$A_N(a) := \max\{u_1 + \pi, (N+1)\pi\}.$$

Solving equation (5.2.3) for u_1 we find that

$$A_N(a) = \max\left\{\frac{a - 2\pi}{(N+1)^{n-1}} + 2\pi, (N+1)\pi\right\}.$$

Optimizing the choice of $N \in \mathbb{N}$, we conclude that the polydisc $P^{2n}(\pi, \ldots, \pi, a)$ symplectically embeds into the cube $C^{2n}(f_{PC}^{2n}(a) + \epsilon)$ for any $\epsilon > 0$, where $f_{PC}^{2n}(a)$ is defined by

$$f_{PC}^{2n}(a) = \min\{A_N(a) \mid N = 1, 2, \ldots\}.$$

This completes the proof of Proposition 5.2.1. $\qquad \square$

Remarks 5.2.2. 1. Arguing as in 4.4.2 we see that for $a \in \,]\pi, 2\pi]$ multiple symplectic folding does not provide a better embedding result then the inclusion $P^{2n}(\pi, \ldots, \pi, a) \hookrightarrow C^{2n}(a)$, and that the procedure proving Proposition 5.2.1 is optimal in the sense that we cannot embed $P^{2n}(\pi, \ldots, \pi, a)$ into a cube smaller than $C^{2n}(f_{PC}^{2n}(a))$ by multiple symplectic folding.

2. The functions $f_{PC}^{2n}(a)$, $n \geq 3$, are compared with the results yielded by Lagrangian folding in 7.2.2.2. $\qquad \diamond$

In view of the proof of the second statement in Theorem 3, which will be completed in Section 6.1, we also prove

Proposition 5.2.3. *Fix $n \geq 2$. For every $a > 3\pi$ there exists a natural number $N(a)$ and a symplectic embedding*

$$\varphi_a \colon R^n(a, \pi) \hookrightarrow R^n((N(a) + 1)\pi)$$

such that the following assertions hold.

(i) *If $u < \pi$,*

$$\varphi_a(u, v, x, y) = (u, v, x, y),$$

and if $u > a - \pi$,

$$\varphi_a(u, v, x, y) = \big(u - a + (N(a) + 1)\pi, v, x, y + N(a)\pi 1_y\big).$$

(ii) $\displaystyle \lim_{a \to \infty} \frac{|\varphi_a(R^n(a, \pi))|}{|R^n((N(a) + 1)\pi)|} = 1.$

Proof. Fix $n \geq 2$ and $N \in 2\mathbb{N}$. We set $u_1 = N\pi$ and

$$\hat{a}_N = 2\pi + (N + 1)^{n-1}(u_1 - \pi) = 2\pi + (N + 1)^{n-1}(N - 1)\pi.$$

We recall that in the previous description of the symplectic embedding $\phi_N^n(\hat{a}_N)$ we have neglected the arbitrarily small δ-terms appearing in the actual construction. Define $\delta_N > 0$ by

$$\delta_N = \frac{\pi}{3\big((N + 1)^{n-1} - 1\big)}.$$

Since $N \geq 2$ we find that $u_1 - 2\delta_N > \pi + 2\delta_N$. In the actual construction associated with $\hat{a}_N$ and N we can therefore achieve the i'th fold as follows. We fold the last floor of the image at $u = u_1 - 2\delta_N$ if i is odd and at $u = \pi + 2\delta_N$ if i is even in such a way that the u-length of the part of the last floor which is mapped to the i'th stairs is equal to δ_N. After folding $(N + 1)^{n-1} - 1$ times we thereby obtain a symplectic embedding

$$\phi_N^n(\hat{a}_N, \delta_N) \colon R^n(a_N, \pi) \hookrightarrow R^n((N + 1)\pi))$$

where

$$\begin{aligned}
a_N &= 2(\pi + 2\delta_N) + (N + 1)^{n-1}(u_1 - \pi - 4\delta_N) + ((N + 1)^{n-1} - 1)\delta_N \\
&= 2\pi + (N + 1)^{n-1}(u_1 - \pi) - 3\big((N + 1)^{n-1} - 1\big)\delta_N \qquad (5.2.4) \\
&= \pi + (N + 1)^{n-1}(N - 1)\pi.
\end{aligned}$$

We abbreviate $\psi_N = \phi_N^n(\hat{a}_N, \delta_N)$. In view of the construction of ψ_N and the inequality $\pi + 2\delta_N < u_1 - 2\delta_N$ we have

$$\psi_N(u, v, x, y) = (u, v, x, y) \qquad\qquad \text{if } u < \pi + 2\delta_N, \qquad (5.2.5)$$

$$\psi_N(u, v, x, y) = (u - a_N + (N + 1)\pi, v, x, y + N\pi 1_y)$$
$$\text{if } u > a_N - \pi - 2\delta_N. \qquad (5.2.6)$$

Notice that

$$a_N < a_{N+2} \text{ for every } N \in 2\mathbb{N} \quad \text{and} \quad a_N \to \infty \text{ as } N \to \infty. \tag{5.2.7}$$

The function $N \colon \,]\pi, \infty[\,\to \mathbb{N}$,

$$N(a) := \min\{\, N \in 2\mathbb{N} \mid a_N \geq a \,\}, \tag{5.2.8}$$

is therefore well-defined, and $N(a_N) = N$. Fix $a > 3\pi$. Since $a \leq a_{N(a)}$, we find a symplectic embedding $\beta_a \colon R(a) \hookrightarrow R(a_{N(a)})$ which is the identity on $\{u < \pi\}$ and the translation $(u, v) \mapsto (u + a_{N(a)} - a, v)$ on $\{u > a - \pi\}$, cf. Figure 3.6 for the case $a < a_{N(a)}$. We define the symplectic embedding

$$\varphi_a \colon R^n(a, \pi) \hookrightarrow R^n\big((N(a) + 1)\pi\big)$$

by

$$\varphi_a = \psi_{N(a)} \circ (\beta_a \times \mathrm{id}_{2n-2}).$$

In view of the formulae (5.2.5) and (5.2.6) and in view of its definition, φ_a meets assertion (i) in Proposition 5.2.3.

In order to verify assertion (ii), we first of all observe that

$$\big|\varphi_a\big(R^n(a, \pi)\big)\big| = \big|R^n(a, \pi)\big| \quad \text{and} \quad \varphi_a\big(R^n(a, \pi)\big) \subset R^n((N(a) + 1)\pi)$$

for all $a > 3\pi$. Therefore,

$$1 \geq \frac{\big|\varphi_a\left(R^n(a, \pi)\right)\big|}{\big|R^n((N(a) + 1)\pi)\big|} = \frac{\big|R^n(a, \pi)\big|}{\big|R^n((N(a) + 1)\pi)\big|} = \frac{a}{(N(a) + 1)^n\,\pi} \tag{5.2.9}$$

for all $a > 3\pi$. Assume now that $a > a_2$. In view of (5.2.7) and the definition (5.2.8) of $N(a)$ we have $a \in \,]a_{N(a)-2}, a_{N(a)}]$. Using this and the formula (5.2.4) for $a_{N(a)-2}$ we can further estimate

$$\frac{a}{(N(a) + 1)^n\,\pi} > \frac{a_{N(a)-2}}{(N(a) + 1)^n\,\pi} = \frac{(N(a) - 1)^{n-1}(N(a) - 3) + 1}{(N(a) + 1)^n\,\pi}. \tag{5.2.10}$$

The definition (5.2.8) and (5.2.7) imply that

$$N(a) \to \infty \quad \text{as } a \to \infty. \tag{5.2.11}$$

Combining the estimates (5.2.9) and (5.2.10) we therefore conclude that

$$\lim_{a \to \infty} \frac{\big|\varphi_a\left(R^n(a, \pi)\right)\big|}{\big|R^n\left((N(a) + 1)\pi\right)\big|} = 1,$$

and so the proof of Proposition 5.2.3 is complete. $\square$

5.3   Embedding ellipsoids into balls

In this section we shall study a problem closely related to symplectically embedding skinny ellipsoids $E^{2n}(\pi, \ldots, \pi, a)$ into balls $B^{2n}(A)$. As in Chapter 3 we start with replacing these sets by symplectomorphic sets which are more convenient to work with. Recall that given a domain $U \subset \mathbb{R}^{2n}$ and $\lambda > 0$ we set

$$\lambda U = \{\lambda z \in \mathbb{R}^{2n} \mid z \in U\}.$$

Reorganizing the coordinates, we consider the Lagrangian splitting $\mathbb{R}^n(x) \times \mathbb{R}^n(y)$ of $\mathbb{R}^{2n}$. We set

$$\triangle(a_1, \ldots, a_n) = \left\{0 < x_1, \ldots, x_n \;\middle|\; \sum_{i=1}^n \frac{x_i}{a_i} < 1\right\} \subset \mathbb{R}^n(x),$$

$$\Box(b_1, \ldots, b_n) = \{0 < y_i < b_i, \; 1 \le i \le n\} \subset \mathbb{R}^n(y),$$

and we abbreviate $\triangle^n(a) = \triangle(a, \ldots, a)$ and $\Box^n(b) = \Box(b, \ldots, b)$.

Lemma 5.3.1. *Assume $\epsilon > 0$. Then*

(i) *the ellipsoid $E(a_1, \ldots, a_n)$ symplectically embeds into the Lagrangian product $((1 + \epsilon)\triangle(a_1, \ldots, a_n)) \times \Box^n(1)$,*

(i) *the Lagrangian product $\triangle(a_1, \ldots, a_n) \times \Box^n(1)$ symplectically embeds into the ellipsoid $(1 + \epsilon)E(a_1, \ldots, a_n)$.*

Proof. (i) Define ϵ' by $\sum_{i=1}^n \frac{\epsilon'}{a_i} = \epsilon$. Replacing the parameter a in the proof of Lemma 3.1.8 (i) by a_i, $1 \le i \le n$, we obtain area and orientation preserving diffeomorphisms $\alpha_i \colon D(a_i) \to R(a_i)$ satisfying

$$x_i(\alpha_i(z_i)) \le \pi |z_i|^2 + \epsilon' \quad \text{for all } z_i \in D(a_i), \; 1 \le i \le n,$$

cf. Figure 3.3. For $(z_1, \ldots, z_n) \in E(a_1, \ldots, a_n)$ we then find

$$\sum_{i=1}^n \frac{x_i(\alpha_i(z_i))}{a_i} \le \sum_{i=1}^n \frac{\pi |z_i|^2}{a_i} + \frac{\epsilon'}{a_i} < 1 + \epsilon.$$

It follows that the restriction of the symplectic embedding

$$\alpha_1 \times \cdots \times \alpha_n \colon D(a_1) \times \cdots \times D(a_n) \hookrightarrow \mathbb{R}^{2n}$$

to $E(a_1, \ldots, a_n)$ is as desired.

(ii) Define ϵ' by $\sum_{i=1}^n \frac{\epsilon'}{a_i} = \epsilon^2$. Replacing the parameters a and ϵ in the proof of Lemma 3.1.8 (ii) by a_i and ϵ', $1 \le i \le n$, we obtain area and orientation preserving embeddings $\omega_i \colon R(a_i) \hookrightarrow D(a_i + \epsilon')$ satisfying

$$\pi |\omega_i(z_i)|^2 \le x_i + \epsilon' \quad \text{for all } z_i = (x_i, y_i) \in R(a_i), \; 1 \le i \le n,$$

cf. Figure 3.3. For $(z_1, \dots, z_n) \in \triangle(a_1, \dots, a_n) \times \square^n(1)$ we then find

$$\sum_{i=1}^{n} \frac{\pi |\omega_i(z_i)|^2}{a_i} \leq \sum_{i=1}^{n} \frac{x_i}{a_i} + \frac{\epsilon'}{a_i} < 1 + \epsilon^2 < (1 + \epsilon)^2.$$

It follows that the restriction of the symplectic embedding

$$\omega_1 \times \cdots \times \omega_n \colon R(a_1) \times \cdots \times R(a_n) \hookrightarrow \mathbb{R}^{2n}$$

to $\triangle(a_1, \dots, a_n) \times \square^n(1)$ is as desired. $\qquad\square$

In view of Lemma 5.3.1 we may think of an ellipsoid as a Lagrangian product of a simplex and a cube. In the setting of symplectic folding, however, we want to work with sets which fibre over a symplectic rectangle. We therefore set again $(u, v) = (x_1, y_1)$ and define the $2n$-dimensional trapezoid $T^n(a, b)$ by

$$T^n(a, b) = \left\{ \begin{array}{l} (u, v, x, y) \in \mathbb{R}^2 \times \mathbb{R}^{n-1}(x) \times \mathbb{R}^{n-1}(y) \mid \\ (u, v) \in R(a), \ (x, y) \in \left(\left(1 - \frac{u}{a}\right)\triangle^{n-1}(b)\right) \times \square^{n-1}(1) \end{array} \right\}.$$

Then

$$T^n(a, b) = \triangle(a, b, \dots, b) \times \square^n(1). \qquad (5.3.1)$$

If $b = a$, we abbreviate $T^n(a) = T^n(a, a)$.

Corollary 5.3.2. *Assume $\epsilon > 0$. Then*

(i) $E^{2n}(\pi, \dots, \pi, a)$ *symplectically embeds into* $T^n(a + \epsilon, \pi + \epsilon)$,

(ii) $T^n(a)$ *symplectically embeds into* $B^{2n}(a + \epsilon)$.

Proof. (i) The ellipsoid $E^{2n}(\pi, \dots, \pi, a)$ is symplectomorphic to the ellipsoid $E^{2n}(a, \pi, \dots, \pi)$, and by Lemma 5.3.1 (i) and the identity (5.3.1) this ellipsoid symplectically embeds into

$$\left((1 + \epsilon')\triangle(a, \pi, \dots, \pi) \right) \times \square^n(1) = T^n(a + \epsilon'a, \pi + \epsilon'\pi)$$

for every $\epsilon' > 0$. The claim thus follows.

(ii) follows from the identity (5.3.1) and Lemma 5.3.1 (ii). $\qquad\square$

By Corollary 5.3.2 the problem of symplectically embedding skinny ellipsoids $E^{2n}(\pi, \dots, \pi, a)$ into balls $B^{2n}(A)$ is equivalent to the problem of symplectically embedding trapezoids $T^n(\pi, a)$ into trapezoids $T^n(A)$. In view of the proof of the first statement in Theorem 3, which will be completed in Section 6.2, we shall, however, consider a somewhat different embedding problem. Instead of embeddings of $T^n(\pi, a)$ we shall study embeddings of the larger set

$$S_a := R(a) \times \triangle^{n-1}(\pi) \times \square^{n-1}(1) \subset \mathbb{R}^2 \times \mathbb{R}^{n-1}(x) \times \mathbb{R}^{n-1}(y)$$

into $T^n(A)$.

Proposition 5.3.3. *Fix $n \geq 2$. For every $a > 3\pi$ there exists a natural number $l(a)$ and a symplectic embedding $\varphi_a \colon S_a \hookrightarrow T^n(l(a)^2)$ such that the following assertions hold.*

(i) *If $u < \pi$,*

$$\varphi_a(u, v, x, y) = \left(u, v, \tfrac{(l(a)-1)l(a)}{\pi} x, \tfrac{\pi}{(l(a)-1)l(a)} y\right),$$

and if $u > a - \pi$,

$$\varphi_a(u, v, x, y) = \left(u - a + (l(a) - 1)l(a), v, \tfrac{l(a)}{\pi} x, \tfrac{\pi}{l(a)} y\right).$$

(ii) $\displaystyle \lim_{a \to \infty} \frac{|\varphi_a(S_a)|}{|T^n(l(a)^2)|} = 1.$

Proof. The proof of Proposition 5.3.3 is more difficult than the proof of the analogous Proposition 5.2.3. The reason is that for $n \geq 4$ it is impossible to fill a large $(n - 1)$-simplex with small $(n - 1)$-simplices. We shall therefore repeatedly rescale the fibres of S_a and fill the cube-factor $\square^{n-1}(1)$ of the fibres of $T^n\left(l(a)^2\right)$ with the small cube-factors of the rescaled fibres.

Fix $n \geq 2$ and $a > 3\pi$. We prove Proposition 5.3.3 in six steps. In the first four steps we construct for each odd number $l > 3n\pi$ a symplectic embedding ψ_l of the unbounded set

$$S_\infty := \{0 < u, \, 0 < v < 1\} \times \Delta^{n-1}(\pi) \times \square^{n-1}(1)$$

into $\mathbb{R}^{2n}$. The basic idea behind the embeddings ψ_l is explained in Figure 5.3. In Step 5 we associate to $a > 3\pi$ an odd number $l(a)$ and use the embeddings ψ_l to construct symplectic embeddings $\varphi_a \colon S_a \hookrightarrow T^n(l(a)^2)$ which meet assertion (i). In Step 6 we verify that these embeddings also meet assertion (ii).

Step 1. Preparations. Fix $l \in 2\mathbb{N} + 1$ with $l > 3n\pi$. We define subsets $P_i = P_i(l)$ of $T^n(l^2)$ by

$$P_i := \left\{ (u, v, x, y) \in T^n(l^2) \mid (i - 1)l < u < il \right\}, \quad 1 \leq i \leq l, \qquad (5.3.2)$$

cf. Figure 5.4. Define real numbers $k_i = k_i(l)$ by

$$k_i := \tfrac{1}{\pi}(l - i)l, \quad 1 \leq i \leq l - 1. \qquad (5.3.3)$$

Since $l > 3n\pi$, we find that $k_i - \frac{k_i}{k_{i+1}} > 3$, $1 \leq i \leq l - 2$, and $k_{l-1} > 3$. We may therefore define even numbers $N_i = N_i(l)$ by

$$N_i := \max\left\{ N \in 2\mathbb{N} \mid N + 1 < k_i - \tfrac{k_i}{k_{i+1}} \right\}, \quad 1 \leq i \leq l - 2, \qquad (5.3.4)$$

$$N_{l-1} := \max\left\{ N \in 2\mathbb{N} \mid N + 1 < k_{l-1} \right\}. \qquad (5.3.5)$$

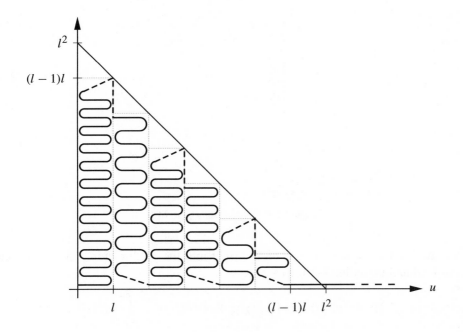

Figure 5.3. The embedding $\psi_l \colon S_\infty \hookrightarrow \mathbb{R}^{2n}$ for $l = 7$.

In view of definition (5.3.4) we have

$$0 < d_i := 1 - \frac{1}{k_{i+1}} - \frac{N_i+1}{k_i} \leq \frac{2}{k_i}, \quad 1 \leq i \leq l-2, \tag{5.3.6}$$

and in view of definition (5.3.3) we have $k_{i+1} < k_i$, $1 \leq i \leq l-2$. Then

$$e_i := \frac{2k_{i+1}}{k_i + k_{i+1}}\, d_i \in\,]0, d_i[, \quad 1 \leq i \leq l-2. \tag{5.3.7}$$

The set S_∞ is symplectomorphic to the set

$$F_1 := \{0 < u,\ 0 < v < 1\} \times \triangle^{n-1}(k_1\pi) \times \square^{n-1}\!\left(\tfrac{1}{k_1}\right)$$

via the linear symplectomorphism

$$\sigma \colon S_\infty \to F_1, \quad (u, v, x, y) \mapsto \left(u, v, k_1 x, \tfrac{1}{k_1} y\right).$$

In view of the definitions of k_1 and P_1, the fibres $\triangle^{n-1}(k_1\pi) \times \square^{n-1}(1/k_1)$ of F_1 are contained in the fibres of P_1, cf. Figure 5.4.

Step 2. Multiple folding in P_1. We symplectically embed a part of F_1 into P_1 by the multiple folding procedure described in Section 5.2. In the following description we shall again neglect the arbitrarily small δ-terms appearing in the actual construction.

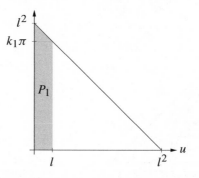

Figure 5.4. The subset P_1 of $T^n(l^2)$.

We first fold F_1 at $u = l - \pi$ on the right into the y_2-direction. The lifting map involved in this folding has the form

$$(u, x_2, v, y_2) \mapsto \left(u, x_2, v + c(u)x_2, y_2 + \int_0^u c(s)\, ds \right)$$

where the cut off function $c \colon \mathbb{R} \to [0, 1/(k_1\pi)]$ has support in $[l - \pi, l]$. We next fold the part of the image on which $y_2 > 1/k_1$ at $u = \pi$ on the left into the y_2-direction. This is possible because $l > 2\pi$. The length of the second floor is $l - 2\pi$. We fold N_1 times alternatingly at $u = l - \pi$ on the right and at $u = \pi$ on the left into the y_2-direction. We then fold once into the y_3-direction. Since N_1 is even, we do this by folding the part of the image on which $y_2 > N_1/k_1$ at $u = l - \pi$ on the right. Going on this way, we altogether fold $(N_1 + 1)^{n-1} - 1$ times, in which we fold N_1 times into the y_n-direction. Denote the multiple folding embedding $F_1 \hookrightarrow \mathbb{R}^{2n}$ thus obtained by μ_1. The image of the projection of $\mu_1(F_1)$ onto $\mathbb{R}^{n-1}(y)$ is contained in the cube $\square^{n-1}((N_1+1)/k_1)$, cf. Figure 5.6. Since N_1 is even, the infinite end of $\mu_1(F_1)$ points into the u-direction. More precisely, the last floor of $\mu_1(F_1)$ is the subset

$$F_1' := \,]\pi, \infty[\, \times\,]0, 1[\, \times\, \Delta^{n-1}(k_1\pi) \times C_1'$$

of $\mathbb{R}^2 \times \mathbb{R}^{n-1}(x) \times \mathbb{R}^{n-1}(y)$ where

$$C_1' = \Big\{ (y_2, \ldots, y_n) \,\Big|\, \tfrac{N_1}{k_1} < y_j < \tfrac{N_1+1}{k_1},\ \ j = 2, \ldots, n \Big\},$$

cf. Figure 5.5 and Figure 5.6.

Define

$$\delta_1 := d_1 - e_1. \tag{5.3.8}$$

We choose the δ-terms in the actual construction of the embedding μ_1 in such a way that the u-length u_1 of the part of F_1 mapped to $\mu_1(F_1) \setminus F_1'$ is equal to

$$u_1 = (l - \pi) + \big((N_1 + 1)^{n-1} - 2\big)(l - 2\pi) - \delta_1. \tag{5.3.9}$$

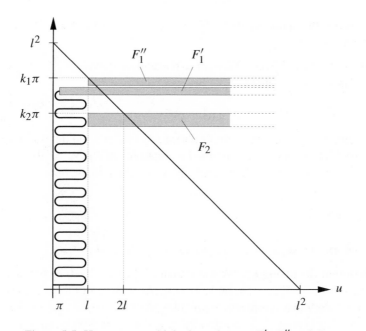

Figure 5.5. How one can think about the sets F_1', F_1'' and F_2.

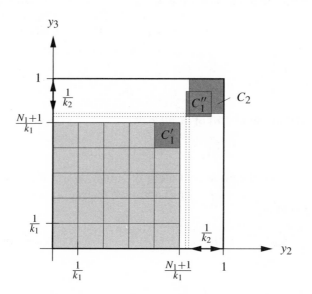

Figure 5.6. The cubes $\Box^{n-1}\left(\frac{N_1+1}{k_1}\right)$, C_1', C_1'' and C_2 for $n = 3$.

By construction, the set $\mu_1(F_1) \setminus F_1'$ is contained in P_1. We next want to pass to P_2 and fill as much of P_2 as possible. The fibres $\triangle^{n-1}(k_1\pi) \times C_1'$ of F_1', however, are not contained in the smallest fibre $\triangle^{n-1}(k_2\pi) \times \square^{n-1}(\pi)$ of P_2. We therefore need to rescale the fibres of F_1'.

Step 3. Rescaling the fibres. We want to rescale the fibres $\triangle^{n-1}(k_1\pi) \times C_1'$ of F_1' to fibres $\triangle^{n-1}(k_2\pi) \times C_2$ where

$$C_2 := \big\{ (y_2, \ldots, y_n) \mid 1 - \tfrac{1}{k_2} < y_j < 1, \ j = 2, \ldots, n \big\},$$

cf. Figure 5.6. In view of the definition (5.3.3) of k_2 and the definition (5.3.2) of P_2 the fibres $\triangle^{n-1}(k_2\pi) \times C_2$ are contained in the fibres of P_2. We shall first separate the fibres $\triangle^{n-1}(k_1\pi) \times C_1'$ from themselves by lifting them to the fibres $\triangle^{n-1}(k_1\pi) \times C_1''$ where

$$C_1'' := \big\{ (y_2, \ldots, y_n) \mid \tfrac{N_1+1}{k_1} + e_1 < y_j < \tfrac{N_1+2}{k_1} + e_1, \ j = 2, \ldots, n \big\}$$

and then deform the separated fibres to the fibres $\triangle^{n-1}(k_2\pi) \times C_2$.

Construction of the lifting λ_1. We shall separate F_1' from itself by lifting its fibres into each y_j-direction, $j = 2, \ldots, n$, by $1/k_1 + e_1$. As in Step 1 of the folding construction in Section 3.2 we find a symplectic embedding $\beta_1 \colon \,]0, \infty[\times]0, 1[\hookrightarrow \,]0, \infty[\times]0, 1[$ which is the identity on $\{u < \pi\}$ and the translation $(u, v) \mapsto (u + l - \pi - \delta_1, v)$ on $\{u > \pi + \delta_1\}$, cf. Figure 3.6. Define $s_1 := \pi + k_1\pi d_1$. In view of the second inequality in (5.3.6) and $l > 3n\pi$ we find

$$\pi + (n-1)s_1 = \pi + (n-1)\pi(1 + k_1 d_1)$$
$$\leq \pi + (n-1)\pi(1 + 2)$$
$$< 3n\pi$$
$$< l.$$

In view of the definition (5.3.8) of δ_1 we have

$$s_1 \tfrac{1}{k_1\pi} = \delta_1 + \big(\tfrac{1}{k_1} + e_1 \big).$$

For $j = 2, \ldots, n$ we therefore find a cut off function $c_j \colon \mathbb{R} \to [0, 1/(k_1\pi)]$ with support $[\pi + (j-2)s_1, \pi + (j-1)s_1]$ and such that $\int_0^\infty c_j(s)\,ds = 1/k_1 + e_1$. The symplectic embedding

$$\varphi_1 \colon \operatorname{Im}\beta_1 \times \triangle^{n-1}(k_1\pi) \times C_1' \hookrightarrow \mathbb{R}^{2n}, \quad (u, v, x, y) \mapsto (u', v', x', y')$$

defined by

$$u' = u, \quad v' = v + \sum_{j=2}^{n} c_j(u)x_j, \quad x_j' = x_j, \quad y_j' = y_j + \int_0^u c_j(s)\,ds, \quad j = 2, \ldots, n,$$

is the identity on $\{u < \pi\}$, maps $\{u < l\}$ to P_1 and translates $\{u > l\}$ to the set

$$F_1'' := \{u > l, \ 0 < v < 1\} \times \triangle^{n-1}(k_1\pi) \times C_1'',$$

cf. Figure 5.5. We restrict the symplectic embedding

$$\varphi_1 \circ (\beta_1 \times \mathrm{id}_{n-2}) \colon \,]0, \infty[\, \times \,]0, 1[\, \times \, \triangle^{n-1}(k_1\pi) \times C_1' \, \hookrightarrow \, \mathbb{R}^{2n}$$

to the intersection of its domain with $\mu_1(F_1)$ and extend this restriction by the identity to the symplectic embedding $\lambda_1 \colon \mu_1(F_1) \hookrightarrow \mathbb{R}^{2n}$.

In view of the construction of the "translation" $\beta_1 \times \mathrm{id}_{2n-2}$ the u-length of the part of F_1' which λ_1 embeds into P_1 is δ_1. In view of the identity (5.3.9) we therefore conclude that the u-length u_1' of the part of F_1 which $\lambda_1 \circ \mu_1$ embeds into P_1 is equal to

$$u_1' = (l - \pi) + \big((N_1 + 1)^{n-1} - 2\big)(l - 2\pi). \tag{5.3.10}$$

Construction of the deformation α_1. The deformation of the fibres $\triangle^{n-1}(k_1\pi) \times C_1''$ of F_1'' to fibres $\triangle^{n-1}(k_2\pi) \times C_2$ is based on the following lemma.

Lemma 5.3.4. *There exists a symplectic embedding*

$$\alpha \colon \,]0, k_1\pi[\, \times \,]0, \tfrac{N_1+2}{k_1} + e_1[\, \hookrightarrow \, \mathbb{R}^2$$

which restricts to the identity on $\{\,(x, y) \mid y \leq (N_1 + 1)/k_1\,\}$, restricts to the affine map

$$(x, y) \mapsto \big(\tfrac{k_2}{k_1}x, \, \tfrac{k_1}{k_2}y + 1 - \tfrac{1}{k_2}(N_1 + 2 + k_1e_1)\big) \tag{5.3.11}$$

on $\{\,(x, y) \mid y \geq (N_1 + 1)/k_1 + e_1\,\}$, and is such that

$$x'(\alpha(x, y)) \leq x \quad \text{and} \quad y'(\alpha(x, y)) < 1 \tag{5.3.12}$$

for all $(x, y) \in \,]0, k_1\pi[\, \times \,]0, (N_1 + 2)/k_1 + e_1[$, cf. Figure 5.7.

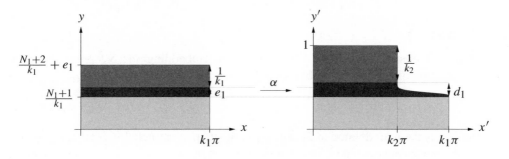

Figure 5.7. The map α.

Proof. Choose a smooth function $h \colon \mathbb{R} \to \mathbb{R}$ such that

(i) $h(w) = 1$ for $w \leq \frac{N_1+1}{k_1}$,

(ii) $h'(w) < 0$ for $w \in \, \big]\frac{N_1+1}{k_1}, \frac{N_1+1}{k_1} + e_1\big[$,

(iii) $h(w) = \frac{k_2}{k_1}$ for $w \geq \frac{N_1+1}{k_1} + e_1$.

In view of the definition (5.3.7) of e_1 and the inequality $k_2 < k_1$ we have

$$e_1 < d_1 < e_1 \frac{k_1}{k_2}.$$

We may therefore further require that

(iv) $\int_{\frac{N_1+1}{k_1}}^{\frac{N_1+1}{k_1}+e_1} \frac{1}{h(w)} \, dw = d_1.$

Then the map

$$\alpha \colon \;]0, k_1\pi[\; \times \;]0, \tfrac{N_1+2}{k_1} + e_1[\; \to \; \mathbb{R}^2, \quad (x, y) \mapsto \left(h(y)x, \int_0^y \frac{1}{h(w)} dw \right)$$

is a symplectic embedding which is as required.   □

Denote by α_1 the restriction to $\lambda_1(\mu_1(F_1))$ of the symplectic embedding

$$]0, \infty[\times]0, 1[\times \square^{n-1}(k_1\pi) \times \square^{n-1}\left(\tfrac{N_1+2}{k_1} + e_1\right) \hookrightarrow \mathbb{R}^2 \times \mathbb{R}^{n-1}(x) \times \mathbb{R}^{n-1}(y),$$

$$(u, v, x_2, \dots, x_n, y_2, \dots, y_n) \mapsto (u, v, x_2', \dots, x_n', y_2', \dots, y_n')$$

where $(x_j', y_j') = \alpha(x_j, y_j)$, $j = 2, \dots, n$. In view of the inequalities (5.3.12), α_1 maps the set $\lambda_1(\mu_1(F_1)) \cap P_1$ into P_1, and in view of (5.3.11), α_1 maps the set $F_1'' = \lambda_1(\mu_1(F_1)) \setminus P_1$ symplectically onto the set

$$F_2 := \{u > l, \, 0 < v < 1\} \times \Delta^{n-1}(k_2\pi) \times C_2,$$

cf. Figure 5.5.

Step 4. Construction of ψ_l. The symplectic embedding $\psi_l \colon S_\infty \hookrightarrow \mathbb{R}^{2n}$ is the composition of symplectic embeddings

$$\lambda_{l-1} \circ \mu_{l-1} \circ (\alpha_{l-2} \circ \lambda_{l-2} \circ \mu_{l-2}) \circ \cdots \circ (\alpha_2 \circ \lambda_2 \circ \mu_2) \circ (\alpha_1 \circ \lambda_1 \circ \mu_1) \circ \sigma.$$

Here, σ is the map defined in Step 1, μ_1 is the map constructed in Step 2, λ_1 and α_1 are the maps constructed in Step 3, and the maps $\mu_i, \lambda_i, \alpha_i$, $i \geq 2$, are constructed in a similar way as $\mu_1, \lambda_1, \alpha_1$. In order to describe the maps $\mu_i, \lambda_i, \alpha_i$, $i \geq 2$, in more detail, we assume by induction that we have already constructed embeddings μ_j, λ_j, α_j, $j = 1, \dots, i - 1$, where $i \leq l - 2$, and that the set

$$\{ (u, v, x, y) \in (\alpha_{i-1} \circ \lambda_{i-1} \circ \mu_{i-1} \circ \cdots \circ \alpha_1 \circ \lambda_1 \circ \mu_1)(F_1) \mid u > (i - 1)l \}$$

is the set

$$F_i := \{u > (i - 1)l, \, 0 < v < 1\} \times \Delta^{n-1}(k_i\pi) \times C_i$$

where

$$
C_i := \begin{cases} \left\{ (y_2, \ldots, y_n) \mid 0 < y_j < \frac{1}{k_i}, \ j = 2, \ldots, n \right\} & \text{if } i \text{ is odd,} \\ \left\{ (y_2, \ldots, y_n) \mid 1 - \frac{1}{k_i} < y_j < 1, \ j = 2, \ldots, n \right\} & \text{if } i \text{ is even.} \end{cases}
$$

The multiple folding map μ_i embeds a part of F_i into P_i. We fold N_i times alternatingly at $u = il - \pi$ on the right and at $u = (i-1)l + \pi$ on the left into the y_2-direction, and so on. Since N_i is even, the last floor of the image of μ_i is

$$
F'_i := \,](i-1)l + \pi, \infty[\, \times \,]0, 1[\, \times \Delta^{n-1}(k_i \pi) \times C'_i
$$

where

$$
C'_i = \begin{cases} \left\{ (y_2, \ldots, y_n) \mid \frac{N_i}{k_i} < y_j < \frac{N_i+1}{k_i}, \ j = 2, \ldots, n \right\} & \text{if } i \text{ is odd,} \\ \left\{ (y_2, \ldots, y_n) \mid 1 - \frac{N_i+1}{k_i} < y_j < 1 - \frac{N_i}{k_i}, \ j = 2, \ldots, n \right\} & \text{if } i \text{ is even.} \end{cases}
$$

We define $\delta_i := d_i - e_i$ and choose the δ-terms in the actual construction of μ_i in such a way that the u-length u_i of the part of F_i mapped to $\mu_i(F_i) \setminus F'_i$ is equal to

$$
u_i = (l - \pi) + \left((N_i + 1)^{n-1} - 2 \right)(l - 2\pi) - \delta_i. \tag{5.3.13}
$$

The maps λ_i and α_i rescale the fibres $\Delta^{n-1}(k_i \pi) \times C'_i$ of the floor F'_i to fibres $\Delta^{n-1}(k_{i+1} \pi) \times C_{i+1}$ where

$$
C_{i+1} := \begin{cases} \left\{ (y_2, \ldots, y_n) \mid 1 - \frac{1}{k_{i+1}} < y_j < 1, \ j = 2, \ldots, n \right\} & \text{if } i \text{ is odd,} \\ \left\{ (y_2, \ldots, y_n) \mid 0 < y_j < \frac{1}{k_{i+1}}, \ j = 2, \ldots, n \right\} & \text{if } i \text{ is even.} \end{cases}
$$

In view of the definition (5.3.3) of k_{i+1} and the definition (5.3.2) of P_{i+1} the fibres $\Delta^{n-1}(k_{i+1} \pi) \times C_{i+1}$ are contained in the fibres of P_{i+1}.

The lifting map λ_i separates F'_i from itself by lifting its fibres into each y_j-direction, $j = 2, \ldots, n$, by $(-1)^{i+1}(1/k_i + e_i)$. More precisely, we define $s_i := \pi + k_i \pi d_i$ and find as in the construction of λ_1 that

$$
(i-1)l + \pi + (n-1)s_i < il.
$$

Proceeding as in the construction of λ_1 we can therefore construct a symplectic embedding $\lambda_i : \operatorname{Im} \mu_i \hookrightarrow \mathbb{R}^{2n}$ which is the identity on $\operatorname{Im} \mu_i \setminus F'_i$ and translates $\{ (u, v, x, y) \in F'_i \mid u > (i-1)l + \pi + \delta_i \}$ to the set

$$
F''_i := \{ u > il, \ 0 < v < 1 \} \times \Delta^{n-1}(k_i \pi) \times C''_i
$$

where

$$
C''_i := \begin{cases} \left\{ (y_2, \ldots, y_n) \mid \frac{N_i+1}{k_i} + e_i < y_j < \frac{N_i+2}{k_i} + e_i, \ j = 2, \ldots, n \right\} \\ \hspace{7cm} \text{if } i \text{ is odd,} \\ \left\{ (y_2, \ldots, y_n) \mid 1 - \frac{N_i+2}{k_i} - e_i < y_j < 1 - \frac{N_i+1}{k_i} - e_i, \ j = 2, \ldots, n \right\} \\ \hspace{7cm} \text{if } i \text{ is even.} \end{cases}
$$

The u-length of the part of F_i' which λ_i embeds into P_i is δ_i. In view of the identity (5.3.13) we therefore conclude that the u-length u_i' of the part of F_i which $\lambda_i \circ \mu_i$ embeds into P_i is equal to

$$u_i' = (l - \pi) + ((N_i + 1)^{n-1} - 2)(l - 2\pi). \qquad (5.3.14)$$

The symplectic embedding $\alpha_i \colon \operatorname{Im} \lambda_i \hookrightarrow \mathbb{R}^{2n}$ maps the set $\operatorname{Im} \lambda_i \cap P_i$ into P_i and maps $F_i'' = \operatorname{Im} \lambda_i \setminus P_i$ onto the set

$$F_{i+1} := \{ u > il, \, 0 < v < 1 \} \times \Delta^{n-1}(k_{i+1}\pi) \times C_{i+1}.$$

The deformation α_i is constructed the same way as α_1 if i is odd, and in a similar way if i is even.

Next, the multiple folding map μ_{l-1} embeds a part of F_{l-1} into P_{l-1}. We fold N_{l-1} times alternatingly at $u = (l - 1)l - \pi$ on the right and at $u = (l - 2)l + \pi$ on the left into the y_2-direction, and so on. Since N_{l-1} is even and l is odd, the last floor of the image of μ_{l-1} is

$$F_{l-1}' := \,](l - 2)l + \pi, \infty[\,\times\,]0, 1[\,\times\, \Delta^{n-1}(k_{l-1}\pi) \times C_{l-1}'$$

where

$$C_{l-1}' = \Big\{ (y_2, \ldots, y_n) \mid 1 - \tfrac{N_{l-1}+1}{k_{l-1}} < y_j < 1 - \tfrac{N_{l-1}}{k_{l-1}}, \; j = 2, \ldots, n \Big\}.$$

We define $\delta_{l-1} := 1/k_{l-1}$ and choose the δ-terms in the actual construction of μ_{l-1} in such a way that the u-length u_{l-1} of the part of F_{l-1} mapped to $\mu_{l-1}(F_{l-1}) \setminus F_{l-1}'$ is equal to

$$u_{l-1} = (l - \pi) + ((N_{l-1} + 1)^{n-1} - 2)(l - 2\pi) - \delta_{l-1}. \qquad (5.3.15)$$

Finally, define $s_{l-1} := k_{l-1}\pi - (N_{l-1} + 1)\pi + \pi$. In view of the definition (5.3.5) of N_{l-1} we have

$$1 - \tfrac{N_{l-1}+1}{k_{l-1}} \le \tfrac{2}{k_{l-1}}.$$

This and $l > 3n\pi$ imply that

$$(l - 2)l + \pi + (n - 1)s_{l-1} < (l - 1)l - \pi.$$

Proceeding as in the construction of λ_i, i even, we therefore find a symplectic embedding $\lambda_{l-1} \colon \operatorname{Im} \mu_{l-1} \hookrightarrow \mathbb{R}^{2n}$ which is the identity on $\operatorname{Im} \mu_{l-1} \setminus F_{l-1}'$ and translates $\{ (u, v, x, y) \in F_{l-1}' \mid u > (l - 2)l + \pi + \delta_{l-1} \}$ to the set

$$F_l := \{ u > (l - 1)l - \pi, \, 0 < v < 1 \} \times \Delta^{n-1}(k_{l-1}\pi) \times C_l$$

where

$$C_l := \Big\{ (y_2, \ldots, y_n) \mid 0 < y_j < \tfrac{1}{k_{l-1}}, \; j = 2, \ldots, n \Big\}.$$

The u-length of the part of F'_{l-1} which λ_{l-1} embeds into P_{l-1} is $\delta_{l-1} + \pi$. In view of the identity (5.3.15) we therefore conclude that the u-length u'_{l-1} of the part of F_{l-1} which $\lambda_{l-1} \circ \mu_{l-1}$ embeds into P_{l-1} is equal to

$$u'_{l-1} = (l - \pi) + \big((N_{l-1} + 1)^{n-1} - 2\big)(l - 2\pi) + \pi. \qquad (5.3.16)$$

This completes the construction of the symplectic embedding

$$\psi_l = \lambda_{l-1} \circ \mu_{l-1} \circ \alpha_{l-2} \circ \lambda_{l-2} \circ \mu_{l-2} \circ \ldots \circ \alpha_1 \circ \lambda_1 \circ \mu_1 \circ \sigma \colon S_\infty \hookrightarrow \mathbb{R}^{2n}.$$

Step 5. Construction of φ_a. In view of the construction of the symplectic embedding $\psi_l \colon S_\infty \hookrightarrow \mathbb{R}^{2n}$ in the previous four steps and in view of the identities (5.3.10), (5.3.14) and (5.3.16), the u-length of the part of F_1 embedded into P_i is equal to

$$u'_i = \begin{cases} (l - \pi) + \big((N_i + 1)^{n-1} - 2\big)(l - 2\pi) & \text{if } i \le l - 2, \\ (l - \pi) + \big((N_i + 1)^{n-1} - 2\big)(l - 2\pi) + \pi & \text{if } i = l - 1. \end{cases}$$

Therefore, the u-length $a_l := \sum_{i=1}^{l-1} u_i$ of the part of F_1 embedded into $T^n(l^2) \setminus P_l$ is

$$a_l = l\pi + \sum_{i=1}^{l-1} \big((N_i + 1)^{n-1} - 2\big)(l - 2\pi). \qquad (5.3.17)$$

Moreover, by construction of ψ_l,

$$\psi_l(u, v, x, y) = \big(u, v, k_1 x, \tfrac{1}{k_1} y\big) \qquad\qquad \text{if } u < l - \pi - \delta_1,$$
$$\psi_l(u, v, x, y) = \big(u - a_l + (l-1)l, v, k_{l-1} x, \tfrac{1}{k_{l-1}} y\big) \quad \text{if } u > a_l - \pi.$$

Using the definition (5.3.8) of δ_1 and (5.3.7), (5.3.6) and (5.3.3) we find that

$$\delta_1 = d_1 - e_1 < d_1 \le \tfrac{2}{k_1} < 2\pi.$$

Since $l > 3n\pi$, we therefore find that $\pi < l - \pi - \delta_1$. This and the definition (5.3.3) of k_1 and k_{l-1} imply that

$$\psi_l(u, v, x, y) = \big(u, v, \tfrac{(l-1)l}{\pi} x, \tfrac{\pi}{(l-1)l} y\big) \qquad \text{if } u < \pi, \qquad (5.3.18)$$
$$\psi_l(u, v, x, y) = \big(u - a_l + (l-1)l, v, \tfrac{l}{\pi} x, \tfrac{\pi}{l} y\big) \quad \text{if } u > a_l - \pi. \qquad (5.3.19)$$

Before defining the embeddings φ_a, we further investigate the sequence (a_l).

Lemma 5.3.5. (i) $a_l < a_{l+2}$ *for every $l \in 2\mathbb{N} + 1$ with $l > 3n\pi$.*

(ii) $a_l \to \infty$ *as* $l \to \infty$.

Proof. (i) Fix $l \in 2\mathbb{N} + 1$ with $l > 3n\pi$. As in Step 1 we abbreviate

$$k_i = k_i(l), \quad N_i = N_i(l), \quad i = 1, \ldots, l - 1.$$

Moreover, we set $l' = l + 2$ and abbreviate

$$k_i' = k_i(l'), \quad N_i' = N_i(l'), \quad i = 1, \ldots, l' - 1.$$

By computation,

$$\frac{l-i}{l-i-1} > \frac{l'-i}{l'-i-1}, \quad i = 1, \ldots, l - 2.$$

Using the definition (5.3.3) of k_i and k_i', we therefore find

$$k_i - \frac{k_i}{k_{i+1}} = \frac{1}{\pi}(l-i)l - \frac{l-i}{l-i-1} < \frac{1}{\pi}(l'-i)l' - \frac{l'-i}{l'-i-1} = k_i' - \frac{k_i'}{k_{i+1}'},$$

$i = 1, \ldots, l - 2$. In view of the definition (5.3.4) of N_i and N_i', we conclude that

$$N_i \le N_i', \quad i = 1, \ldots, l - 2. \tag{5.3.20}$$

Moreover, we read off from definition (5.3.3) that $k_{l-1} = \frac{1}{\pi}l < \frac{1}{\pi}l' = k_{l'-1}$, and so, in view of definition (5.3.5),

$$N_{l-1} \le N_{l'-1}. \tag{5.3.21}$$

Using equation (5.3.17) and the inequalities (5.3.20) and (5.3.21), we can now estimate

$$a_l = l\pi + \sum_{i=1}^{l-2} \left((N_i + 1)^{n-1} - 2 \right)(l - 2\pi) + \left((N_{l-1} + 1)^{n-1} - 2 \right)(l - 2\pi)$$

$$< l'\pi + \sum_{i=1}^{l-2} \left((N_i' + 1)^{n-1} - 2 \right)(l' - 2\pi) + \left((N_{l'-1} + 1)^{n-1} - 2 \right)(l' - 2\pi)$$

$$< l'\pi + \sum_{i=1}^{l'-1} \left((N_i' + 1)^{n-1} - 2 \right)(l' - 2\pi)$$

$$= a_{l'}.$$

This proves (i).

(ii) follows from equation (5.3.17). $\qquad\qquad\square$

In view of Lemma 5.3.5 (ii) the function $l \colon]\pi, \infty[\to \mathbb{N}$,

$$l(a) := \min \{ l \in 2\mathbb{N} + 1 \mid l > 3n\pi, \, a_l \ge a \}, \tag{5.3.22}$$

is well-defined. Lemma 5.3.5 (i) shows that $l(a_l) = l$. Fix $a > 3\pi$. Since $a \le a_{l(a)}$, we find a symplectic embedding $\beta_a \colon R(a) \hookrightarrow R(a_{l(a)})$ which is the identity on $\{u < \pi\}$ and the translation $(u, v) \mapsto (u + a_{l(a)} - a, v)$ on $\{u > a - \pi\}$, cf. Figure 3.6 for the case $a < a_{l(a)}$. We finally define the symplectic embedding $\varphi_a \colon S_a \hookrightarrow T^n(l(a)^2)$ by

$$\varphi_a = \psi_{l(a)} \circ (\beta_a \times \mathrm{id}_{2n-2}).$$

In view of the formulae (5.3.18) and (5.3.19) and in view of its definition, φ_a meets assertion (i) in Proposition 5.3.3.

Step 6. Verification of assertion (ii) in Proposition 5.3.3. Recall that assertion (ii) in Proposition 5.3.3 claims that

$$\frac{|\varphi_a(S_a)|}{|T^n(l(a)^2)|} \to 1 \quad \text{as } a \to \infty. \tag{5.3.23}$$

Lemma 5.3.6. *Assertion (5.3.23) is a consequence of*

$$\frac{|T^n(l^2) \setminus \psi_l(S_{a_l})|}{|T^n(l^2)|} \to 0 \quad \text{as } l \to \infty. \tag{5.3.24}$$

Proof. Since $|\psi_l(S_{a_l})| = |S_{a_l}|$ and $\psi_l(S_{a_l}) \subset T^n(l^2)$, we have

$$\frac{|S_{a_l}|}{|T^n(l^2)|} = 1 - \frac{|T^n(l^2)| - |\psi_l(S_{a_l})|}{|T^n(l^2)|} = 1 - \frac{|T^n(l^2) \setminus \psi_l(S_{a_l})|}{|T^n(l^2)|}. \tag{5.3.25}$$

Fix $\epsilon \in {]}0, 1{[}$. The assumption (5.3.24) and (5.3.25) imply that there exists $l_0 \in 2\mathbb{N}+1$ such that

$$\frac{|S_{a_l}|}{|T^n(l^2)|} > \sqrt{1 - \epsilon} \quad \text{for all } l \in 2\mathbb{N} + 1 \text{ with } l \geq l_0. \tag{5.3.26}$$

Choosing l_0 larger if necessary, we may assume that

$$\frac{l}{l+2} > \sqrt[4n]{1 - \epsilon} \quad \text{for all } l \in 2\mathbb{N} + 1 \text{ with } l \geq l_0. \tag{5.3.27}$$

Assume now that $a > a_{l_0}$. In view of Lemma 5.3.5 we have $a \in {]}a_l, a_{l+2}]$ for some $l \in 2\mathbb{N} + 1$ with $l \geq l_0$. The definition (5.3.22) of $l(a)$ implies $l(a) = l + 2$. Using the estimates (5.3.26) and (5.3.27), we can therefore estimate

$$\frac{|S_a|}{|T^n(l(a)^2)|} > \frac{|S_{a_l}|}{|T^n((l+2)^2)|}$$

$$= \frac{|T^n(l^2)|}{|T^n((l+2)^2)|} \frac{|S_{a_l}|}{|T^n(l^2)|}$$

$$= \frac{l^{2n}}{(l+2)^{2n}} \frac{|S_{a_l}|}{|T^n(l^2)|}$$

$$> \sqrt{1 - \epsilon} \, \sqrt{1 - \epsilon}$$

$$= 1 - \epsilon.$$

Since $\varphi_a(S_a) \subset T^n(l(a)^2)$ and $|\varphi_a(S_a)| = |S_a|$, we therefore find

$$1 \geq \frac{|\varphi_a(S_a)|}{|T^n(l(a)^2)|} = \frac{|S_a|}{|T^n(l(a)^2)|} > 1 - \epsilon.$$

Since $\epsilon \in {]}0, 1{[}$ was arbitrary, Lemma 5.3.6 follows.   □

In order to prove assertion (5.3.24), we fix $l \in 2\mathbb{N} + 1$ with $l > 3n\pi$ and introduce several subsets of $T^n(l^2)$. We define the subsets $Q_i(l)$ of $P_i(l)$ by

$$Q_i(l) := \{\, (u, v, x, y) \in P_i(l) \mid x \in \Delta^{n-1}((l-1)l) \,\}, \qquad i = 1, \ldots, l,$$

and we define subsets $X_i(l)$, $Y_i(l)$ and $Z_i(l)$ of $P_i(l)$ by

$$\begin{aligned}
X_i(l) &:= P_i(l) \setminus Q_i(l), & i &= 1, \ldots, l, \\
Y_i(l) &:= \{\, (u, v, x, y) \in Q_i(l) \mid u \notin {]}(i-1)l + \pi, il - \pi{[} \,\}, & i &= 1, \ldots, l-1, \\
Z_i(l) &:= \{\, (u, v, x, y) \in Q_i(l) \mid y \notin \square^{n-1}\big(\tfrac{N_i+1}{k_i}\big) \,\}, & i &= 1, \ldots, l-1.
\end{aligned}$$

We also set

$$X(l) = \coprod_{i=1}^{l} X_i(l), \quad Y(l) = \coprod_{i=1}^{l-1} Y_i(l), \quad Z(l) = \coprod_{i=1}^{l-1} Z_i(l).$$

The sets $X(l)$ and $Y(l)$ are illustrated in Figure 5.8, and for $Z_1(l)$ we refer to Figure 5.6.

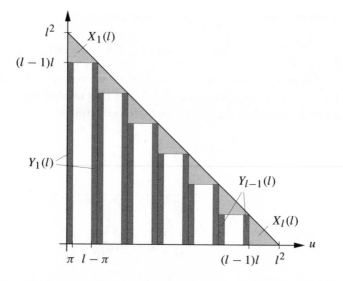

Figure 5.8. The subsets $X(l) = \coprod_{i=1}^{l} X_i(l)$ and $Y(l) = \coprod_{i=1}^{l-1} Y_i(l)$ of $T^n(l^2)$.

We recall from the construction of the embedding ψ_l that the unfilled space $X_i(l)$ in $P_i(l) \setminus \psi_l(S_{a_l})$ is caused by the fact that the size of the fibres of F_i is constant, that the space $Y(l)$ was needed for folding and contains all stairs, and that the space $Z_i(l)$ is the union of the space needed to deform the fibres of F_i' and the space caused by the fact that the N_i have to be even integers. Denote the closure of $\psi_l(S_{a_l})$ in $\mathbb{R}^{2n}$ by $\overline{\psi_l(S_{a_l})}$. By construction of ψ_l we have $|\psi_l(S_{a_l})| = \left|\overline{\psi_l(S_{a_l})}\right|$ and

$$T^n(l^2) \setminus \overline{\psi_l(S_{a_l})} \subset X(l) \cup Y(l) \cup Z(l).$$

We conclude that

$$\left|T^n(l^2) \setminus \psi_l(S_{a_l})\right| = \left|T^n(l^2) \setminus \overline{\psi_l(S_{a_l})}\right| \leq |X(l)| + |Y(l)| + |Z(l)|. \qquad (5.3.28)$$

Lemma 5.3.7.

(i) $|X(l)| < \frac{n}{l} \left|T^n(l^2)\right|$.

(ii) $|Y(l)| < \frac{2\pi}{l} \left|T^n(l^2)\right|$.

(iii) $|Z(l)| < \frac{2\pi n}{l} \left|T^n(l^2)\right|$.

Proof. (i) Notice that $X(l) \subset T^n(l^2) \setminus T^n((l-1)l)$, cf. Figure 5.8. Since $(l-1)^n > l^n - n\,l^{n-1}$, we can therefore estimate

$$
\begin{aligned}
|X(l)| &\leq \left|T^n(l^2)\right| - |T^n((l-1)l)| \\
&= \tfrac{1}{n!}\,(l^{2n} - (l-1)^n l^n) \\
&< \tfrac{1}{n!}\,n\,l^{2n-1} \\
&= \tfrac{n}{l}|T^n(l^2)|.
\end{aligned}
$$

(ii) The definitions of $Y_i(l)$ and $Q_i(l)$ yield

$$\frac{|Y_i(l)|}{|Q_i(l)|} = \frac{2\pi}{l}, \quad i = 1, \ldots, l-1,$$

cf. Figure 5.8. Therefore,

$$|Y(l)| = \sum_{i=1}^{l-1} |Y_i(l)| = \frac{2\pi}{l} \sum_{i=1}^{l-1} |Q_i(l)| < \frac{2\pi}{l} \sum_{i=1}^{l-1} |P_i(l)| < \frac{2\pi}{l} \left|T^n(l^2)\right|.$$

(iii) Assume first that $i \leq l - 2$. In view of definition (5.3.3) we then find

$$\frac{k_i}{k_{i+1}} = \frac{l-i}{l-i-1} \leq 2,$$

and so, in view of definition (5.3.4), $N_i + 1 \geq k_i - 4$, i.e.,

$$\frac{N_i + 1}{k_i} \geq 1 - \frac{4}{k_i}.$$

Since $i \leq l - 2$ and $l > 3n\pi$ we have

$$\frac{4}{k_i} = \frac{4\pi}{(l-i)l} \leq \frac{4\pi}{2l} = \frac{2\pi}{l} < 1.$$

Applying the formula $(1 - r)^{n-1} > 1 - (n - 1)r$ valid for all $r \in]0, 1[$ we can therefore estimate

$$\left(\frac{N_i + 1}{k_i}\right)^{n-1} \geq \left(1 - \frac{4}{k_i}\right)^{n-1} > 1 - (n - 1)\frac{4}{k_i} \geq 1 - (n - 1)\frac{2\pi}{l}. \qquad (5.3.29)$$

Similarly, the definition (5.3.5) shows that $N_{l-1} + 1 \geq k_{l-1} - 2$, whence

$$\frac{N_{l-1} + 1}{k_{l-1}} \geq 1 - \frac{2}{k_{l-1}}.$$

Estimating as before, we find

$$\left(\frac{N_{l-1} + 1}{k_{l-1}}\right)^{n-1} \geq \left(1 - \frac{2}{k_{l-1}}\right)^{n-1} > 1 - (n - 1)\frac{2}{k_{l-1}} = 1 - (n - 1)\frac{2\pi}{l}. \qquad (5.3.30)$$

The definitions of $Z_i(l)$ and $Q_i(l)$ and the estimates (5.3.29) and (5.3.30) now yield

$$\frac{|Z_i(l)|}{|Q_i(l)|} = \frac{\left|\Box^{n-1}(1) \setminus \Box^{n-1}\left(\frac{N_i + 1}{k_i}\right)\right|}{\left|\Box^{n-1}(1)\right|} = 1 - \left(\frac{N_i + 1}{k_i}\right)^{n-1} < (n - 1)\frac{2\pi}{l} < \frac{2\pi n}{l},$$

$i = 1, \ldots, l - 1$, and so

$$|Z(l)| = \sum_{i=1}^{l-1} |Z_i(l)| < \frac{2\pi n}{l} \sum_{i=1}^{l-1} |Q_i(l)| < \frac{2\pi n}{l} \left|T^n(l^2)\right|.$$

This completes the proof of Lemma 5.3.7. $\qquad\qquad\square$

In view of the estimate (5.3.28) and Lemma 5.3.7 we find

$$\frac{\left|T^n(l^2) \setminus \psi_l(S_{a_l})\right|}{\left|T^n(l^2)\right|} \leq \frac{|X(l)|}{\left|T^n(l^2)\right|} + \frac{|Y(l)|}{\left|T^n(l^2)\right|} + \frac{|Z(l)|}{\left|T^n(l^2)\right|} \leq \frac{n}{l} + \frac{2\pi}{l} + \frac{2\pi n}{l}.$$

Taking the limit $l \to \infty$, we see that assertion (5.3.24) holds true, and so the proof of assertion (ii) in Proposition 5.3.3 is complete.

The proof of Proposition 5.3.3 is accomplished. $\qquad\qquad\square$

Chapter 6

Proof of Theorem 3

Throughout this chapter (M, ω) is a given connected $2n$-dimensional symplectic manifold of finite volume $\text{Vol}(M, \omega) = \frac{1}{n!} \int_M \omega^n$. We denote by μ the measure on M induced by the volume form $\frac{1}{n!} \omega^n$. As before, $|S|$ denotes the Lebesgue measure of a measurable subset S of $\mathbb{R}^{2n}$. For $w \in \mathbb{R}^{2n}$ we denote the translation $z \mapsto z + w$ of $\mathbb{R}^{2n}$ by τ_w. Of course, τ_w is a symplectomorphism of $(\mathbb{R}^{2n}, \omega_0)$. In this chapter the symplectic coordinates will again be denoted by

$$(x_1, y_1, x_2, y_2, \ldots, x_n, y_n) = (u, v, x_2, y_2, \ldots, x_n, y_n) \in \mathbb{R}^{2n},$$

and we again abbreviate $x = (x_2, \ldots, x_n)$ and $y = (y_2, \ldots, y_n)$ as well as

$$1_y = (1, \ldots, 1) \in \mathbb{R}^{n-1}(y).$$

6.1 Proof of $\lim\limits_{a \to \infty} p_a^P(M, \omega) = 1$

We recall from the introduction that for every $a \geq \pi$ the real number $p_a^P(M, \omega)$ is defined by

$$p_a^P(M, \omega) = \sup_\lambda \frac{|\lambda P(\pi, \ldots, \pi, a)|}{\text{Vol}(M, \omega)}$$

where the supremum is taken over all those λ for which $\lambda P^{2n}(\pi, \ldots, \pi, a)$ symplectically embeds into (M, ω). Moreover, we recall from Section 5.2 that the polydisc $P^{2n}(\pi, \ldots, \pi, a)$ is symplectomorphic to the set $R^n(a, \pi)$. We conclude that the second statement in Theorem 3 in Section 1.3.2 can be reformulated as

Theorem 6.1.1. *For every $\epsilon > 0$ there exists a number $a_0 = a_0(\epsilon) > \pi$ having the following property. For every $a \geq a_0$ there exist a number $\lambda(a) > 0$ and a symplectic embedding $\Phi_a \colon \lambda(a) R^n(a, \pi) \hookrightarrow M$ such that*

$$\mu\left(M \setminus \Phi_a\left(\lambda(a) R^n(a, \pi)\right)\right) < \epsilon.$$

Proof. We shall proceed along the following lines. We shall first fill almost all of M with finitely many symplectically embedded cubes whose closures are disjoint, and connect these cubes by neighbourhoods of lines. In view of Proposition 5.2.3 we can then almost fill the cubes with symplectically embedded thin polydiscs, and we shall use the neighbourhoods of the lines to pass from one cube to another, cf. Figure 6.1.

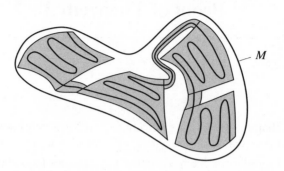

Figure 6.1. Filling M with a thin polydisc.

Step 1. Filling M by cubes. We denote by $C(s)$ the $2n$-dimensional open cube.

$$C(s) = \big\{(x_1, y_1, \ldots, x_n, y_n) \in \mathbb{R}^{2n} \mid 0 < x_i < s,\ 0 < y_i < s,\ i = 1, \ldots, n\big\}.$$

Lemma 6.1.2. *For every $\epsilon > 0$ there exist $s \in\ ]0, 1[$, an integer k and a symplectic embedding*

$$\gamma \colon \coprod_{i=1}^{k} C_i(s) \hookrightarrow M$$

of a disjoint union of k translates $C_i(s)$ of $C(s)$ in $\mathbb{R}^{2n}$ such that

$$\mu\Big(M \setminus \gamma\Big(\coprod C_i(s)\Big)\Big) < \epsilon.$$

Proof. We choose for each point $p \in M$ a Darboux chart $\chi_p \colon U_p \to V_p \subset M$. We can assume that the sets U_p are bounded. As every manifold, M satisfies the second axiom of countability, and so M is Lindelöf, i.e., every open covering of M has a countable subcovering. We therefore find a countable subcovering $\{V_{p_i}\}$ of the open covering $\{V_p\}$ of M. Since the sets $U_{p_i} \subset \mathbb{R}^{2n}$ are bounded, we find points $w_i \in \mathbb{R}^{2n}$ such that the translates $U_i := \tau_{w_i}(U_{p_i})$, $i \geq 1$, are disjoint. We abbreviate $\chi_i = \chi_{p_i} \circ \tau_{-w_i}$ and $V_i = V_{p_i}$. We have constructed countably many disjoint Darboux charts $\chi_i \colon U_i \to V_i$ which cover M.

We define subsets V_i' of M by

$$V_1' = V_1 \quad\text{and}\quad V_i' = V_i \setminus \bigcup_{j=1}^{i-1} V_j,\ i \geq 2.$$

Then

$$\coprod_{i \geq 1} V_i' = \bigcup_{i \geq 1} V_i = M.$$

Since the open sets V_i are μ-measurable, the sets V_i' are also μ-measurable. It follows that

$$\sum_{i \geq 1} \mu(V_i') = \mu\left(\coprod_{i \geq 1} V_i'\right) = \mu(M).$$

Since $\mu(M) < \infty$ we therefore find $m \in M$ such that

$$\sum_{i=1}^{m} \mu(V_i') > \mu(M) - \frac{\epsilon}{2}. \tag{6.1.1}$$

Set $U_i' = \chi_i^{-1}(V_i') \subset U_i, i = 1, \ldots, m$. Since V_i' is μ-measurable and χ_i^{-1} is smooth, U_i' is Lebesgue-measurable, $i = 1, \ldots, m$. We therefore find $s \in]0, 1[$ and finitely many disjoint translates

$$C_{i,j}(s) \subset U_i', \quad j = 1, \ldots, j_i,$$

of the cube $C(s)$ such that

$$\sum_{j=1}^{j_i} |C_{i,j}(s)| > |U_i'| - \frac{\epsilon}{2m}, \quad i = 1, \ldots, m. \tag{6.1.2}$$

We set $k = \sum_{i=1}^{m} j_i$. Since the sets U_i' are disjoint, the k cubes $C_{i,j}(s)$ are disjoint. Moreover, the embeddings $\chi_i \colon U_i' \hookrightarrow M$ are symplectic, and so the embedding γ defined by

$$\gamma = \coprod_{i,j} \chi_i|_{C_{i,j}(s)} \colon \coprod_{i,j} C_{i,j}(s) \hookrightarrow M$$

is symplectic, and

$$\mu(\chi_i(C_{i,j}(s))) = |C_{i,j}(s)| \quad \text{and} \quad \mu(V_i') = |U_i'|.$$

In view of the estimates (6.1.1) and (6.1.2) we therefore find

$$\mu\left(M \setminus \gamma\left(\coprod C_{i,j}(s)\right)\right) = \mu(M) - \sum_{i,j} \mu(\chi_i(C_{i,j}(s)))$$

$$= \mu(M) - \sum_{i,j} |C_{i,j}(s)|$$

$$= \mu(M) - \sum_{i=1}^{m} \mu(V_i') + \sum_{i=1}^{m} \left(|U_i'| - \sum_{j=1}^{i_j} |C_{i,j}(s)|\right)$$

$$< \frac{\epsilon}{2} + \sum_{i=1}^{m} \frac{\epsilon}{2m}$$

$$= \epsilon,$$

and so the proof of Lemma 6.1.2 is complete.   □

Let $\epsilon > 0$ be as in Theorem 6.1.1 and set $\epsilon' = \epsilon/3$. In view of Lemma 6.1.2 we find $s' \in \,]0, 1[$ and a symplectic embedding

$$\hat{\gamma} = \coprod_{i=1}^{k} \hat{\gamma}_i \colon \coprod_{i=1}^{k} \hat{C}_i(s') \hookrightarrow M$$

of a disjoint union of k translates $\hat{C}_i(s')$ of $C(s')$ such that

$$\mu\left(M \setminus \hat{\gamma}\left(\coprod \hat{C}_i(s')\right)\right) < \epsilon'. \tag{6.1.3}$$

We choose $s \in \,]0, s'[$ so large that

$$k\left((s')^{2n} - s^{2n}\right) < \epsilon'. \tag{6.1.4}$$

We abbreviate $d := (s' - s)/2$. For each $\delta \in [0, d]$ we define

$$C_i(\delta) = \{z + ((i-1)s - \delta, -\delta, \dots, -\delta) \mid z \in C(s + 2\delta)\},$$

and we abbreviate $C_i = C_i(0)$ and $C_i' = C_i(d)$, $i = 1, \dots, k$, cf. Figure 6.2. After choosing $s \in \,]0, s'[$ larger if necessary we can assume that $2d \leq s$ so that the cubes C_i' are disjoint. We define $w_i \in \mathbb{R}^{2n}$ through the identity $\tau_{w_i}\left(\hat{C}_i(s')\right) = C_i'$, and we define the symplectic embedding $\gamma_i' \colon C_i' \hookrightarrow M$ by

$$\gamma_i' = \hat{\gamma}_i \circ \tau_{-w_i} \colon C_i' \hookrightarrow M.$$

We denote the restriction of γ_i' to C_i by γ_i, and we write

$$\gamma = \coprod_{i=1}^{k} \gamma_i \colon \coprod_{i=1}^{k} C_i \hookrightarrow M \quad \text{and} \quad \gamma' = \coprod_{i=1}^{k} \gamma_i' \colon \coprod_{i=1}^{k} C_i' \hookrightarrow M.$$

Since γ and γ' are symplectic, we have $\mu\left(\gamma\left(\coprod C_i\right)\right) = \sum |C_i| = ks^{2n}$ and

$$\mu\left(\gamma'\left(\coprod C_i'\right) \setminus \gamma\left(\coprod C_i\right)\right) = \sum |C_i' \setminus C_i| = k\left((s')^{2n} - s^{2n}\right).$$

In view of the inequalities (6.1.3) and (6.1.4) we can therefore estimate

$$\begin{aligned}
\mu(M) &= \mu\left(\gamma\left(\coprod C_i\right)\right) + \mu\left(\gamma'\left(\coprod C_i'\right) \setminus \gamma\left(\coprod C_i\right)\right) + \mu\left(M \setminus \gamma'\left(\coprod C_i'\right)\right) \\
&= ks^{2n} + k\left((s')^{2n} - s^{2n}\right) + \mu\left(M \setminus \hat{\gamma}\left(\coprod C_i(s')\right)\right) \\
&< ks^{2n} + \epsilon' + \epsilon' \\
&= ks^{2n} + 2\epsilon'.
\end{aligned} \tag{6.1.5}$$

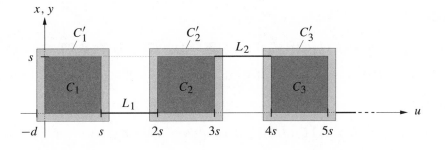

Figure 6.2. The cubes C_i and C_i', $i = 1, 2, 3$, and the lines L_i, $i = 1, 2$.

Step 2. Connecting the cubes. In this step we extend the embedding $\gamma \colon \coprod C_i \hookrightarrow M$ to a symplectic embedding of a connected domain. For $i = 1, \ldots, k-1$ we abbreviate

$$I_i = [\,(2i-1)s, \, 2is\,],$$

and we define straight lines $L_i(t)\colon I_i \hookrightarrow \mathbb{R}^2 \times \mathbb{R}^{n-1}(x) \times \mathbb{R}^{n-1}(y)$ by

$$L_i(t) = \begin{cases} (t, 0, 0, 0) & \text{if } i \text{ is odd,} \\ (t, 0, 0, s1_y) & \text{if } i \text{ is even,} \end{cases}$$

cf. Figure 6.2. Then

$$\left.\begin{aligned} L_i(t) \in C_i' & \quad \text{if } t \in [\,(2i-1)s, \, (2i-1)s+d\,[, \\ L_i(t) \in C_{i+1}' & \quad \text{if } t \in \,]\,2is-d, \, 2is\,]. \end{aligned}\right\} \tag{6.1.6}$$

For $\delta > 0$ we define the "δ-neighbourhood" $N_i(\delta)$ of L_i in $\mathbb{R}^{2n}$ by

$$N_i(\delta) = \begin{cases} I_i \times \,]-\delta, \delta\,[\times (\,]-\delta, \delta\,[\times \,]-\delta, \delta\,[)^{n-1} & \text{if } i \text{ is odd,} \\ I_i \times \,]-\delta, \delta\,[\times (\,]-\delta, \delta\,[\times \,]s-\delta, s+\delta\,[)^{n-1} & \text{if } i \text{ is even.} \end{cases}$$

Proposition 6.1.3. *There exist $\delta \in \,]0, d/8[$ and a symplectic embedding*

$$\rho \colon \coprod_{i=1}^{k} C_i(d/3) \cup \coprod_{i=1}^{k-1} N_i(\delta) \hookrightarrow M.$$

Proof. By construction of the map γ', the set $M \setminus \gamma' \left(\coprod C_i'\right)$ is connected. Using this and the inclusions (6.1.6) we find a smooth embedding

$$\coprod_{i=1}^{k-1} \lambda_i \colon \coprod_{i=1}^{k-1} L_i \hookrightarrow M \setminus \gamma \left(\coprod C_i\right)$$

such that

$$\lambda_i(L_i(t)) = \gamma_i'(L_i(t)) \qquad \text{if } t \in [\,(2i-1)s,\ (2i-1)s + d/2\,], \qquad (6.1.7)$$

$$\lambda_i(L_i(t)) = \gamma_{i+1}'(L_i(t)) \quad \text{if } t \in [\,2is - d/2,\ 2is\,], \qquad\qquad (6.1.8)$$

and such that

$$\lambda_i(L_i(t)) \in M \setminus \gamma'\left(\coprod C_i(d/2)\right) \quad \text{if } t \in [\,(2i-1)s + d/2,\ 2is - d/2\,]. \quad (6.1.9)$$

We are now going to construct a symplectic extension of λ_1 to a neighbourhood of L_1. A symplectic extension of λ_i to a neighbourhood of L_i for $i \geq 2$ can be constructed the same way. We denote by

$$\{\,e_1(t), e_2(t), \ldots, e_{2n-1}(t), e_{2n}(t)\,\} = \left\{ \tfrac{\partial}{\partial x_1}, \tfrac{\partial}{\partial y_1}, \ldots, \tfrac{\partial}{\partial x_n}, \tfrac{\partial}{\partial y_n} \right\}$$

the standard symplectic frame of the tangent space $T_{L_1(t)}\mathbb{R}^{2n}$.

Lemma 6.1.4. *There exists a smooth 1-parameter family of symplectic frames $\{f_j(t)\}$, $j = 1, \ldots, 2n$, along the curve $\lambda_1(t) = \lambda_1(L_1(t))$ such that*

$$f_1(t) = \tfrac{d}{dt}\lambda_1(t) \quad \text{for all } t \in [s, 2s]$$

and such that for all $j = 1, \ldots, 2n$,

$$f_j(t) = \left(T_{L_1(t)}\gamma_1'\right)(e_j(t)) \quad \text{if } t \in [\,s,\ s + d/2\,], \qquad (6.1.10)$$

$$f_j(t) = \left(T_{L_1(t)}\gamma_2'\right)(e_j(t)) \quad \text{if } t \in [\,2s - d/2,\ 2s\,]. \qquad (6.1.11)$$

Proof. For $t \in [s, 2s]$ we define $f_1(t) = \tfrac{d}{dt}\lambda_1(t)$. In view of the identities (6.1.7) and (6.1.8) for $i = 1$ the assertions (6.1.10) and (6.1.11) for $j = 1$ are met. Using (6.1.10) for $j = 1$ and that γ_1' is symplectic we find a smooth vector field $f_2'(t)$ along $\lambda_1(t)$ such that

$$f_2'(t) = \left(T_{L_1(t)}\gamma_1'\right)(e_2(t)) \quad \text{if } t \in [s, s + d/2] \qquad (6.1.12)$$

and such that $\omega\left(f_1(t), f_2'(t)\right) = 1$ for all $t \in [s, 2s]$. Choose a smooth function $c(t) \colon [s, 2s] \to [0, 1]$ such that

$$c(t) = \begin{cases} 0 & \text{if } t \leq 2s - d/2, \\ 1 & \text{if } t \geq 2s - d/3, \end{cases}$$

and define a second smooth vector field $f_2''(t)$ along $\lambda_1(t)$ by

$$f_2''(t) = \begin{cases} 0 & \text{if } t \leq 2s - d/2, \\ c(t)\left(T_{L_1(t)}\gamma_2'\right)(e_2(t)) & \text{if } t > 2s - d/2. \end{cases}$$

We define the smooth vector field $f_2(t)$ along $\lambda_1(t)$ by

$$f_2(t) = c(t) f_2''(t) + (1 - c(t)) f_2'(t).$$

In view of the formula (6.1.12) and the definition of $f_2''(t)$ the assertions (6.1.10) and (6.1.11) for $j = 2$ are met, and since γ_2' is symplectic, we find $\omega(f_1(t), f_2(t)) = 1$ for all $t \in [s, 2s]$. The linear subspace V_t of $T_{\lambda_1(t)} M$ spanned by $f_1(t)$ and $f_2(t)$ is therefore a symplectic subspace. Denote the symplectic complement of V_t in $T_{\lambda_1(t)} M$ by $V_t^\perp$. Since the linear symplectic group $\mathrm{Sp}(n - 1; \mathbb{R})$ is path-connected, we find a smooth 1-parameter family of symplectic frames $\{f_3(t), \ldots, f_{2n}(t)\}$ of $V_t^\perp$, $t \in [s, 2s]$, such that the assertions (6.1.10) and (6.1.11) are met for all $j = 3, \ldots, 2n$. The family of symplectic frames $\{f_1(t), f_2(t), f_3(t), \ldots, f_{2n}(t)\}$ of $V_t \oplus V_t^\perp = T_{\lambda_1(t)} M$ is as desired. $\qquad\square$

Lemma 6.1.5. *There exist $\kappa_1 > 0$ and a smooth embedding*

$$\sigma_1 : C_1(d/3) \cup N_1(\kappa_1) \cup C_2(d/3) \hookrightarrow M$$

such that

$$\sigma_1\big|_{C_1(d/3)} = \gamma_1', \quad \sigma_1\big|_{L_1} = \lambda_1, \quad \sigma_1\big|_{C_2(d/3)} = \gamma_2',$$

and such that $\left(T_{L_1(t)}\sigma_1\right)(e_j(t)) = f_j(t)$ *for all* $t \in [s, 2s]$ *and* $j = 1, \ldots, 2n$.

Proof. Define the smooth embedding $\sigma : C_1(d/2) \cup L_1 \cup C_2(d/2) \hookrightarrow M$ by

$$\sigma\big|_{C_1(d/2)} = \gamma_1', \quad \sigma\big|_{L_1} = \lambda_1, \quad \sigma\big|_{C_2(d/2)} = \gamma_2'.$$

Applying the Whitney extension theorem to the restriction of σ to the closed set $\overline{C_1(d/3)} \cup L_1 \cup \overline{C_2(d/3)}$ we find $\hat\kappa > 0$ and a smooth map

$$\hat\sigma : C_1(d/3) \cup N_1(\hat\kappa) \cup C_2(d/3) \to M$$

which extends the restriction of σ to $C_1(d/3) \cup L_1 \cup C_2(d/3)$. We read off from the identities (6.1.7) and (6.1.8) and the inclusions (6.1.9) for $i = 1$ that the restriction of σ to the compact set $\overline{C_1(d/3)} \cup L_1 \cup \overline{C_2(d/3)}$ is an embedding. After choosing $\hat\kappa > 0$ smaller if necessary, we can therefore assume that $\hat\sigma$ is an embedding.

For each $t \in [s, 2s]$ we define the frame $\{\hat e_j(t)\}$ of $T_{L_1(t)}\mathbb{R}^{2n}$ by

$$\hat e_j(t) = \hat\sigma^* f_j(t), \quad j = 1, \ldots, 2n. \tag{6.1.13}$$

Since $f_1(t) = \frac{d}{dt}\lambda_1(t)$ and $\hat\sigma\big|_{L_1} = \lambda_1$ we can compute

$$\begin{aligned}
\hat e_1(t) &= \hat\sigma^* f_1(t) \\
&= \hat\sigma^*\left(\tfrac{d}{dt}\lambda_1(t)\right) \\
&= \hat\sigma^*\left(T_{L_1(t)}\lambda_1\right)(e_1(t)) \\
&= \hat\sigma^*\left(T_{L_1(t)}\hat\sigma\right)(e_1(t)) \\
&= e_1(t) \tag{6.1.14}
\end{aligned}$$

for all $t \in [s, 2s]$. Similarly, the identities (6.1.10), (6.1.11) and $\hat{\sigma}|_{C_1(d/3)} = \gamma_1'$, $\hat{\sigma}|_{C_2(d/3)} = \gamma_2'$ imply that for all $j = 1, \ldots, 2n$,

$$\hat{e}_j(t) = e_j(t) \quad \text{if } t \in [s, s + d/3] \cup [2s - d/3, 2s]. \tag{6.1.15}$$

Define the smooth map $\alpha \colon [s, 2s] \times \mathbb{R}^{2n-1} \to \mathbb{R}^{2n}$ by

$$\alpha\left(L_1(t) + \sum_{j=2}^{2n} b_i e_i(t)\right) = L_1(t) + \sum_{j=2}^{2n} b_i \hat{e}_i(t). \tag{6.1.16}$$

In view of the identities (6.1.14) and (6.1.15) the map α restricts to the identity on $\left(C_1(d/3) \cap N_1(\hat{\kappa})\right) \cup L_1 \cup \left(C_2(d/3) \cap N_1(\hat{\kappa})\right)$. We choose $\kappa_1 > 0$ so small that the restriction of α to $N_1(\kappa_1)$ is an embedding whose image is contained in $N_1(\hat{\kappa})$. We can now define a smooth embedding

$$\sigma_1 \colon C_1(d/3) \cup N_1(\kappa_1) \cup C_2(d/3) \hookrightarrow M$$

by

$$\sigma_1|_{C_1(d/3)} = \gamma_1', \quad \sigma_1|_{N_1(\kappa_1)} = \hat{\sigma} \circ \alpha, \quad \sigma_1|_{C_2(d/3)} = \gamma_2'.$$

For $z \in L_1$ we then have $\sigma_1(z) = \hat{\sigma}(\alpha(z)) = \hat{\sigma}(z) = \lambda_1(z)$. Moreover, we read off from the identity (6.1.14) and the definition (6.1.16) of α that $\left(T_{L_1(t)}\alpha\right)(e_j(t)) = \hat{e}_j(t)$. Together with the definitions (6.1.13) we therefore conclude that

$$\left(T_{L_1(t)}\sigma_1\right)(e_j(t)) = \left(T_{L_1(t)}\hat{\sigma}\right)\left(T_{L_1(t)}\alpha\right)(e_j(t)) = \left(T_{L_1(t)}\hat{\sigma}\right)(\hat{e}_j(t)) = f_j(t)$$

for all $t \in [s, 2s]$ and $j = 1, \ldots, 2n$. The proof of Lemma 6.1.5 is complete. $\quad\square$

Since the maps γ_1' and γ_2' are symplectic and since the frames $\{f_j(t)\}$ along $\lambda_1(t)$ are symplectic, the map σ_1 guaranteed by Lemma 6.1.5 is symplectic on the set $C_1(d/3) \cup L_1 \cup C_2(d/3)$. Applying the proof of Lemma 3.14 in [62] to $L_1 \subset N_1(\kappa_1)$ and to the symplectic forms ω_0 and $\sigma_1^*\omega$, we therefore find $\delta_1 > 0$ and a smooth embedding $\psi_1 \colon N_1(\delta_1) \hookrightarrow N_1(\kappa_1)$ such that

$$\psi_1|_{(N_1(\delta) \cap C_1(d/3)) \cup L_1 \cup (N_1(\delta) \cap C_2(d/3))} = \text{id} \quad \text{and} \quad \psi_1^*\left(\sigma_1^*\omega\right) = \omega_0. \tag{6.1.17}$$

Define the embedding $\hat{\rho}_1 \colon N_1(\delta_1) \hookrightarrow M$ by $\hat{\rho}_1 = \sigma_1 \circ \psi_1$. In view of the first statement in (6.1.17) and the identities $\sigma_1|_{C_1(d/3)} = \gamma_1'$ and $\sigma_1|_{C_2(d/3)} = \gamma_2'$ we have

$$\hat{\rho}_1|_{N_1(\delta_1) \cap C_1(d/3)} = \gamma_1' \quad \text{and} \quad \hat{\rho}_1|_{N_1(\delta_1) \cap C_2(d/3)} = \gamma_2',$$

and in view of the second statement in (6.1.17) we have

$$\hat{\rho}_1^*\omega = (\sigma_1 \circ \psi_1)^*\omega = \psi_1^*(\sigma_1^*\omega) = \omega_0,$$

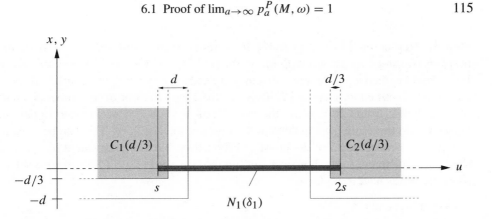

Figure 6.3. The domain $C_1(d/3) \cup N_1(\delta_1) \cup C_2(d/3)$ of ρ_1.

i.e., $\hat{\rho}_1$ is symplectic. We can therefore define a smooth symplectic embedding

$$\rho_1 : C_1(d/3) \cup N_1(\delta_1) \cup C_2(d/3) \hookrightarrow M$$

by

$$\rho_1\big|_{C_1(d/3)} = \gamma_1', \quad \rho_1\big|_{N_1(\delta_1)} = \hat{\rho}_1, \quad \rho_1\big|_{C_2(d/3)} = \gamma_2'. \tag{6.1.18}$$

Proceeding as in the construction of ρ_1 we find $\delta_i > 0$ and symplectic embeddings

$$\rho_i : C_i(d/3) \cup N_i(\delta_i) \cup C_{i+1}(d/3) \hookrightarrow M$$

such that

$$\rho_i\big|_{C_i(d/3)} = \gamma_i', \quad \rho_i\big|_{N_i(\delta_i)} = \hat{\rho}_i, \quad \rho_i\big|_{C_{i+1}(d/3)} = \gamma_{i+1}' \tag{6.1.19}$$

where $\hat{\rho}_i : N_i(\delta_i) \hookrightarrow M$ is a symplectic extension of λ_i, $i = 1, \ldots, k-1$. In view of the identities (6.1.18) and (6.1.19) the map

$$\rho : \coprod_{i=1}^{k} C_i(d/3) \cup \coprod_{i=1}^{k-1} N_i(\delta_i) \to M$$

defined by

$$\rho\big|_{C_i(d/3) \cup N_i(\delta_i)} = \rho_i\big|_{C_i(d/3) \cup N_i(\delta_i)}, \quad i = 1, \ldots, k-1, \quad \rho\big|_{C_k(d/3)} = \rho_{k-1}\big|_{C_k(d/3)}$$

is smooth and symplectic. In view of the inclusions (6.1.9) we finally find some $\delta \in \,]0, \min\{\delta_1, \ldots, \delta_k, d/8\}[$ such that the restriction of ρ to $\coprod C_i(d/3) \cup \coprod N_i(\delta)$ is an embedding. The proof of Proposition 6.1.3 is thus complete. $\qquad \square$

Step 3.   Replacing $\coprod C_i \cup \coprod N_i(\delta)$ by a more convenient set.   In view of the previous two steps we are left with filling the set $\coprod C_i \cup \coprod N_i(\delta)$ with thin polydiscs. If we would embed a part of a polydisc into the cube C_1 by using the multiple folding technique described in Section 5.2, the x-width of the last floor of the embedded part of the polydisc would be s, while the x-width of $N_1(\delta)$ is $\delta < s$. In order to pass to C_2 we would therefore have to deform the fibres of the last floor. This can be done in a similar way as in Step 3 of the proof of Proposition 5.3.3. For technical reasons we shall take a different route, however, and replace the set $\coprod C_i \cup \coprod N_i(\delta)$ by a set all of whose fibres have x-width δ.

We abbreviate

$$\nu = \frac{s^2}{\delta}.$$

We define the subset K of $\mathbb{R}^{2n}$ by

$$K = \left\{ (x_1, y_1, \ldots, x_n, y_n) \in \mathbb{R}^{2n} \mid 0 < x_i < \delta,\ 0 < y_i < \nu,\ i = 1, \ldots, n \right\},$$

and for $i = 1, \ldots, k$ we define subsets K_i of $\mathbb{R}^{2n}$ by

$$K_i = \{ z + ((i-1)(\nu + s), 0, \ldots, 0) \mid z \in K \}, \quad i = 1, \ldots, k.$$

Notice that the sets K_i are symplectomorphic to the cubes C_i. We abbreviate

$$I_i^s = [\, (i-1)(\nu + s) + \nu,\ i(\nu + s) \,], \quad i = 1, \ldots, k-1,$$

and define the "δ-halfneighbourhood" H_i in $\mathbb{R}^{2n}$ by

$$H_i = \begin{cases} I_i^s \times {]0, \delta[} \times ({]0, \delta[} \times {]0, \delta[})^{n-1} & \text{if } i \text{ is odd,} \\ I_i^s \times {]0, \delta[} \times ({]0, \delta[} \times {]\nu - \delta, \nu[})^{n-1} & \text{if } i \text{ is even,} \end{cases}$$

cf. Figure 6.7.

Proposition 6.1.6. *There exists a symplectic embedding*

$$\xi : \coprod_{i=1}^{k} K_i \cup \coprod_{i=1}^{k-1} H_i \ \hookrightarrow\ \coprod_{i=1}^{k} C_i(d/3) \cup \coprod_{i=1}^{k-1} N_i(\delta).$$

Proof. We start with

Lemma 6.1.7. *There exists a symplectic embedding*

$$\alpha : {]0, \delta[} \times {]0, \nu[} \ \hookrightarrow\ {]0, s + d/3[} \times {]0, s[}$$

which restricts to the identity on $\{ (x, y) \mid y < \delta \}$ and to the translation $(x, y) \mapsto (x, y - \nu + s)$ on $\{ (x, y) \mid y > \nu - \delta \}$, cf. Figure 6.4.

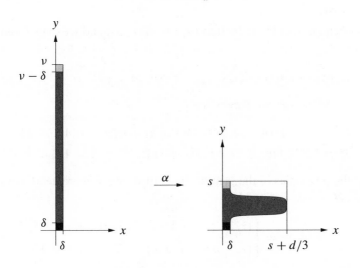

Figure 6.4. The embedding α.

Proof. Choose a smooth function $h\colon]0, v[\to [1, (s+d/3)/\delta]$ such that

(i) $h(w) = 1$ if $w \in]0, \delta[\cup]v - \delta/3, v[$.

Since $\delta < d/8$ and $d < s$ we find by computation that

$$2\delta + \frac{v-2\delta}{(s+d/3)/\delta} < s < v.$$

We may therefore further require that

(ii) $\displaystyle\int_0^v \frac{1}{h(w)}\, dw = s.$

Then the map

$$\alpha\colon]0, \delta[\times]0, v[\to \mathbb{R}^2, \quad (x, y) \mapsto \left(h(y)x, \int_0^y \frac{1}{h(w)}\, dw\right)$$

is the desired symplectic embedding. $\square$

The $(n-1)$-fold product $\alpha \times \cdots \times \alpha$ embeds the fibres $(]0, \delta[\times]0, v[)^{n-1}$ of K_i into the fibres of $C_i(d/3)$ in such a way that the fibres of H_i are embedded into the fibres of $N_i(\delta)$. We next embed the base of $\coprod K_i \cup \coprod H_i$ into the base of $\coprod C_i(d/3) \cup \coprod N_i(\delta)$. We denote by $\mathcal{K}$ and $\mathcal{C}$ the projections of the sets $\coprod K_i \cup \coprod H_i$ and $\coprod C_i(d/3) \cup \coprod N_i(\delta)$ onto the (u, v)-plane, cf. Figure 6.5.

Lemma 6.1.8. *There exists a symplectic embedding $\bar{\alpha}\colon \mathcal{K} \hookrightarrow \mathcal{C}$, cf. Figure 6.5.*

Proof. We denote by ζ the reflection $(u, v) \mapsto (v, u)$, and we define translations τ_i^+ and τ_i^- by

$$\tau_i^+(u, v) = (u + (i - 1)2s, v), \quad \tau_i^-(u, v) = (u - (i - 1)(v + s), v),$$

$i = 1, \ldots, k$. Moreover, we abbreviate

$$I_i^v = \,] (i - 1)(v + s), (i - 1)(v + s) + v\,[, \quad i = 1, \ldots, k,$$
$$I_i^s = [\,(i - 1)(v + s) + v, i(v + s)\,], \quad\quad i = 1, \ldots, k - 1.$$

In view of the properties of the symplectic embedding α guaranteed by Lemma 6.1.7 the map $\bar{\alpha} \colon \mathcal{K} \to \mathbb{R}^2$ defined by

$$\bar{\alpha}(u, v) = \begin{cases} \left(\tau_i^+ \circ \zeta \circ \alpha \circ \zeta \circ \tau_i^-\right)(u, v) & \text{if } u \in I_i^v, \\ \left(\tau_{i+1}^+ \circ \tau_{i+1}^-\right)(u, v) & \text{if } u \in I_i^s, \end{cases}$$

is a smooth symplectic embedding as desired, cf. Figure 6.5.   □

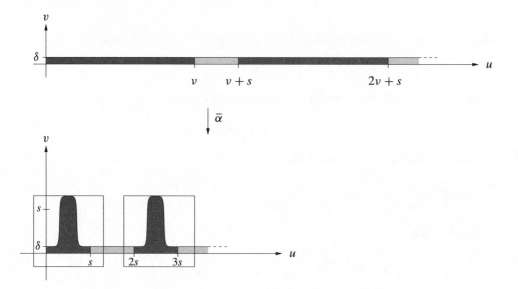

Figure 6.5. The embedding $\bar{\alpha} \colon \mathcal{K} \hookrightarrow \mathcal{C}$.

We finally define ξ to be the restriction to $\coprod K_i \cup \coprod H_i$ of the symplectic embedding

$$\bar{\alpha} \times \alpha \times \cdots \times \alpha \colon \mathcal{K} \times (\,]0, \delta[\times \,]0, v[)^{n-1} \hookrightarrow \mathcal{C} \times (\,]0, s + d/3[\times \,]0, s[)^{n-1}.$$

By construction, $\xi\left(\coprod K_i \cup \coprod H_i\right) \subset \coprod C_i(d/3) \cup \coprod N_i(\delta)$, and so the proof of Proposition 6.1.6 is complete.   □

Step 4. Filling $\coprod K_i \cup \coprod H_i$ with thin polydiscs,. Let k, s, d and δ be the numbers found in the previous three steps, and recall that $\epsilon' = \epsilon/3$ and $\nu = s^2/\delta$. For each $\hat{a} > 3\pi$ we let $N(\hat{a})$ and

$$\varphi_{\hat{a}} \colon R^n(\hat{a}, \pi) \hookrightarrow R^n((N(\hat{a}) + 1)\pi)$$

be the natural number and the symplectic embedding found in Proposition 5.2.3. By Proposition 5.2.3 (ii) there exists $\hat{a}_0 > 3\pi$ such that for all $\hat{a} \geq \hat{a}_0$,

$$\frac{\left| \varphi_{\hat{a}} \left(R^n(\hat{a}, \pi) \right) \right|}{\left| R^n \left((N(\hat{a}) + 1)\pi \right) \right|} > 1 - \frac{\epsilon'}{k \, s^{2n}}. \tag{6.1.20}$$

In view of the definition (5.2.8) we have $N(\hat{a}) \geq N(\hat{a}_0)$ whenever $\hat{a} \geq \hat{a}_0$, and in view of (5.2.11) we have $N(\hat{a}) \to \infty$ as $\hat{a} \to \infty$. Choosing $\hat{a}_0$ larger if necessary, we can therefore assume that

$$\frac{\nu}{N(\hat{a}) + 1} \leq \delta \quad \text{for all } \hat{a} \geq \hat{a}_0. \tag{6.1.21}$$

We set $a_0 = k\hat{a}_0$, and we define the function $\lambda \colon [a_0, \infty[\to \mathbb{R}$ by

$$\lambda(a) = \frac{s}{\sqrt{\left(N\left(\frac{a}{k}\right) + 1\right)\pi}}. \tag{6.1.22}$$

Proposition 6.1.9. *For each $a \geq a_0$ there exists a symplectic embedding*

$$\Psi_a \colon \lambda(a) R^n(a, \pi) \hookrightarrow \coprod_{i=1}^{k} K_i \cup \coprod_{i=1}^{k-1} H_i.$$

Proof. Fix $a \geq a_0$. We set $\hat{a} = a/k$, and we abbreviate $N = N(\hat{a})$ and $\lambda = \lambda(a)$. In abuse of notation we denote the dilatation $z \mapsto \lambda z$, $z \in \mathbb{R}^{2n}$, also by λ. Moreover, we define the linear symplectomorphism σ of $\mathbb{R}^{2n}$ by

$$\sigma(u, v, x, y) = \left(\tfrac{\lambda}{\delta} u, \tfrac{\delta}{\lambda} v, \tfrac{\delta}{\lambda} x, \tfrac{\lambda}{\delta} y\right).$$

Then $\sigma\left(\lambda R^n((N+1)\pi)\right) = K$, and so the map $\psi_{\hat{a}} \colon \lambda R^n(\hat{a}, \pi) \to \mathbb{R}^{2n}$ defined by

$$\psi_{\hat{a}} = \sigma \circ \lambda \circ \varphi_{\hat{a}} \circ \lambda^{-1}$$

symplectically embeds $\lambda R^n(\hat{a}, \pi)$ into K. Using Proposition 5.2.3 (i), the definitions of $\psi_{\hat{a}}$, σ and λ, and $\nu = s^2/\delta$, we find that

$$\psi_{\hat{a}}(u, v, x, y) = \sigma(u, v, x, y) \qquad \text{if } u < \lambda\pi, \tag{6.1.23}$$

$$\psi_{\hat{a}}(u, v, x, y) = \sigma(u, v, x, y) + \left(\nu - \tfrac{\nu}{N+1}\tfrac{\hat{a}}{\pi}, 0, 0, \left(\nu - \tfrac{\nu}{N+1}\right)1_y\right) \tag{6.1.24}$$
$$\text{if } u > \lambda(\hat{a} - \pi),$$

cf. the left picture in Figure 6.6.

Denote the reflection $(u, v, x, y) \mapsto (u, v, x, -y)$ by ζ_y. Since $\psi_{\hat{a}}$ symplectically embeds $\lambda R^n(\hat{a}, \pi)$ into K, the map $\overline{\psi_{\hat{a}}}$ defined by

$$\overline{\psi_{\hat{a}}} = \tau_{v 1_y} \circ \zeta_y \circ \psi_{\hat{a}} \circ \tau_{\lambda \pi 1 y} \circ \zeta_y$$

symplectically embeds $\lambda R^n(\hat{a}, \pi)$ into K as well, cf. Figure 6.6. We read off from the identities (6.1.23) and (6.1.24) that

$$\overline{\psi_{\hat{a}}}(u, v, x, y) = \sigma(u, v, x, y) + \left(0, 0, 0, \left(v - \tfrac{v}{N+1}\right)1_y\right) \quad \text{if } u < \lambda \pi, \qquad (6.1.25)$$

$$\overline{\psi_{\hat{a}}}(u, v, x, y) = \sigma(u, v, x, y) + \left(v - \tfrac{v}{N+1}\tfrac{\hat{a}}{\pi}, 0, 0, 0\right) \quad \text{if } u > \lambda(\hat{a} - \pi), \quad (6.1.26)$$

cf. the right picture in Figure 6.6.

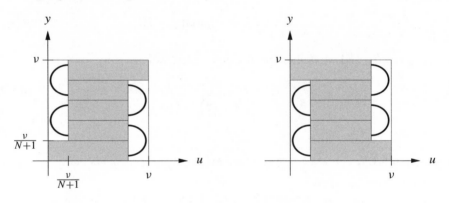

Figure 6.6. The embeddings $\psi_{\hat{a}}$ and $\overline{\psi_{\hat{a}}} \colon \lambda R^n(\hat{a}, \pi) \hookrightarrow K$.

We are now going to use the embeddings $\psi_{\hat{a}}$ and $\overline{\psi_{\hat{a}}}$ of $\lambda R^n(\hat{a}, \pi)$ into K to construct a symplectic embedding Ψ_a of $\lambda R^n(a, \pi)$ into $\coprod K_i \cup \coprod H_i$, cf. Figure 6.7. As in Step 1 of Section 3.2 we find a symplectic embedding

$$\beta \colon \,]0, v[\times]0, \delta[\, \hookrightarrow \,]0, v + s[\times]0, \delta[$$

which restricts to the identity on $\left\{u \le v - \tfrac{v}{N+1}\right\}$ and to the translation $(u, v) \mapsto (u + s, v)$ on $\left\{u \ge v - \tfrac{1}{2}\tfrac{v}{N+1}\right\}$. For $i = 1, \ldots, k$ we define

$$R_i := \left\{ (u, v, x, y) \in \lambda R^n(a, \pi) \mid (i - 1)\lambda \hat{a} < u \le i \lambda \hat{a} \right\}.$$

Then

$$\lambda R^n(a, \pi) = \coprod_{i=1}^{k} R_i.$$

We denote the translation $(u, v, x, y) \mapsto (u + (i - 1)(v + s), v, x, y)$ of $\mathbb{R}^{2n}$ by τ_i. For each odd $i \in \{1, \ldots, k - 1\}$ we define the map $\psi_i \colon R_i \to \mathbb{R}^{2n}$ by

$$\psi_i(u, v, x, y) = \begin{cases} \left(\tau_i \circ \overline{\psi_{\hat{a}}} \circ \tau_i^{-1}\right)(u, v, x, y) & \text{if } u < \lambda(i\hat{a} - \pi), \\ \left(\tau_i \circ (\beta \times \mathrm{id}_{2n-2}) \circ \overline{\psi_{\hat{a}}} \circ \tau_i^{-1}\right)(u, v, x, y) & \text{if } u \geq \lambda(i\hat{a} - \pi). \end{cases}$$

The estimate (6.1.21) and formula (6.1.26) imply that ψ_i is a smooth symplectic embedding of R_i into $K_i \cup H_i$ for which

$$\psi_i(u, v, x, y) = \sigma(u, v, x, y) + (u_i, 0, 0, 0) \quad \text{if } u \geq \lambda\left(i\hat{a} - \tfrac{\pi}{2}\right) \qquad (6.1.27)$$

where u_i is such that the right end of R_i is mapped to the right end of H_i, cf. Figure 6.7. For each even $i \in \{2, \ldots, k - 1\}$ we define the map $\psi_i \colon R_i \to \mathbb{R}^{2n}$ by

$$\psi_i(u, v, x, y) = \begin{cases} \left(\tau_i \circ \psi_{\hat{a}} \circ \tau_i^{-1}\right)(u, v, x, y) & \text{if } u < \lambda(i\hat{a} - \pi), \\ \left(\tau_i \circ (\beta \times \mathrm{id}_{2n-2}) \circ \psi_{\hat{a}} \circ \tau_i^{-1}\right)(u, v, x, y) & \text{if } u \geq \lambda(i\hat{a} - \pi). \end{cases}$$

The estimate (6.1.21) and formula (6.1.24) imply that ψ_i is a smooth symplectic embedding of R_i into $K_i \cup H_i$ for which

$$\psi_i(u, v, x, y) = \sigma(u, v, x, y) + \left(u_i, 0, 0, \left(v - \tfrac{v}{N+1}\right)1_y\right) \quad \text{if } u \geq \lambda\left(i\hat{a} - \tfrac{\pi}{2}\right) \qquad (6.1.28)$$

where u_i is such that the right end of R_i is mapped to the right end of H_i, cf. Figure 6.7. We finally define the symplectic embedding $\psi_k \colon R_k \hookrightarrow K_k$ by

$$\psi_k(u, v, x, y) = \begin{cases} \left(\tau_i \circ \overline{\psi_{\hat{a}}} \circ \tau_i^{-1}\right)(u, v, x, y) & \text{if } k \text{ is odd}, \\ \left(\tau_i \circ \psi_{\hat{a}} \circ \tau_i^{-1}\right)(u, v, x, y) & \text{if } k \text{ is even}. \end{cases}$$

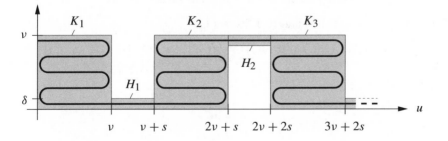

Figure 6.7. The embedding $\Psi_a \colon \lambda R^n(a, \pi) \hookrightarrow \coprod_{i=1}^k K_i \cup \coprod_{i=1}^{k-1} H_i$.

In view of the identities (6.1.23), (6.1.25), (6.1.27) and (6.1.28) the embedding

$$\Psi_a: \lambda R^n(a, \pi) \hookrightarrow \coprod_{i=1}^{k} K_i \cup \coprod_{i=1}^{k-1} H_i$$

defined by

$$\Psi_a|_{R_i} = \psi_i, \quad i = 1, \ldots, k,$$

is a smooth symplectic embedding. The proof of Proposition 6.1.9 is complete. $\square$

Step 5. End of the proof of Theorem 6.1.1. We let $\epsilon > 0$ be as in Theorem 6.1.1, set $\epsilon' = \epsilon/3$ and let k and s be as introduced after the proof of Lemma 6.1.2. We choose $a_0 = a_0(\epsilon)$ as before Proposition 6.1.9, fix $a \geq a_0$ and define $\lambda(a)$ as in (6.1.22).

Lemma 6.1.10. *We have*

$$\left| \lambda(a) R^n(a, \pi) \right| > ks^{2n} - \epsilon'. \tag{6.1.29}$$

Proof. We set again $\hat{a} = a/k$, $N = N(\hat{a})$ and $\lambda = \lambda(a)$. Since the embedding $\varphi_{\hat{a}}: R^n(\hat{a}, \pi) \hookrightarrow R^n((N+1)\pi)$ is volume preserving, we have

$$\left| \lambda R^n(a, \pi) \right| = \lambda^{2n} k \left| R^n(\hat{a}, \pi) \right| = k\lambda^{2n} \left| \varphi_{\hat{a}} \left(R^n(\hat{a}, \pi) \right) \right|, \tag{6.1.30}$$

and multiplying the inequality (6.1.20) by

$$k\lambda^{2n} \left| R^n \left((N+1)\pi \right) \right| = k\lambda^{2n} \left((N+1)\pi \right)^n = ks^{2n}$$

we find that

$$k\lambda^{2n} \left| \varphi_{\hat{a}}(R^n(\hat{a}, \pi)) \right| > ks^{2n} - \epsilon'. \tag{6.1.31}$$

Lemma 6.1.10 now follows from combining the identity (6.1.30) with the estimate (6.1.31). $\square$

Composing the symplectic embeddings Ψ_a, ξ and ρ guaranteed by Proposition 6.1.9, Proposition 6.1.6 and Proposition 6.1.3 we obtain the symplectic embedding

$$\Phi_a := \rho \circ \xi \circ \Psi_a: \lambda(a) R^n(a, \pi) \hookrightarrow M.$$

Using the estimates (6.1.5) and (6.1.29) we find

$$\begin{aligned} \mu \left(M \setminus \Phi_a \left(\lambda(a) R^n(a, \pi) \right) \right) &= \mu(M) - \mu \left(\Phi_a \left(\lambda(a) R^n(a, \pi) \right) \right) \\ &= \mu(M) - \left| \lambda(a) R^n(a, \pi) \right| \\ &< \left(ks^{2n} + 2\epsilon' \right) - \left(ks^{2n} - \epsilon' \right) \\ &= 3\epsilon' \\ &= \epsilon. \end{aligned}$$

This is the required estimate in Theorem 6.1.1 and so the proof of Theorem 6.1.1 is complete. $\square$

6.2 Proof of $\displaystyle\lim_{a \to \infty} p_a^E(M, \omega) = 1$

We recall from the introduction that for every $a \geq \pi$ the real number $p_a^E(M, \omega)$ is defined by

$$p_a^E(M, \omega) = \sup_\lambda \frac{|\lambda E(\pi, \ldots, \pi, a)|}{\mathrm{Vol}(M, \omega)}$$

where the supremum is taken over all those λ for which $\lambda E^{2n}(\pi, \ldots, \pi, a)$ symplectically embeds into (M, ω). Corollary 5.3.2 (i) implies that the first statement in Theorem 3 in Section 1.3.2 is a consequence of

Theorem 6.2.1. *For every $\epsilon > 0$ there exists a number $a_0 = a_0(\epsilon) > \pi$ having the following property. For every $a \geq a_0$ there exist a number $\lambda(a) > 0$ and a symplectic embedding $\Phi_a \colon \lambda(a) T^n(a, \pi) \hookrightarrow M$ such that*

$$\mu\left(M \setminus \Phi_a\left(\lambda(a) T^n(a, \pi)\right)\right) < \epsilon.$$

Proof. We shall proceed along the same lines as in the proof of Theorem 6.1.1. We shall first fill almost all of M with finitely many symplectically embedded balls whose closures are disjoint, and connect these balls by neighbourhoods of lines. Using Corollary 5.3.2 (ii) and Proposition 5.3.3 we can then almost fill the balls with symplectically embedded parts of $\lambda(a) T^n(a, \pi)$, and we shall use the neighbourhoods of the lines to pass from one ball to another. The proof of Theorem 6.2.1 is substantially more difficult than the proof of Theorem 6.1.1, however. The first reason is that for $n \geq 3$ there is no elementary method of symplectically filling M with balls. We shall overcome this difficulty by using a result of McDuff and Polterovich in [61].

The second reason is that symplectically filling a ball with a part of $\lambda(a) T^n(a, \pi)$ is more difficult than symplectically filling a cube with a polydisc. We have overcome this difficulty in Corollary 5.3.2 (ii) and Proposition 5.3.3. The third reason is that the fibres of $\lambda(a) T^n(a, \pi)$ are not constant. This will merely cause technical complications.

Step 1. Filling M by balls. We denote by $B(r)$ the $2n$-dimensional open ball

$$B(r) = \{z \in \mathbb{R}^{2n} \mid |z| < r\}.$$

Lemma 6.2.2. *For every $\epsilon > 0$ there exists an integer k_0 with the following property. For each integer $k > k_0$ there exist $r = r(k) > 0$ and a symplectic embedding*

$$\eta_k \colon \coprod_{i=1}^k B_i(r) \hookrightarrow M$$

of a disjoint union of k translates $B_i(r)$ of $B(r)$ in $\mathbb{R}^{2n}$ such that

$$\mu\left(M \setminus \eta_k\left(\coprod B_i(r)\right)\right) < \epsilon. \tag{6.2.1}$$

Proof. We follow [61, Remark 1.5.G]. Fix $\epsilon > 0$. In view of Lemma 6.1.2 we find $s > 0$ and a symplectic embedding $\gamma \colon \coprod C_i(s) \hookrightarrow M$ of a disjoint union of m translates of $C(s)$ such that

$$\mu\left(M \setminus \gamma\left(\coprod C_i(s)\right)\right) < \frac{\epsilon}{3}. \tag{6.2.2}$$

Choose $l_0 \in \mathbb{N}$ so large that for all $l \geq l_0$,

$$\frac{(l-1)^n}{l^n} > 1 - \frac{\epsilon}{3ms^{2n}}. \tag{6.2.3}$$

We define the integer k_0 by $k_0 = m\,n!\,l_0^n$. We fix $k > k_0$ and define the integer $l > l_0$ through the inequalities

$$m\,n!\,(l-1)^n < k \leq m\,n!\,l^n. \tag{6.2.4}$$

The crucial ingredient of the proof of Lemma 6.2.2 is the following result which is proved in [61, Corollary 1.5.F]. We abbreviate $\varsigma = s/\sqrt{\pi}$.

Lemma 6.2.3 (Mc Duff–Polterovich). *There exist $r_l > 0$ and a symplectic embedding*

$$\beta_l \colon \coprod_{j=1}^{n!\,l^n} \hat{B}_j(r_l) \hookrightarrow C^{2n}(\pi) = D(\pi) \times \cdots \times D(\pi)$$

of a disjoint union of $n!\,l^n$ translates $\hat{B}_j(r_l)$ of $B(r_l)$ such that

$$\left| C^{2n}(\pi) \setminus \beta_l\left(\coprod \hat{B}_j(r_l)\right) \right| < \frac{\epsilon}{3m\varsigma^{2n}}.$$

Remark 6.2.4. If $n = 2$, Lemma 6.2.3 follows from Lemma 5.3.1 (i) and Lemma 3.1.5, see also 9.3.2 in Chapter 9 and [81]. For general n, however, the only known proof in [61] uses non-elementary, algebro-geometric methods. $\diamond$

Continuing with the proof of Lemma 6.2.2 we define $r = \varsigma r_l$ and $\hat{B}_j(r) = \varsigma \hat{B}_j(r_l)$. For each $i = 1, \ldots, m$ we choose $w_i \in \mathbb{R}^{2n}$ such that the $m\,n!\,l^n$ balls $B_{i,j}(r) = \tau_{-w_i}\left(\hat{B}_j(r)\right)$ are disjoint. In view of Lemma 3.1.5 the disc $D(\pi)$ is symplectomorphic to the square $]0, \sqrt{\pi}\,[\times\,]0, \sqrt{\pi}\,[$. We therefore find a symplectomorphism

$$\sigma \colon C^{2n}(\pi) \to C(\sqrt{\pi}).$$

We finally denote the dilatation $z \mapsto \varsigma z$ of $\mathbb{R}^{2n}$ also by ς and the translation $C(s) \to C_i(s)$ by τ_i. We can now define a symplectic embedding

$$\beta \colon \coprod_{i,j} B_{i,j}(r) \hookrightarrow \coprod_i C_i(s)$$

by

$$\beta\big|_{B_{i,j}(r)} = \left(\tau_i \circ \varsigma \circ \sigma \circ \beta_l \circ \varsigma^{-1} \circ \tau_{w_i}\right)\big|_{B_{i,j}(r)}.$$

In view of the second inequality in (6.2.4) we can choose k members $B_1(r), \ldots, B_k(r)$ of the family $\{B_{i,j}(r)\}$. We define the symplectic embedding

$$\eta_k : \coprod_{i=1}^{k} B_i(r) \hookrightarrow M$$

by $\eta_k = \gamma \circ \beta$. In order to verify the estimate (6.2.1) we first use Lemma 6.2.3 to estimate

$$
\begin{aligned}
\left| \coprod_i C_i(s) \setminus \beta\Big(\coprod_{i,j} B_{i,j}(r) \Big) \right| &= \sum_{i=1}^{m} \left| (\varsigma \circ \sigma)(C^{2n}(\pi)) \setminus \coprod_j (\varsigma \circ \sigma \circ \beta_l)\big(\hat{B}_j(r_l)\big) \right| \\
&= \sum_{i=1}^{m} \varsigma^{2n} \left| C^{2n}(\pi) \setminus \coprod_j \beta_l\big(\hat{B}_j(r_l)\big) \right| \\
&< m \varsigma^{2n} \frac{\epsilon}{3m\varsigma^{2n}} \qquad\qquad (6.2.5) \\
&= \frac{\epsilon}{3}.
\end{aligned}
$$

Moreover, since β is volume preserving, the inclusion $\beta\left(\coprod B_{i,j}(r) \right) \subset \coprod C_i(s)$ implies that

$$m \, n! \, l^n \, |B(r)| \le ms^{2n}. \qquad\qquad (6.2.6)$$

The left inequality in (6.2.4) and the estimates (6.2.6) and (6.2.3) now yield

$$
\begin{aligned}
\left| \coprod B_{i,j}(r) \setminus \coprod B_i(r) \right| &= (m \, n! \, l^n - k) \, |B(r)| \\
&< m \, n! \, (l^n - (l-1)^n) \, |B(r)| \\
&\le ms^{2n} \left(1 - \frac{(l-1)^n}{l^n} \right) \qquad\qquad (6.2.7) \\
&< \frac{\epsilon}{3}.
\end{aligned}
$$

Using the fact that β and γ are both volume preserving and using the estimates (6.2.2), (6.2.5) and (6.2.7) we finally find

$$
\begin{aligned}
&\mu\left(M \setminus \eta_k\big(\coprod B_i(r) \big) \right) \\
&\quad = \mu\left(M \setminus \gamma\big(\coprod C_i(s) \big) \right) + \left| \coprod C_i(s) \setminus \beta\big(\coprod B_{i,j}(r) \big) \right| + \left| \coprod B_{i,j}(r) \setminus \coprod B_i(r) \right| \\
&\quad < \frac{\epsilon}{3} + \frac{\epsilon}{3} + \frac{\epsilon}{3} \\
&\quad = \epsilon.
\end{aligned}
$$

The proof of Lemma 6.2.2 is complete.   □

Let $\epsilon > 0$ be as in Theorem 6.2.1 and set $\epsilon' = \epsilon/5$. In view of Lemma 6.2.2 we find an integer k_0 such that for each integer $k > k_0$ there exists $r_k > 0$ and a symplectic embedding

$$\eta_k : \coprod_{i=1}^{k} B_i(r_k) \hookrightarrow M$$

of a disjoint union of k translates $B_i(r_k)$ of $B(r_k)$ such that

$$\mu\left(M \setminus \eta_k\left(\coprod B_i(r_k)\right)\right) < \epsilon'. \tag{6.2.8}$$

Fix $k > k_0$. In order to bring Proposition 5.3.3 into play we shall next replace the balls $B_i(r_k)$ by trapezoids. We choose $s' \in \,]0, \sqrt{\pi}\, r_k[$ so large that

$$k \tfrac{1}{n!}\left(\pi^n r_k^{2n} - (s')^{2n}\right) < \epsilon', \tag{6.2.9}$$

and we choose $s \in \,]0, s'[$ so large that

$$k \tfrac{1}{n!}\left((s')^{2n} - s^{2n}\right) < \epsilon'. \tag{6.2.10}$$

We abbreviate $d := (s' - s)/(2n)$. For each $\delta \in [0, d]$ we define the trapezoid $T(\delta)$ by

$$T(\delta) = \Delta^n(s + 2n\delta) \times \square^n(s + 2n\delta) \subset \mathbb{R}^n(x) \times \mathbb{R}^n(y).$$

Moreover, we define the translates $T_i(\delta)$ by

$$T_i(\delta) = \{ z + ((i - 1)s - \delta, -\delta, \ldots, -\delta) \mid z \in T(\delta) \},$$

and we abbreviate $T_i = T_i(0)$ and $T_i' = T_i(d)$, $i = 1, \ldots, k$, cf. Figure 6.8 and Figure 6.9. Notice that $T_i(\delta) \subset T_i(\delta')$ whenever $\delta \leq \delta'$. After choosing $s \in \,]0, s'[$ larger if necessary we can assume that $2nd \leq s$ so that the trapezoids T_i' are disjoint.

Composing the linear symplectomorphism

$$T(d) = \Delta^n(s') \times \square^n(s') \to \Delta^n\left((s')^2\right) \times \square^n(1) = T^n\left((s')^2\right), \quad (x, y) \mapsto \left(s'x, \tfrac{1}{s'}y\right)$$

with the symplectic embedding $T^n((s')^2) \hookrightarrow B^{2n}\left(\pi r_k^2\right) = B(r_k)$ guaranteed by Corollary 5.3.2 (ii), we obtain a symplectic embedding

$$\sigma : T(d) \hookrightarrow B(r_k).$$

We define the points v_i and w_i in $\mathbb{R}^{2n}$ through the identities $\tau_{v_i}\left(T_i'\right) = T(d)$ and $\tau_{w_i}\left(B(r_k)\right) = B_i(r_k)$. We can now define the symplectic embedding $\vartheta_i' : T_i' \hookrightarrow M$ by

$$\vartheta_i' = \eta_k \circ \tau_{w_i} \circ \sigma \circ \tau_{v_i} : T_i' \hookrightarrow M.$$

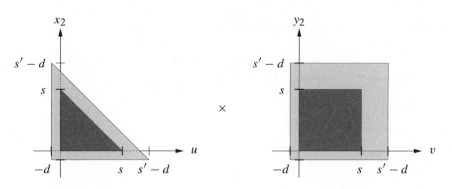

Figure 6.8. The trapezoids T_1 and T_1' for $n = 2$.

We denote the restriction of ϑ_i' to T_i by ϑ_i, and we write

$$\vartheta = \coprod_{i=1}^{k} \vartheta_i : \coprod_{i=1}^{k} T_i \hookrightarrow M \quad \text{and} \quad \vartheta' = \coprod_{i=1}^{k} \vartheta_i' : \coprod_{i=1}^{k} T_i' \hookrightarrow M.$$

Since ϑ, ϑ' and η_k are symplectic, we have $\mu\left(\vartheta\left(\coprod T_i\right)\right) = \sum |T_i| = k\frac{1}{n!}s^{2n}$,

$$\mu\left(\vartheta'\left(\coprod T_i'\right) \setminus \vartheta\left(\coprod T_i\right)\right) = \sum |T_i' \setminus T_i| = k\frac{1}{n!}\left((s')^{2n} - s^{2n}\right)$$

and

$$\mu\left(\eta_k\left(\coprod B_i(r_k)\right) \setminus \vartheta'\left(\coprod T_i'\right)\right) = k\left(|B(r_k)| - |T(d)|\right) = k\frac{1}{n!}\left(\pi^n r_k^{2n} - (s')^{2n}\right).$$

In view of the inequalities (6.2.8), (6.2.9) (6.2.10) we can therefore estimate

$$\begin{aligned}
\mu(M) &= \mu\left(\vartheta\left(\coprod T_i\right)\right) + \mu\left(\vartheta'\left(\coprod T_i'\right) \setminus \vartheta\left(\coprod T_i\right)\right) \\
&\quad + \mu\left(\eta_k\left(\coprod B_i(r_k)\right) \setminus \vartheta'\left(\coprod T_i'\right)\right) + \mu\left(M \setminus \eta_k\left(\coprod B_i(r_k)\right)\right) \\
&= k\frac{1}{n!}s^{2n} + k\frac{1}{n!}\left((s')^{2n} - s^{2n}\right) \\
&\quad + k\frac{1}{n!}\left(\pi^n r_k^{2n} - (s')^{2n}\right) + \mu\left(M \setminus \eta_k\left(\coprod B_i(r_k)\right)\right) \\
&< k\frac{1}{n!}s^{2n} + \epsilon' + \epsilon' + \epsilon' \\
&= k\frac{1}{n!}s^{2n} + 3\epsilon'.
\end{aligned}$$ (6.2.11)

Step 2. Connecting the trapezoids. We next extend the embedding $\vartheta : \coprod T_i \hookrightarrow M$ to a symplectic embedding of a connected domain. For $i = 1, \ldots, k - 1$ we define straight lines

$$L_i(t) : [\,(2i - 1)s, \, 2is\,] \hookrightarrow \mathbb{R}^2 \times \mathbb{R}^{n-1}(x) \times \mathbb{R}^{n-1}(y)$$

by $L_i(t) = (t, 0, 0, 0)$, cf. Figure 6.9. Then

$$\left.\begin{array}{ll} L_i(t) \in T_i' & \text{if } t \in [\,(2i-1)s,\ (2i-1)s + (2n-2)d\,[\,, \\ L_i(t) \in T_{i+1}' & \text{if } t \in \,]\,2is - d,\ 2is\,]. \end{array}\right\} \qquad (6.2.12)$$

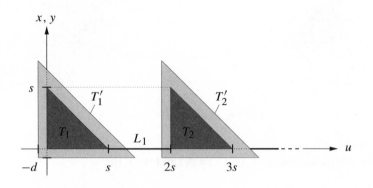

Figure 6.9. The trapezoids T_i and T_i', $i = 1, 2$, and the line L_1.

For $\delta > 0$ we define the "δ-neighbourhood" $N_i(\delta)$ of L_i in $\mathbb{R}^{2n}$ by

$$N_i(\delta) = [\,(2i-1)s,\ 2is\,] \times \,]-\delta,\ \delta\,[^{2n-1}.$$

For later use we shall verify that the left end of $N_i(\delta)$ and the right end of $T_i(\delta)$ fit together nicely. In the sequel we denote by $p_{u,v}$, $p_{u,x}$ and $p_{v,y}$ the projections of $\mathbb{R}^{2n}$ onto $\mathbb{R}^2(u, v)$, $\mathbb{R}^n(u, x)$ and $\mathbb{R}^n(v, y)$.

Lemma 6.2.5. *For each $\delta \in \,]0, d\,]$ and each $i \in \{1, \dots, k-1\}$ we have*

$$\{\,(u, v, x, y) \in N_i(\delta) \mid u = (2i-1)s + \delta\,\} \subset T_i(\delta).$$

Proof. We may assume that $i = 1$. We compute that

$$\{\,(u, x_2, \dots, x_n) \mid u = s + \delta,\ -\delta < x_2, \dots, x_n < \delta\,\}$$

$$\subset \{\,(u, x_2, \dots, x_n) \mid u = s + \delta,\ -\delta < x_2, \dots, x_n \text{ and } x_2 + \cdots + x_n < (n-1)\delta\,\}$$

$$= \{\,(u - \delta, x_2 - \delta, \dots, x_n - \delta) \mid u - \delta = s + \delta,\ 0 < x_2, \dots, x_n \text{ and}$$
$$u + x_2 + \cdots + x_n < s + 2n\delta\,\}$$

$$\subset \{\,(u - \delta, x_2 - \delta, \dots, x_n - \delta) \mid 0 < u, x_2, \dots, x_n \text{ and}$$
$$u + x_2 + \cdots + x_n < s + 2n\delta\,\},$$

i.e.,

$$p_{u,x}\big(\{\,(u, v, x, y) \in N_1(\delta) \mid u = s + \delta\,\}\big) \subset p_{u,x}\,(T_1(\delta)).$$

Moreover, $\,]-\delta, \delta\,[^n \subset \,]-\delta, s + (2n-1)\delta\,[^n$, i.e.,

$$p_{v,y}\big(\{\,(u, v, x, y) \in N_1(\delta) \mid u = s + \delta\,\}\big) \subset p_{v,y}(T_1(\delta)).$$

Lemma 6.2.5 thus follows. $\qquad\qquad\qquad\qquad\qquad\qquad\qquad\qquad\quad$ □

The following proposition parallels Proposition 6.1.3.

Proposition 6.2.6. *There exist $\delta \in\,]0, d\,[$ and a symplectic embedding*

$$\rho: \coprod_{i=1}^{k} T_i(\delta) \,\cup\, \coprod_{i=1}^{k-1} N_i(\delta) \;\hookrightarrow\; M.$$

Proof. By construction of the map ϑ' the set $M \setminus \vartheta'(\coprod T_i')$ is connected. Using this and the inclusions (6.2.12) we find a smooth embedding

$$\coprod_{i=1}^{k-1} \lambda_i: \coprod_{i=1}^{k-1} L_i \;\hookrightarrow\; M \setminus \vartheta\left(\coprod T_i\right)$$

such that

$$\begin{aligned}
\lambda_i(L_i(t)) &= \vartheta_i'(L_i(t)) &&\text{if } t \in [\,(2i-1)s,\; (2i-1)s + (2n-1)d/2\,], \\
\lambda_i(L_i(t)) &= \vartheta_{i+1}'(L_i(t)) &&\text{if } t \in [\,2is - d/2,\; 2is\,],
\end{aligned}$$

and such that

$$\lambda_i(L_i(t)) \in M \setminus \vartheta'\big(\coprod T_i(d/2)\big) \quad \text{if } t \in [\,(2i-1)s + (2n-1)d/2,\; 2is - d/2\,].$$

For $i = 1, \dots, k-1$ and $\delta \in [0, d]$ we define the truncated trapezoid

$$\check{T}_i(\delta) = \{\, z \in T_i(\delta) \mid u < (2i-1)s + \delta\,\},$$

cf. Figure 6.10. Replacing the maps $\gamma_1', \dots, \gamma_k'$ and the sets

$$C_1(d/3),\; C_1(d/2),\; \dots,\; C_{k-1}(d/3),\; C_{k-1}(d/2),\; C_k(d/3),\; C_k(d/2)$$

in the proof of Proposition 6.1.3 by the maps $\vartheta_1', \dots, \vartheta_k'$ and the sets

$$\check{T}_1(d/3),\; \check{T}_1(d/2),\; \dots,\; \check{T}_{k-1}(d/3),\; \check{T}_{k-1}(d/2),\; T_k(d/3),\; T_k(d/2)$$

we find $\hat{\delta} > 0$ and a symplectic embedding

$$\rho: \coprod_{i=1}^{k-1} \Big(\check{T}_i(d/3) \cup N_i\big(\hat{\delta}\big) \Big) \,\cup\, T_k(d/3) \;\hookrightarrow\; M,$$

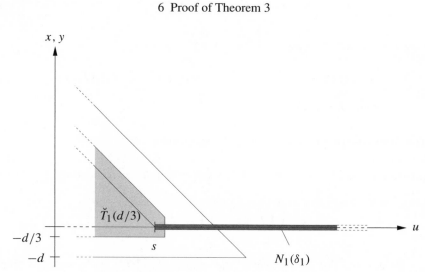

Figure 6.10. The set $\check{T}_1(d/3) \cup N_1(\delta_1)$.

cf. Figure 6.10. We define δ by

$$\delta = \min\left\{\hat{\delta}, \frac{d/3}{2n-1}\right\}.$$

Then $\delta < d$ and $T_i(\delta) \subset \check{T}_i(d/3)$ for all $i = 1, \ldots, k-1$. The restriction of ρ to the domain

$$\coprod_{i=1}^{k} T_i(\delta) \cup \coprod_{i=1}^{k-1} N_i(\delta)$$

is as desired. $\hfill\square$

Step 3. The choice of k and of a_0. We recall that $\epsilon' = \epsilon/5$, and we choose the integer k_0 as after the proof of Lemma 6.2.2. We now choose the integer k such that $k \geq 2$ and $k > k_0$ and

$$\left(1 + \frac{1}{\sqrt[n]{k}}\right)^n - 1 \leq \frac{1}{\mu(M)}\epsilon'. \tag{6.2.13}$$

The inequality (6.2.13) will be crucial for the proof of Lemma 6.2.9 below. Let s and d be the numbers associated with k after the proof of Lemma 6.2.2, and let δ be as in Proposition 6.2.6. For each $\hat{a} > 3\pi$ we let $l(\hat{a})$ and

$$\varphi_{\hat{a}} \colon S_{\hat{a}} \hookrightarrow T^n\big(l(\hat{a})^2\big)$$

be the natural number and the symplectic embedding found in Proposition 5.3.3. We abbreviate

$$q = 1 - \frac{\epsilon'}{k}\frac{n!}{s^{2n}}. \tag{6.2.14}$$

By Proposition 5.3.3 (ii) there exists $\hat{a}_0 > 3\pi$ such that for all $\hat{a} \geq \hat{a}_0$,

$$\frac{|\varphi_{\hat{a}}(S_{\hat{a}})|}{|T^n(l(\hat{a})^2)|} > q. \tag{6.2.15}$$

In view of Lemma 5.3.5 and the definition (5.3.22) we have that $l(\hat{a}) \geq l(\hat{a}_0)$ whenever $\hat{a} \geq \hat{a}_0$ and that $l(\hat{a}) \to \infty$ as $\hat{a} \to \infty$. Choosing $\hat{a}_0$ larger if necessary, we can therefore assume that

$$l(\hat{a}) \geq \pi \frac{s}{\delta} \quad \text{for all } \hat{a} \geq \hat{a}_0. \tag{6.2.16}$$

We set $a_0 = kn\hat{a}_0$. In the sequel we fix $a \geq a_0$. We set $\hat{a} = \frac{a}{kn}$.

Step 4. The set $\coprod_{i=1}^k S_i$ and the choice of $\lambda(a)$. With each k-tuple $(u_1, \ldots, u_k) \in \mathbb{R}^k$ we associate the k-tuple $(v_1, \ldots, v_k)$ defined by $v_i = u_1 + \cdots + u_i$. We say that a k-tuple $(u_1, \ldots, u_k)$ is admissible if $u_i > 0$ for all i and if $v_k = a$.

Lemma 6.2.7. *There exists a unique admissible k-tuple $(u_1, \ldots, u_k)$ such that*

$$u_i \left(1 - \frac{v_{i-1}}{a}\right)^{n-1} = u_1, \quad i = 2, \ldots, k. \tag{6.2.17}$$

Proof. Fix $u_1 \in \,]0, a[$. We inductively associate with u_1 numbers $u_2, \ldots, u_k$ as follows. Assume that $i \in \{2, \ldots, k\}$ and that we have already constructed $u_2, \ldots, u_{i-1}$. We set $v_j = u_1 + \cdots + u_j$, $j = 1, \ldots, i-1$. We define u_i by

$$u_i \left(1 - \frac{v_{j_0}}{a}\right)^{n-1} = u_1$$

where $j_0 = \max\{\, j \mid v_j < a \,\}$. The function $f \colon \,]0, a[\,\to\, \mathbb{R}$ defined by

$$f(u_1) = v_k$$

is then continuous and strictly increasing. Since $f(u_1) \to 0$ as $u_1 \to 0$ and $f(u_1) \to \infty$ as $u_1 \to a$ we conclude that there exists a unique $u_1 \in \,]0, a[$ with $f(u_1) = a$. By construction, the k-tuple $(u_1, \ldots, u_k)$ associated with this u_1 is admissible and meets the identities (6.2.17). The proof of Lemma 6.2.7 is complete. $\square$

Let $(u_1, \ldots, u_k)$ be the k-tuple guaranteed by Lemma 6.2.7. As before we abbreviate $v_i = u_1 + \cdots + u_i$, and we set $v_0 = 0$. We consider the subsets

$$S_1 = \,]0, v_1[\,\times\,]0, 1[\,\times\, \triangle^{n-1}(\pi) \times \square^{n-1}(1),$$

$$S_i = [v_{i-1}, v_i[\,\times\,]0, 1[\,\times\, \triangle^{n-1}\left(\pi - \frac{\pi}{a} v_{i-1}\right) \times \square^{n-1}(1), \quad 2 \leq i \leq k,$$

of $\mathbb{R}^2 \times \mathbb{R}^{n-1}(x) \times \mathbb{R}^{n-1}(y)$. Then $T^n(a, \pi) \subset \coprod S_i$, cf. Figure 6.11. Notice that

$$|S_i| = u_i \frac{\pi^{n-1}}{(n-1)!} \left(1 - \frac{v_{i-1}}{a}\right)^{n-1}, \quad 1 \leq i \leq k.$$

The identities (6.2.17) therefore imply that

$$|S_1| = \cdots = |S_k|.\tag{6.2.18}$$

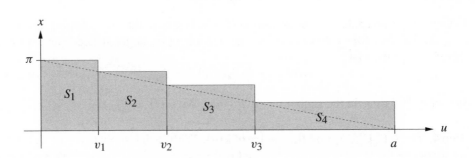

Figure 6.11. The set $\coprod_{i=1}^{k} S_i$ for $k = 4$.

Lemma 6.2.8. *We have*

$$\hat{a}_0 \le \frac{a}{kn} \le u_1 < u_2 < \cdots < u_k \le \frac{a}{\sqrt[n]{k}}.\tag{6.2.19}$$

Proof. The first inequality in (6.2.19) is equivalent to our assumption $a_0 \le a$. The inclusion $T^n(a, \pi) \subset \coprod S_i$ and the identities (6.2.18) yield

$$\frac{a\pi^{n-1}}{n!} = \left| T^n(a, \pi) \right| \le \left| \coprod S_i \right| = k\,|S_1| = k\frac{u_1\pi^{n-1}}{(n-1)!}$$

and so $\frac{a}{kn} \le u_1$. The inequalities $u_1 < u_2 < \cdots < u_k$ follow from the identities (6.2.17). Finally, the identities (6.2.18) and the inclusion $\coprod S_i \subset S_a$ yield

$$k u_k \frac{\left(\frac{\pi}{a} u_k\right)^{n-1}}{(n-1)!} = k\,|S_k| = \left| \coprod S_i \right| \le |S_a| = a\frac{\pi^{n-1}}{(n-1)!}$$

and so $u_k \le \frac{a}{\sqrt[n]{k}}$. $\qquad\square$

We now define the real number $\lambda = \lambda(a)$ by

$$\frac{|\lambda S_1|}{|T_1|} = q.\tag{6.2.20}$$

Lemma 6.2.9. $\left| \lambda \coprod_{i=1}^{k} S_i \setminus \lambda T^n(a, \pi) \right| < \epsilon'.$

Proof. In view of Lemma 6.2.8 we have $u_1 < u_2 < \cdots < u_k$, and so

$$\coprod_{i=1}^{k} S_i \subset T^n\left(a + u_k, \pi + \tfrac{\pi}{a}u_k\right),\qquad (6.2.21)$$

cf. Figure 6.12.

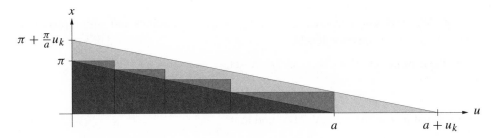

Figure 6.12. $T^n(a,\pi) \subset \coprod_{i=1}^{k} S_i \subset T^n\left(a + u_k, \pi + \tfrac{\pi}{a}u_k\right)$.

Since the embedding $\vartheta : \coprod T_i \hookrightarrow M$ is symplectic, we have

$$\mu(M) \geq \mu\left(\vartheta\left(\coprod T_i\right)\right) = \sum |T_i| = k\tfrac{1}{n!}s^{2n}.\qquad (6.2.22)$$

In view of the inclusion (6.2.21), the last inequality in (6.2.19) and the estimates (6.2.13) and (6.2.22) we can now estimate

$$
\begin{aligned}
\left|\coprod S_i \setminus T^n(a,\pi)\right| &\leq \left|T^n\left(a + u_k, \pi + \tfrac{\pi}{a}u_k\right) \setminus T^n(a,\pi)\right| \\
&= \tfrac{\pi^{n-1}}{n!}\left((a + u_k)\left(1 + \tfrac{u_k}{a}\right)^{n-1} - a\right) \\
&= \tfrac{\pi^{n-1}}{n!}\left(\left(1 + \tfrac{u_k}{a}\right)^n - 1\right)a \\
&\leq \tfrac{\pi^{n-1}}{n!}\left(\left(1 + \tfrac{1}{\sqrt[n]{k}}\right)^n - 1\right)a \\
&\leq \pi^{n-1}\tfrac{1}{n!\,\mu(M)}\epsilon' a \\
&\leq \pi^{n-1}\tfrac{1}{ks^{2n}}\epsilon' a.
\end{aligned}
$$

Using the definitions (6.2.20) and (6.2.14) and the second inequality in (6.2.19) we find that

$$\lambda^{2n} = q\,\frac{|T_1|}{|S_1|} < \frac{|T_1|}{|S_1|} = \frac{\tfrac{s^{2n}}{n!}}{u_1\tfrac{\pi^{n-1}}{(n-1)!}} = \frac{s^{2n}}{nu_1\pi^{n-1}} \leq \frac{ks^{2n}}{a\pi^{n-1}}.$$

We conclude that

$$\left|\lambda\coprod S_i \setminus \lambda T^n(a,\pi)\right| = \lambda^{2n}\left|\coprod S_i \setminus T^n(a,\pi)\right| < \epsilon'$$

and so the proof of Lemma 6.2.9 is complete. $\square$

Step 5. Embedding $\lambda \coprod S_i$ **into** $\coprod T_i(\delta) \cup \coprod N_i(\delta)$

Proposition 6.2.10. *There exists a symplectic embedding*

$$\Psi_a : \lambda \coprod_{i=1}^{k} S_i \;\hookrightarrow\; \coprod_{i=1}^{k} T_i(\delta) \;\cup\; \coprod_{i=1}^{k-1} N_i(\delta).$$

Proof. We start with introducing several geometric quantities and with replacing the sets λS_i by more convenient sets.

5.A. Preliminaries. For $i = 1, \ldots, k$ we set

$$h_i = 1 - \frac{v_{i-1}}{a}.$$

Notice that $h_i \pi$ is the x-width of S_i and that

$$1 = h_1 > h_2 > \cdots > h_k.$$

We define a_i by

$$a_i = \frac{u_i}{h_i}.$$

Then $a_1 = u_1$, and in view of Lemma 6.2.8 we have $a_i \geq u_i \geq u_1 > \hat{a}_0$. We denote the natural number $l(a_i)$ associated with a_i in Proposition 5.3.3 by

$$l_i = l(a_i).$$

Since $a_i > \hat{a}_0$ the estimate (6.2.16) shows that

$$l_i \geq \pi \frac{s}{\delta}. \tag{6.2.23}$$

We finally define λ_i by

$$\lambda_i = \frac{s}{l_i}. \tag{6.2.24}$$

Lemma 6.2.11. $\lambda_i > \sqrt{h_i}\,\lambda$ *and* $\delta > \sqrt{h_i}\,\pi\lambda.$

Proof. Since $|T_i| = \frac{s^{2n}}{n!}$ and $\left| T^n\left(l_i^2\right) \right| = \frac{l_i^{2n}}{n!}$ we have

$$|T_i| = \lambda_i^{2n} \left| T^n\left(l_i^2\right) \right|,$$

and since $a_i \geq \hat{a}_0$, the estimate (6.2.15) shows that

$$\frac{|S_{a_i}|}{\left| T^n\left(l_i^2\right) \right|} > q.$$

Together with the definition (6.2.20) of λ and the identities (6.2.18) we can now estimate

$$\frac{h_i^{-n}\,|S_i|}{\lambda_i^{-2n}\,|T_i|} = \frac{|S_{a_i}|}{\left| T^n\left(l_i^2\right) \right|} > q = \frac{|\lambda S_1|}{|T_1|} = \frac{\lambda^{2n}\,|S_i|}{|T_i|},$$

and so the first statement in Lemma 6.2.11 follows. The second statement follows from the first one, from the definition of λ_i and from the estimate (6.2.23). $\square$

We define the linear symplectomorphism β of $\mathbb{R}^2$ by

$$\beta(u, v) = \left(\tfrac{\lambda}{s} u, \tfrac{s}{\lambda} v \right). \tag{6.2.25}$$

Moreover, we set $\tilde{u}_i = \tfrac{\lambda^2}{s} u_i$ and $\tilde{v}_i = \tilde{u}_1 + \cdots + \tilde{u}_i$, and we abbreviate

$$\widetilde{S}_i := (\beta \times \mathrm{id}_{2n-2})(\lambda S_i), \quad 1 \le i \le k.$$

Then

$$\widetilde{S}_1 = \,]0, \tilde{v}_1[\, \times \,]0, s[\, \times \Delta^{n-1}(\pi\lambda) \times \square^{n-1}(\lambda),$$

$$\widetilde{S}_i = [\tilde{v}_{i-1}, \tilde{v}_i[\, \times \,]0, s[\, \times \Delta^{n-1}(h_i \pi\lambda) \times \square^{n-1}(\lambda), \quad 2 \le i \le k.$$

We shall next embed $\widetilde{S}_1$ into T_1 and shall then successively embed $\widetilde{S}_i$ into $N_{i-1}(\delta) \cup T_i(\delta), i = 2, \ldots, k$.

5.B. The embedding $\psi_1 \colon \widetilde{S}_1 \hookrightarrow T_1$. We recall that $a_1 = u_1$ and $s = \lambda_1 l_1$, and we read off from the first statement in Lemma 6.2.11 that $\lambda_1 > \lambda$. Therefore,

$$t_1 := \tfrac{\lambda_1}{l_1} a_1 - \tilde{u}_1 > 0,$$

and so the affine symplectomorphism β_1 of $\mathbb{R}^2$ defined by

$$\beta_1(u, v) = \left(l_1(u + t_1), \tfrac{1}{l_1} v \right)$$

embeds $p_{u,v}(\widetilde{S}_1)$ into $]0, \lambda_1 a_1[\, \times \,]0, \lambda_1[$. Using once more that $\lambda_1 > \lambda$ we conclude that

$$\left(\lambda_1^{-1} \circ (\beta_1 \times \mathrm{id}_{2n-2}) \right)(\widetilde{S}_1) \subset S_{a_1}.$$

We define the linear symplectomorphism σ_1 of $\mathbb{R}^{2n}$ by

$$\sigma_1(u, v, x, y) = \left(\tfrac{1}{l_1} u, l_1 v, \tfrac{1}{l_1} x, l_1 y \right).$$

Composing the affine embedding $\lambda_1^{-1} \circ (\beta_1 \times \mathrm{id}_{2n-2}) \colon \widetilde{S}_1 \hookrightarrow S_{a_1}$ with the symplectic embedding

$$\varphi_{a_1} \colon S_{a_1} \hookrightarrow T^n(l_1^2)$$

guaranteed by Proposition 5.3.3 and with the linear diffeomorphism

$$\sigma_1 \circ \lambda_1 \colon T^n(l_1^2) \to T_1$$

we obtain the symplectic embedding

$$\psi_1 := \sigma_1 \circ \lambda_1 \circ \varphi_{a_1} \circ \lambda_1^{-1} \circ (\beta_1 \times \mathrm{id}_{2n-2}) \colon \widetilde{S}_1 \hookrightarrow T_1.$$

We abbreviate $s_1 = s - \lambda_1$. Using Proposition 5.3.3 (i) and the definitions of β_1, σ_1 and λ_1 we find that

$$\psi_1(u, v, x, y) = \left(u - \tilde{v}_1 + s_1, v, \tfrac{1}{\pi}x, \pi y\right) \quad \text{if } u > \tilde{v}_1 - \tfrac{\lambda_1}{l_1}\pi.$$

5.C. The embedding $\psi_j \colon \coprod_{i=1}^{j} \tilde{S}_i \hookrightarrow \coprod_{i=1}^{j} T_i(\delta) \cup \coprod_{i=1}^{j-1} N_i(\delta)$, $j = 2, \ldots, k$.

For $i = 1, \ldots, k$ we abbreviate

$$s_i = (2i - 1)s - \lambda_i, \quad d_i = \min\left\{\tfrac{\tilde{u}_i}{2}, \delta\right\}, \quad \varepsilon_i = \min\left\{\tilde{u}_i - d_i, \tfrac{\lambda_i}{l_i}\pi\right\}. \quad (6.2.26)$$

Proceeding by induction we fix $j \in \{1, \ldots, k - 1\}$ and assume that we have already constructed a symplectic embedding

$$\psi_j \colon \coprod_{i=1}^{j} \tilde{S}_i \hookrightarrow \coprod_{i=1}^{j} T_i(\delta) \cup \coprod_{i=1}^{j-1} N_i(\delta) \qquad (6.2.27)$$

which is such that $\operatorname{Im} \psi_j \subset \left\{(u, v, x, y) \mid u < s_j\right\}$ and

$$\psi_j(u, v, x, y) = \left(u - \tilde{v}_j + s_j, v, \tfrac{1}{\sqrt{h_j}}\tfrac{1}{\pi}x, \sqrt{h_j}\pi y\right) \quad \text{if } u > \tilde{v}_j - \varepsilon_j, \quad (6.2.28)$$

cf. Figure 6.13. We are going to construct a symplectic extension

$$\psi_{j+1} \colon \coprod_{i=1}^{j+1} \tilde{S}_i \hookrightarrow \coprod_{i=1}^{j+1} T_i(\delta) \cup \coprod_{i=1}^{j} N_i(\delta) \qquad (6.2.29)$$

of ψ_j which is such that $\operatorname{Im} \psi_{j+1} \subset \left\{(u, v, x, y) \mid u < s_{j+1}\right\}$ and

$$\psi_{j+1}(u, v, x, y) = \left(u - \tilde{v}_{j+1} + s_{j+1}, v, \tfrac{1}{\sqrt{h_{j+1}}}\tfrac{1}{\pi}x, \sqrt{h_{j+1}}\pi y\right)$$
$$\text{if } u > \tilde{v}_{j+1} - \varepsilon_{j+1}. \qquad (6.2.30)$$

We extend ψ_j by formula (6.2.28) to the smooth symplectic embedding

$$\tilde{\psi}_j \colon \coprod_{i=1}^{j+1} \tilde{S}_i \hookrightarrow \mathbb{R}^{2n},$$

and we denote by F'_j the end

$$F'_j = \tilde{\psi}_j(\tilde{S}_{j+1}) = \left\{(u, v, x, y) \in \operatorname{Im} \tilde{\psi}_j \mid s_j \leq u < s_j + \tilde{u}_{j+1}\right\},$$

cf. Figure 6.13.

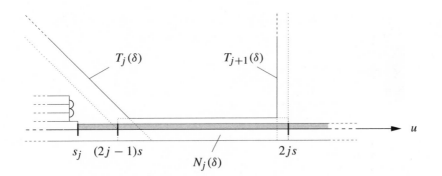

Figure 6.13. The end F_j'.

5.C.1. The map β_j. We recall that $u_{j+1} = h_{j+1}a_{j+1}$ and $s = \lambda_{j+1}l_{j+1}$, and we read off from the first statement in Lemma 6.2.11 that $\lambda_{j+1} > \sqrt{h_{j+1}}\lambda$. Therefore,

$$t_{j+1} := \frac{\lambda_{j+1}}{l_{j+1}}a_{j+1} - \tilde{u}_{j+1} > 0. \tag{6.2.31}$$

As in Step 1 of the folding construction described in Section 3.2 we find a symplectic embedding

$$\beta_j \colon \; [s_j, s_j + \tilde{u}_{j+1}[\, \times \,]0, s[\; \hookrightarrow \; [s_j, 2js + \tfrac{\lambda_{j+1}}{l_{j+1}}a_{j+1}[\times \,]0, s[$$

which restricts to the identity on $\left\{ (u, v) \mid s_j \leq u \leq s_j + \frac{d_{j+1}}{2} \right\}$, restricts to the translation

$$(u, v) \mapsto (u + 2js - s_j + t_{j+1}, v) \quad \text{on } \{ (u, v) \mid u \geq s_j + d_{j+1} \}, \tag{6.2.32}$$

and is such that

$$\operatorname{Im}\beta_j \cap \{(2j - 1)s + \delta \leq u \leq 2js - \delta\} \subset \{0 < v < \delta\}, \tag{6.2.33}$$

$$\operatorname{Im}\beta_j \cap \left\{2js - \delta + \tfrac{\delta}{2n} \leq u \leq 2js - \tfrac{\delta}{2}\right\} \subset \{s - \delta < v < s\}, \tag{6.2.34}$$

$$\operatorname{Im}\beta_j \cap \left\{2js - \tfrac{\delta}{2} + \tfrac{\delta}{2n} \leq u \leq 2js\right\} \subset \{0 < v < \delta\}, \tag{6.2.35}$$

cf. Figure 6.14.

Lemma 6.2.12. *We have*

$$\left(\beta_j \times \operatorname{id}_{2n-2}\right)\left(F_j'\right) \cap \left\{ (u, v, x, y) \mid s_j \leq u \leq 2js - \delta \right\} \subset T_j(\delta) \cup N_j(\delta).$$

Proof. In view of the inclusion (6.2.33) we have

$$p_{u,v}\left(\left(\beta_j \times \operatorname{id}_{2n-2}\right)\left(F_j'\right) \cap \left\{ s_j \leq u \leq 2js - \delta \right\}\right) \subset p_{u,v}\left(T_j(\delta) \cup N_j(\delta)\right).$$

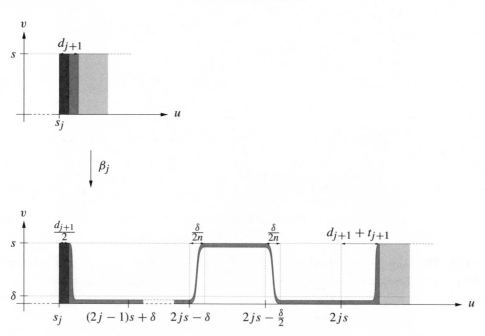

Figure 6.14. The map β_j.

Moreover, the formula (6.2.28) implies that the fibres of F_j' are equal to

$$\left(h_{j+1}\frac{1}{\sqrt{h_j}}\frac{1}{\pi}\lambda\triangle^{n-1}(\pi)\right) \times \sqrt{h_j}\pi\lambda\square^{n-1}(1) = \triangle^{n-1}(w_j) \times \square^{n-1}\left(\sqrt{h_j}\pi\lambda\right)$$

where we abbreviated

$$w_j = \frac{h_{j+1}}{\sqrt{h_j}}\lambda. \tag{6.2.36}$$

In view of the inequality $h_{j+1} < h_j$ and the second estimate in Lemma 6.2.11 these fibres are contained in the fibres of $N_j(\delta)$. Lemma 6.2.12 thus follows in view of Lemma 6.2.5.   □

5.C.2. Rescaling the fibres. We define ν_j by

$$\nu_j = (l_{j+1} - 1)\sqrt{\frac{h_j}{h_{j+1}}}. \tag{6.2.37}$$

In view of the inequalities (6.2.23) we have $l_{j+1} > 3$, and so $\nu_j > 2$. We abbreviate

$$e_j = \frac{1}{\nu_j}\sqrt{h_j}\pi\lambda \tag{6.2.38}$$

and we define

$$C_j' = \left\{ (y_2, \ldots, y_n) \mid 0 < y_i < \sqrt{h_j}\,\pi\lambda, \ i = 2, \ldots, n \right\},$$
$$C_j'' = \left\{ (y_2, \ldots, y_n) \mid -\sqrt{h_j}\,\pi\lambda - 2e_j < y_i < -2e_j, \ i = 2, \ldots, n \right\},$$
$$C_j''' = \left\{ (y_2, \ldots, y_n) \mid -2e_j < y_i < -e_j, \ i = 2, \ldots, n \right\},$$
$$C_{j+1} = \left\{ (y_2, \ldots, y_n) \mid 0 < y_i < e_j, \ i = 2, \ldots, n \right\},$$

cf. Figure 6.15.

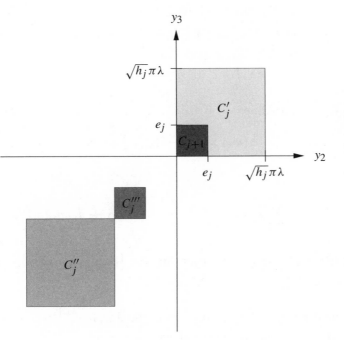

Figure 6.15. The cubes C_j', C_j'', C_j''' and C_{j+1} for $n = 3$.

Our next goal is to rescale the fibres $\triangle^{n-1}(w_j) \times C_j'$ of $(\beta_j \times \mathrm{id}_{2n-2})(F_j')$ over $\{u > 2js - \delta\}$ to the fibres $\triangle^{n-1}(v_j w_j) \times C_{j+1}$. We shall use the same method as in Step 3 of the proof of Proposition 5.3.3 and first lower the fibres $\triangle^{n-1}(w_j) \times C_j'$ at $u = 2js - \delta + \frac{\delta}{2n}$ to the fibres $\triangle^{n-1}(w_j) \times C_j''$, then deform these fibres to the fibres $\triangle^{n-1}(v_j w_j) \times C_j'''$, and finally lift these fibres at $u = 2js - \frac{\delta}{2} + \frac{\delta}{2n}$ to the fibres $\triangle^{n-1}(v_j w_j) \times C_{j+1}$.

The lowering map φ_j^-. Using the definition of e_j and the inequality $v_j > 2$ we estimate

$$2e_j + \sqrt{h_j}\,\pi\lambda \le 2e_j v_j, \tag{6.2.39}$$

and using the definitions of e_j and w_j, the second statement in Lemma 6.2.11 and $\delta < d \le \frac{s}{2n}$ we estimate

$$2e_j v_j w_j = 2h_{j+1}\pi\lambda^2 < \left(\sqrt{h_j}\pi\lambda\right)^2 < \delta^2 < \frac{\delta}{2n}s. \qquad (6.2.40)$$

Combining the estimates (6.2.39) and (6.2.40) we find that

$$\left(2e_j + \sqrt{h_j}\pi\lambda\right)w_j < \frac{\delta}{2n}s.$$

For $i = 2, \ldots, n$ we therefore find a cut off function $c_i^- : \mathbb{R} \to \left[0, \frac{s}{w_j}\right]$ with support $\left[2js - \delta + (i-1)\frac{\delta}{2n}, \, 2js - \delta + i\frac{\delta}{2n}\right]$ and such that

$$\int_0^\infty c_i^-(t)\,dt = 2e_j + \sqrt{h_j}\pi\lambda.$$

The symplectic embedding

$$\varphi_j^- : \operatorname{Im}\beta_j \times \triangle^{n-1}\left(w_j\right) \times C_j' \hookrightarrow \mathbb{R}^{2n}, \quad (u, v, x, y) \mapsto (u', v', x', y')$$

defined by

$$u' = u, \quad v' = v - \sum_{i=2}^n c_i^-(u)x_i, \quad x_i' = x_i, \quad y_i' = y_i - \int_0^u c_i^-(t)\,dt, \quad i = 2, \ldots, n,$$

maps the fibres $\triangle^{n-1}(w_j) \times C_j'$ over the base $\{(u, v) \mid u > 2js - \frac{\delta}{2}\}$ to the fibres $\triangle^{n-1}(w_j) \times C_j''$.

The deformation α_j. The deformation of the fibres $\triangle^{n-1}(w_j) \times C_j''$ to the fibres $\triangle^{n-1}(v_j w_j) \times C_j'''$ is based on the following lemma.

Lemma 6.2.13. *There exists a symplectic embedding*

$$\alpha : \,]0, w_j[\,\times\,] - 2e_j - \sqrt{h_j}\pi\lambda, \infty[\,\hookrightarrow \mathbb{R}^2$$

which restricts to the identity on $\{(x, y) \mid y \ge 0\}$, restricts to the affine map

$$(x, y) \mapsto \left(v_j x, \tfrac{1}{v_j}y + \tfrac{1}{v_j}2e_j - e_j\right) \qquad (6.2.41)$$

on $\{(x, y) \mid y \le -2e_j\}$, and is such that

$$x'(\alpha(x, y)) \le v_j x \quad and \quad y'(\alpha(x, y)) > -\delta \qquad (6.2.42)$$

for all $(x, y) \in]0, w_j[\,\times\,] - 2e_j - \sqrt{h_j}\,\pi\lambda, \infty[$, cf. Figure 6.16.

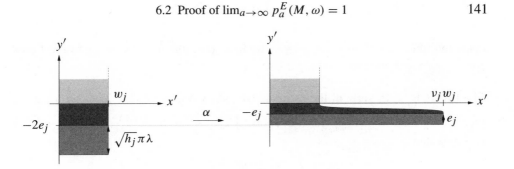

Figure 6.16. The map α.

Proof. Choose a smooth function $f: \mathbb{R} \to [1, \nu_j]$ such that

(i) $f(w) = \nu_j$ for $w \leq -2e_j$,

(ii) $f(w) = 1$ for $w \geq 0$.

Since $\nu_j > 2$ we have

$$2e_j \frac{1}{\nu_j} < e_j < 2e_j.$$

We may therefore further require that

(iii) $\displaystyle\int_{-2e_j}^{0} \frac{1}{f(w)}\, dw = e_j.$

Using (i) and (iii), the definition (6.2.38) of e_j, the inequality $\nu_j > 2$ and the second inequality in Lemma 6.2.11 we can estimate

$$\int_{-2e_j - \sqrt{h_j} \pi \lambda}^{0} \frac{1}{f(w)}\, dw = \frac{1}{\nu_j} \sqrt{h_j} \pi \lambda + e_j = \frac{2}{\nu_j} \sqrt{h_j} \pi \lambda < \delta.$$

We conclude that the map

$$\alpha: \;]0, w_j[\times] - 2e_j - \sqrt{h_j}\pi\lambda, \infty[\; \to \mathbb{R}^2, \quad (x, y) \mapsto \left(f(y)x, \int_0^y \frac{1}{f(w)} dw \right)$$

is a symplectic embedding which is as required. $\square$

In view of (6.2.41) the symplectic embedding

$$\alpha_j: \left(\varphi_j^- \circ \left(\beta_j \times \mathrm{id}_{2n-2} \right) \right)\left(F_j' \right) \hookrightarrow \mathbb{R}^{2n}$$

defined by

$$\alpha_j(u, v, x_2, y_2, \ldots, x_n, y_n) = (u, v, \alpha(x_2, y_2), \ldots, \alpha(x_n, y_n))$$

maps the fibres $\Delta^{n-1}(w_j) \times C''_j$ over the base $\{(u, v) \mid u > 2js - \frac{\delta}{2}\}$ to the fibres $\Delta^{n-1}(v_j w_j) \times C'''_j$.

The lifting φ_j^+. In view of the estimate (6.2.40) we find for $i = 2, \ldots, n$ a cut off function $c_i^+ : \mathbb{R} \to \left[0, \frac{s}{v_j w_j}\right]$ with support $\left[2js - \frac{\delta}{2} + (i-1)\frac{\delta}{2n}, \, 2js - \frac{\delta}{2} + i\frac{\delta}{2n}\right]$ and such that

$$\int_0^\infty c_i^+(t)\,dt = 2e_j.$$

The symplectic embedding

$$\varphi_j^+ : \left(\alpha_j \circ \varphi_j^- \circ \left(\beta_j \times \mathrm{id}_{2n-2}\right)\right)\left(F'_j\right) \hookrightarrow \mathbb{R}^{2n}, \quad (u, v, x, y) \mapsto (u', v', x', y')$$

defined by

$$u' = u, \quad v' = v + \sum_{i=2}^n c_i^+(u)x_i, \quad x'_i = x_i, \quad y'_i = y_i + \int_0^u c_i^+(t)\,dt, \quad i = 2, \ldots, n,$$

maps the fibres $\Delta^{n-1}(v_j w_j) \times C'''_j$ over the base $\{(u, v) \mid u > 2js\}$ to the fibres $\Delta^{n-1}(v_j w_j) \times C_{j+1}$.

We abbreviate the composition

$$\phi_j = \varphi_j^+ \circ \alpha_j \circ \varphi_j^- \circ \left(\beta_j \times \mathrm{id}_{2n-2}\right) : F'_j \hookrightarrow \mathbb{R}^{2n}.$$

Lemma 6.2.14. *We have*

$$\phi_j\left(F'_j\right) \cap \left\{(u, v, x, y) \mid s_j \leq u \leq 2js\right\} \subset T_j(\delta) \cup N_j(\delta) \cup T_{j+1}(\delta).$$

Proof. We fix $(u, v, x, y) \in \left(\beta_j \times \mathrm{id}_{2n-2}\right)\left(F'_j\right) \cap \{s_j \leq u \leq 2js\}$, and we set

$$(u', v', x', y') = \left(\varphi_j^+ \circ \alpha_j \circ \varphi_j^-\right)(u, v, x, y).$$

Then $u' = u$.

Assume first that $u \leq 2js - \delta$. Then $(u', v', x', y') = (u, v, x, y)$, and so Lemma 6.2.12 shows that

$$(u', v', x', y') \in T_j(\delta) \cup N_j(\delta).$$

Assume now that $2js - \delta < u \leq 2js$. We read off from the inclusions (6.2.34) and (6.2.35) and from the definitions of φ_j^-, α_j and φ_j^+ that

$$v' \in \,]-\delta, s + \delta[,$$

cf. Figure 6.17.

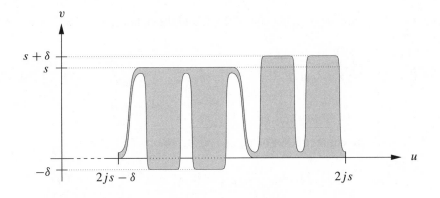

Figure 6.17. The set $p_{u,v}\big(\phi_j\big(F_j'\big) \cap \{2js - \delta < u \leq 2js\}\big)$ for $n = 3$.

Moreover, we have $x \in \triangle^{n-1}(w_j)$, and so the first inequality in (6.2.42) implies that $x' \in \triangle^{n-1}(v_j w_j)$. Using the definitions (6.2.37) and (6.2.36) of v_j and w_j, the first estimate in Lemma 6.2.11 and the definition (6.2.24) of λ_{j+1} we compute

$$v_j w_j = (l_{j+1} - 1)\sqrt{\frac{h_j}{h_{j+1}}} \frac{h_{j+1}}{\sqrt{h_j}}\lambda = (l_{j+1} - 1)\sqrt{h_{j+1}}\lambda < l_{j+1}\lambda_{j+1} = s,$$

and so

$$x' \in \triangle^{n-1}(s).$$

Finally, we have $y \in C_j' = \square^{n-1}\big(\sqrt{h_j}\pi\lambda\big)$, and so the definitions of φ_j^- and φ_j^+ and the second inequality in (6.2.42) imply that $y' \in \,]-\delta, \sqrt{h_j}\pi\lambda[^{n-1}$. Using the second estimate in Lemma 6.2.11 we find $\sqrt{h_j}\pi\lambda < \delta < s$, and so

$$y' \in \,]-\delta, s[^{n-1}.$$

We conclude that

$$(u', v', x', y') \in \,]2js - \delta, 2js] \times \,]-\delta, s+\delta[\times \triangle^{n-1}(s) \times \,]-\delta, s[^{n-1} \subset T_{j+1}(\delta).$$

The proof of Lemma 6.2.14 is thus complete.   □

Recall that β_j is the identity on $\big\{(u, v) \mid s_j \leq u \leq s_j + \frac{d_{j+1}}{2}\big\}$. We can therefore extend the symplectic embedding $\phi_j \colon F_j' = \widetilde{\psi}_j\big(\widetilde{S}_{j+1}\big) \hookrightarrow \mathbb{R}^{2n}$ by the identity to the symplectic embedding

$$\widetilde{\phi}_j \colon \widetilde{\psi}_j\Big(\coprod_{i=1}^{j+1} \widetilde{S}_i\Big) \hookrightarrow \mathbb{R}^{2n}, \qquad \widetilde{\phi}_j = \begin{cases} \mathrm{id} & \text{on } \widetilde{\psi}_j\Big(\coprod_{i=1}^{j} \widetilde{S}_i\Big), \\ \phi_j & \text{on } \widetilde{\psi}_j\big(\widetilde{S}_{j+1}\big). \end{cases}$$

In view of the formulae (6.2.28) and (6.2.32) and the definition (6.2.37) of v_j we find

$$\left(\widetilde{\phi}_j \circ \widetilde{\psi}_j\right)(u, v, x, y)$$
$$= \left(u - \tilde{v}_j + 2js + t_{j+1},\, v,\, (l_{j+1} - 1)\frac{1}{\sqrt{h_{j+1}}}\frac{1}{\pi}x,\, \frac{1}{l_{j+1}-1}\sqrt{h_{j+1}}\pi y\right) \qquad (6.2.43)$$
$$\text{if } u \geq \tilde{v}_j + d_{j+1}.$$

5.C.3. End of the construction of ψ_{j+1}. We denote by F_{j+1} the set

$$F_{j+1} = \left]2js, 2js + \tfrac{\lambda_{j+1}}{l_{j+1}}a_{j+1}\right[\times\,]0, s[\, \times \Delta^{n-1}(v_j w_j) \times C_{j+1}.$$

In view of the construction of ϕ_j we have that

$$\phi_j\left(F_j'\right) \cap \{(u, v, x, y) \mid u > 2js\} \subset F_{j+1}$$

and that the right face of $\phi_j\left(F_j'\right)$ is equal to the right face of F_{j+1}. We are going to embed F_{j+1} into T_{j+1}. We denote by τ_j the translation $(u, v, x, y) \mapsto (u + 2js, v, x, y)$. Moreover, we define the linear symplectomorphisms σ_{j+1} and ξ_{j+1} of $\mathbb{R}^{2n}$ by

$$\sigma_{j+1}(u, v, x, y) = \left(\tfrac{1}{l_{j+1}}u,\, l_{j+1}v,\, \tfrac{1}{l_{j+1}}x,\, l_{j+1}y\right)$$

and

$$\xi_{j+1}(u, v, x, y) = \left(u,\, v,\, \tfrac{\pi}{(l_{j+1}-1)l_{j+1}}x,\, \tfrac{(l_{j+1}-1)l_{j+1}}{\pi}y\right).$$

Using the definitions (6.2.37), (6.2.36), (6.2.38) and (6.2.24) of v_j, w_j, e_j and λ_{j+1} and the first inequality in Lemma 6.2.11 we find that

$$\left(\xi_{j+1} \circ \lambda_{j+1}^{-1} \circ \sigma_{j+1}^{-1} \circ \tau_j^{-1}\right)(F_{j+1})$$
$$= \,]0, a_{j+1}[\, \times\,]0, 1[\, \times \Delta^{n-1}\left(\pi \frac{\sqrt{h_{j+1}}\lambda}{\lambda_{j+1}}\right) \times \square^{n-1}\left(\frac{\sqrt{h_{j+1}}\lambda}{\lambda_{j+1}}\right)$$
$$\subset \,]0, a_{j+1}[\, \times\,]0, 1[\, \times \Delta^{n-1}(\pi) \times \square^{n-1}(1)$$
$$= S_{a_{j+1}}.$$

Composing the affine diffeomorphism $\xi_{j+1} \circ \lambda_{j+1}^{-1} \circ \sigma_{j+1}^{-1} \circ \tau_j^{-1}$ with the symplectic embedding

$$\varphi_{a_{j+1}}\colon S_{a_{j+1}} \hookrightarrow T^n(l_{j+1}^2)$$

guaranteed by Proposition 5.3.3 and with the affine diffeomorphism

$$\tau_j \circ \sigma_{j+1} \circ \lambda_{j+1}\colon T^n(l_{j+1}^2) \to T_{j+1}$$

we obtain the symplectic embedding

$$\rho_j := \tau_j \circ \sigma_{j+1} \circ \lambda_{j+1} \circ \varphi_{a_{j+1}} \circ \xi_{j+1} \circ \lambda_{j+1}^{-1} \circ \sigma_{j+1}^{-1} \circ \tau_j^{-1}\colon F_{j+1} \hookrightarrow T_{j+1}.$$

Using Proposition 5.3.3 (i) we find

$$\rho_j(u, v, x, y) = (u, v, x, y) \quad \text{if } u < 2js + \tfrac{\lambda_{j+1}}{l_{j+1}}\pi \tag{6.2.44}$$

and

$$\rho_j(u, v, x, y) = \left(u + s - \lambda_{j+1} - \tfrac{\lambda_{j+1}}{l_{j+1}}a_{j+1}, v, \tfrac{1}{l_{j+1}-1}x, (l_{j+1}-1)y\right)$$
$$\text{if } u > 2js + \tfrac{\lambda_{j+1}}{l_{j+1}}(a_{j+1} - \pi). \tag{6.2.45}$$

In view of the identity (6.2.44) we can extend the restriction of ρ_j to the subset $\phi_j(F_j') \cap \{(u, v, x, y) \mid u > 2js\}$ by the identity to the symplectic embedding

$$\widetilde{\rho}_j : (\widetilde{\phi}_j \circ \widetilde{\psi}_j)\left(\coprod_{i=1}^{j+1} \widetilde{S}_i\right) \hookrightarrow \mathbb{R}^{2n}, \qquad \widetilde{\rho}_j = \begin{cases} \text{id} & \text{on } \widetilde{\phi}_j \circ \widetilde{\psi}_j\left(\coprod_{i=1}^{j} \widetilde{S}_i\right), \\[2mm] \rho_j & \text{on } \widetilde{\phi}_j \circ \widetilde{\psi}_j\left(\widetilde{S}_{j+1}\right). \end{cases}$$

We finally define the symplectic embedding ψ_{j+1} by

$$\psi_{j+1} := \widetilde{\rho}_j \circ \widetilde{\phi}_j \circ \widetilde{\psi}_j : \coprod_{i=1}^{j+1} \widetilde{S}_i \hookrightarrow \mathbb{R}^{2n}.$$

In view of the induction hypothesis (6.2.27) and Lemma 6.2.14 and the inclusion $\rho_j(F_{j+1}) \subset T_{j+1}$ we have

$$\psi_{j+1} : \coprod_{i=1}^{j+1} \widetilde{S}_i \hookrightarrow \coprod_{i=1}^{j+1} T_i(\delta) \cup \coprod_{i=1}^{j} N_i(\delta),$$

i.e., (6.2.29) holds true. Moreover, we deduce from formulae (6.2.43) and (6.2.45), from $\widetilde{v}_{j+1} = \widetilde{v}_j + \widetilde{u}_{j+1}$ and from the definitions of s_{j+1}, t_{j+1} and ε_{j+1} given in (6.2.26) and (6.2.31) that

$$\psi_{j+1}(u, v, x, y) = \left(u - \widetilde{v}_{j+1} + s_{j+1}, v, \tfrac{1}{\sqrt{h_{j+1}}}\tfrac{1}{\pi}x, \sqrt{h_{j+1}}\pi y\right) \quad \text{if } u > \widetilde{v}_{j+1} - \varepsilon_{j+1},$$

i.e., formula (6.2.30) holds true. This completes the inductive construction of the symplectic embedding ψ_{j+1}.

Composing the linear symplectomorphism

$$\beta \times \text{id}_{2n-2} : \lambda \coprod_{i=1}^{k} S_i \to \coprod_{i=1}^{k} \widetilde{S}_i$$

given by formula (6.2.25) with the inductively constructed symplectic embedding

$$\psi_k : \coprod_{i=1}^{k} \widetilde{S}_i \hookrightarrow \coprod_{i=1}^{k} T_i(\delta) \cup \coprod_{i=1}^{k-1} N_i(\delta)$$

we finally obtain the symplectic embedding

$$\Psi_a := \psi_k \circ (\beta \times \mathrm{id}_{2n-2}) \colon \lambda \coprod_{i=1}^{k} S_i \hookrightarrow \coprod_{i=1}^{k} T_i(\delta) \cup \coprod_{i=1}^{k-1} N_i(\delta).$$

The proof of Proposition 6.2.10 is complete.   □

Step 6. End of the proof of Theorem 6.2.1. We let $\epsilon > 0$ be as in Theorem 6.2.1, set $\epsilon' = \epsilon/5$, choose k and $a_0 = a_0(\epsilon)$ as in Step 3 and let s be the number associated with k and ϵ' after the proof of Lemma 6.2.2. We fix $a \geq a_0$ and define $\lambda = \lambda(a)$ as in (6.2.20).

Using Lemma 6.2.9, the identities (6.2.18) and the definitions (6.2.20) and (6.2.14) we can estimate

$$\begin{aligned}
|\lambda\, T^n(a, \pi)| &> \left|\lambda \coprod S_i\right| - \epsilon' \\
&= k\, |\lambda S_1| - \epsilon' \\
&= kq\, |T_1| - \epsilon' \\
&= k \frac{1}{n!} s^{2n} - 2\epsilon'.
\end{aligned} \tag{6.2.46}$$

Composing the inclusion

$$\iota_a \colon \lambda\, T^n(a, \pi) \hookrightarrow \lambda \coprod_{i=1}^{k} S_i$$

with the symplectic embeddings Ψ_a and ρ guaranteed by Proposition 6.2.10 and Proposition 6.2.6 we obtain the symplectic embedding

$$\Phi_a := \rho \circ \Psi_a \circ \iota_a \colon \lambda\, T^n(a, \pi) \hookrightarrow M.$$

In view of the estimates (6.2.11) and (6.2.46) we find

$$\begin{aligned}
\mu\left(M \setminus \Phi_a\left(\lambda\, T^n(a, \pi)\right)\right) &= \mu(M) - \mu\left(\Phi_a\left(\lambda\, T^n(a, \pi)\right)\right) \\
&= \mu(M) - |\lambda\, T^n(a, \pi)| \\
&< \left(k \frac{1}{n!} s^{2n} + 3\epsilon'\right) - \left(k \frac{1}{n!} s^{2n} - 2\epsilon'\right) \\
&= 5\epsilon' \\
&= \epsilon.
\end{aligned}$$

This is the required estimate in Theorem 6.2.1 an so the proof of Theorem 6.2.1 is complete.   □

6.3 Asymptotic embedding invariants

Consider again a connected $2n$-dimensional symplectic manifold (M, ω) of finite volume $\mathrm{Vol}(M, \omega)$. In view of Theorem 3 the asymptotic symplectic invariants

$$\lim_{a \to \infty} p_a^E(M, \omega) = 1 \quad \text{and} \quad \lim_{a \to \infty} p_a^P(M, \omega) = 1 \tag{6.3.1}$$

are uninteresting. In order to recapture some information on the geometry of (M, ω) one can try to study the convergence speeds in (6.3.1). We define the symplectic invariants $\sigma_E(M, \omega)$ and $\sigma_P(M, \omega)$ in $[0, \infty]$ by

$$\sigma_E(M, \omega) = \sup \left\{ s \; \middle| \; \begin{array}{l} \text{there exists a constant } C < \infty \\ \text{such that } 1 - p_a^E(M, \omega) \le Ca^{-s} \text{ for all } a > \pi \end{array} \right\},$$

$$\sigma_P(M, \omega) = \sup \left\{ s \; \middle| \; \begin{array}{l} \text{there exists a constant } C < \infty \\ \text{such that } 1 - p_a^P(M, \omega) \le Ca^{-s} \text{ for all } a > \pi \end{array} \right\}.$$

In this section we notice that $\sigma_E(M, \omega) \ge \frac{1}{n}$ or $\sigma_P(M, \omega) \ge \frac{1}{n}$ for large classes of $2n$-dimensional symplectic manifolds and thereby improve Theorem 3 for many symplectic manifolds.

Consider a domain U in $\mathbb{R}^{2n}$. The distance $d(p, \partial U)$ between a point $p \in U$ and the boundary ∂U is

$$d(p, \partial U) = \inf \{ d(p, q) \mid q \in \partial U \}.$$

We say that U is *very connected* if there exists $\epsilon > 0$ such that the set

$$U \setminus \{ p \in U \mid d(p, \partial U) < \epsilon \}$$

is connected.

Theorem 6.3.1. *Assume that U is a very connected bounded domain in $\mathbb{R}^{2n}$ with piecewise smooth boundary and that (M, ω) is a compact connected $2n$-dimensional symplectic manifold.*

(i)$_E$ $\sigma_E(U) \ge \frac{1}{n}$ *if $n \le 3$ or if U is a ball.*

(ii)$_E$ $\sigma_E(M, \omega) \ge \frac{1}{n}$ *if $n \le 3$.*

(i)$_P$ $\sigma_P(U) \ge \frac{1}{n}$.

(ii)$_P$ $\sigma_P(M, \omega) \ge \frac{1}{n}$.

Rough outline of the proof of Theorem 6.3.1. Assertion (i)$_E$ for a $2n$-ball follows from Corollary 7.1.6 (i) below which is proved by symplectic wrapping. The other assertions can be proved by the symplectic folding methods previously described.

Assume first that U is a $2n$-cube. If $n = 2$, assertion (i)$_E$ follows from Proposition 4.4.2, and if $n = 3$, assertion (i)$_E$ can be proved by using that in dimension 2 a cube can be filled with small simplices. Assertion (i)$_P$ for a cube follows from Proposition 5.2.1.

Assume next that U is a very connected bounded domain in $\mathbb{R}^{2n}$ with piecewise smooth boundary. Then assertions (i)$_E$ and (i)$_P$ can be proved by first exhausting U with an increasing sequence $U_1 \subset U_2 \subset \cdots$ of connected unions of equal cubes and then extending the embedding techniques used to fill a cube to the sets U_i.

Assume finally that (M, ω) is a compact connected $2n$-dimensional symplectic manifold. Then assertions (ii)$_E$ and (ii)$_P$ can be proved by first choosing finitely many Darboux charts $\varphi_i : U_i \to V_i \subset M$ such that the U_i are very connected bounded domains with piecewise smooth boundary and such that the V_i are disjoint and $\bigcup \overline{V_i} = M$, then connecting the V_i by lines, and finally applying the technique used to prove (i)$_E$ and (i)$_P$ to the sets $V_i = \varphi_i(U_i)$ and using thinner and thinner neighbourhoods of the lines to pass from one V_i to another. $\square$

Chapter 7

Symplectic wrapping

In this chapter we first describe the symplectic wrapping construction invented by Traynor in [81], and then compare the embedding results obtained by symplectic folding and symplectic wrapping.

7.1 The wrapping construction

Symplectic wrapping views the whole ellipsoid or the whole polydisc as a Lagrangian product of a cube and a simplex or of a cube and a cuboid, and then wraps the cube around the base of the cotangent bundle of the torus via a linear map. Symplectic wrapping has therefore a more algebraic flavour than symplectic folding.

For the sake of brevity we shall only study symplectic wrapping embeddings of skinny ellipsoids into balls and of skinny polydiscs into cubes.

Theorem 7.1.1. *Assume that $a > 0$ and that $k_1 < \cdots < k_{n-1}$ are relatively prime numbers.*

(i) *The ellipsoid $E^{2n}(\pi, \ldots, \pi, a)$ symplectically embeds into the ball*

$$B^{2n}\left(\max\left\{(k_{n-1}+1)\,\pi, \frac{a}{k_1 \cdots k_{n-1}}\right\} + \epsilon\right)$$

for any $\epsilon > 0$.

(ii) *The polydisc $P^{2n}(\pi, \ldots, \pi, a)$ symplectically embeds into the cube*

$$C^{2n}\left(\max\left\{k_{n-1}\pi, (n-1)\pi + \frac{a}{k_1 \cdots k_{n-1}}\right\}\right).$$

Proof. Theorem 7.1.1 (i) for $n = 2$ is Theorem 6.4 in [81]. We shall closely follow [81]. We consider again the Lagrangian splitting $\mathbb{R}^n(x) \times \mathbb{R}^n(y)$ of $\mathbb{R}^{2n}$, set

$$\square(a_1, \ldots, a_n) = \{0 < x_i < a_i, \ 1 \leq i \leq n\} \subset \mathbb{R}^n(x),$$

$$\triangle(b_1, \ldots, b_n) = \left\{0 < y_1, \ldots, y_n \ \Big| \ \sum_{i=1}^{n} \frac{y_i}{b_i} < 1\right\} \subset \mathbb{R}^n(y),$$

and abbreviate $\square^n(a) = \square(a, \ldots, a)$ and $\triangle^n(b) = \triangle(b, \ldots, b)$. We also set

$$T^n = \mathbb{R}^n(x)/\pi\mathbb{Z}^n$$

and abbreviate

$$\kappa = k_1 \cdots k_{n-1}$$

and

$$A_E = \max\left\{(k_{n-1} + 1)\pi, \frac{a}{\kappa}\right\},$$
$$A_P = \max\left\{k_{n-1}\pi, (n-1)\pi + \frac{a}{\kappa}\right\}.$$

Choose $\epsilon > 0$. We set $\epsilon' = \epsilon/A_E$. The embeddings provided by symplectic wrapping are compositions of symplectic embeddings

$$E(\pi, \ldots, \pi, a) \xrightarrow{\alpha_E} \square^n(1) \times (1 + \epsilon')\triangle(\pi, \ldots, \pi, a)$$
$$\xrightarrow{\beta} \square\left(\tfrac{\pi}{k_1}, \ldots, \tfrac{\pi}{k_{n-1}}, \kappa\pi\right) \times (1 + \epsilon')\triangle\left(k_1, \ldots, k_{n-1}, \tfrac{a}{\kappa\pi}\right)$$
$$\xrightarrow{\gamma} T^n \times \triangle^n\left(\tfrac{A_E + \epsilon}{\pi}\right)$$
$$\xrightarrow{\delta_E} B^{2n}(A_E + \epsilon)$$

respectively

$$P(\pi, \ldots, \pi, a) \xrightarrow{\alpha_P} \square^n(1) \times \square(\pi, \ldots, \pi, a)$$
$$\xrightarrow{\beta} \square\left(\tfrac{\pi}{k_1}, \ldots, \tfrac{\pi}{k_{n-1}}, \kappa\pi\right) \times \square\left(k_1, \ldots, k_{n-1}, \tfrac{a}{\kappa\pi}\right)$$
$$\xrightarrow{\gamma} T^n \times \square^n\left(\tfrac{A_P}{\pi}\right)$$
$$\xrightarrow{\delta_P} C^{2n}(A_P).$$

1. The maps α_E and α_P. Recall from the proof of Lemma 5.3.1 (i) that there exists a symplectomorphism

$$\alpha_1 \times \cdots \times \alpha_n \colon P(\pi, \ldots, \pi, a) \rightarrow \square(\pi, \ldots, \pi, a) \times \square^n(1)$$

which embeds the subset $E(\pi, \ldots, \pi, a)$ into $\left((1 + \epsilon')\triangle(\pi, \ldots, \pi, a)\right) \times \square^n(1)$. We denote the reflection $(x, y) \mapsto (x, -y)$ of $\mathbb{R}^n(x) \times \mathbb{R}^n(y)$ by ρ and the permutation $(x, y) \mapsto (y, x)$ by σ. The composition

$$\alpha_P := \sigma \circ (\alpha_1 \times \cdots \times \alpha_n) \circ \rho$$

symplectomorphically maps $P(\pi, \ldots, \pi, a)$ to $\square^n(1) \times \square(\pi, \ldots, \pi, a)$, and its restriction to $E(\pi, \ldots, \pi, a)$, which we denote by α_E, symplectically embeds $E(\pi, \ldots, \pi, a)$ into $\square^n(1) \times (1 + \epsilon')\triangle(\pi, \ldots, \pi, a)$.

2. The map β. The map β is the linear symplectomorphism of $\mathbb{R}^n(x) \times \mathbb{R}^n(y)$ given by the diagonal matrix

$$\text{diag}\left[\frac{\pi}{k_1}, \ldots, \frac{\pi}{k_{n-1}}, \kappa\pi, \frac{k_1}{\pi}, \ldots, \frac{k_{n-1}}{\pi}, \frac{1}{\kappa\pi}\right].$$

3. The map γ. In order to describe the wrapping map γ we need an elementary lemma.

Lemma 7.1.2. *Let $M: \mathbb{R}^n(x) \to \mathbb{R}^n(x)$ be the linear map given by the matrix*

$$\begin{pmatrix} 1 & & & & -\frac{1}{k_1} \\ & 1 & & 0 & -\frac{1}{k_2} \\ & & \ddots & & \vdots \\ & 0 & & 1 & -\frac{1}{k_{n-1}} \\ & & & & 1 \end{pmatrix}$$

and let $p: \mathbb{R}^n(x) \to \mathbb{R}^n(x)/\pi\mathbb{Z}^n = T^n$ be the projection. Then the composition $p \circ M: \mathbb{R}^n(x) \to T^n$ embeds $\square\left(\frac{\pi}{k_1}, \ldots, \frac{\pi}{k_{n-1}}, \kappa\pi\right)$ into T^n.

Proof. Let x and x' be points in $\square\left(\frac{\pi}{k_1}, \ldots, \frac{\pi}{k_{n-1}}, \kappa\pi\right)$ for which $(p \circ M)(x) = (p \circ M)(x')$. Then

$$\left|x_i' - x_i\right| < \tfrac{\pi}{k_i}, \quad i = 1, \ldots, n-1, \tag{7.1.1}$$

and

$$\left|x_n' - x_n\right| < \kappa\pi, \tag{7.1.2}$$

and there exist integers $l_1, \ldots, l_{n-1}$ such that

$$x_i' - \tfrac{x_n'}{k_i} = x_i - \tfrac{x_n}{k_i} + l_i\pi \tag{7.1.3}$$

and an integer l_n such that

$$x_n' = x_n + l_n\pi. \tag{7.1.4}$$

Inserting the identity (7.1.4) into the estimate (7.1.2) we obtain

$$|l_n| < \kappa \tag{7.1.5}$$

and inserting (7.1.4) into the identities (7.1.3) we obtain

$$x_i' - x_i = \left(\tfrac{l_n}{k_i} + l_i\right)\pi, \quad i = 1, \ldots, n-1. \tag{7.1.6}$$

The identities (7.1.6) and the estimates (7.1.1) yield

$$\left|\tfrac{l_n}{k_i} + l_i\right| < \tfrac{1}{k_i}, \quad i = 1, \ldots, n-1. \tag{7.1.7}$$

Assume that $l_n \neq 0$. Then the estimates (7.1.7) imply that $l_n + k_i l_i = 0$ for $i = 1, \ldots, n-1$. Since the numbers $k_1, \ldots, k_{n-1}$ are relatively prime, we conclude that $l_n = m k_1 \cdots k_{n-1}$ for some integer $m \neq 0$. In particular,

$$|l_n| \geq k_1 \cdots k_{n-1} = \kappa$$

in contradiction to (7.1.5). Therefore $l_n = 0$. The estimates (7.1.7) now imply that also $l_1 = \cdots = l_{n-1} = 0$, and so $x = x'$ in view of the identities (7.1.4) and (7.1.3). The restriction of $p \circ M$ to $\square\left(\frac{\pi}{k_1}, \ldots, \frac{\pi}{k_{n-1}}, \kappa\pi\right)$ is therefore injective, as claimed. $\square$

We denote by M^* the transpose of the inverse of M. Then the map

$$M \times M^* \colon \mathbb{R}^n(x) \times \mathbb{R}^n(y) \to \mathbb{R}^n(x) \times \mathbb{R}^n(y)$$

is a linear symplectomorphism. In view of Lemma 7.1.2 the map

$$\gamma \colon \square\left(\frac{\pi}{k_1}, \ldots, \frac{\pi}{k_{n-1}}, \kappa\pi\right) \times \mathbb{R}^n(y) \to T^n \times \mathbb{R}^n(y)$$

defined by $\gamma = (p \circ M) \times M^*$ is a symplectic embedding.

Lemma 7.1.3. *We have*

$$\gamma\left(\square\left(\frac{\pi}{k_1}, \ldots, \frac{\pi}{k_{n-1}}, \kappa\pi\right) \times (1+\epsilon')\Delta\left(k_1, \ldots, k_{n-1}, \frac{a}{\kappa\pi}\right)\right) \subset T^n \times \Delta^n\left(\frac{A_E + \epsilon}{\pi}\right)$$

and

$$\gamma\left(\square\left(\frac{\pi}{k_1}, \ldots, \frac{\pi}{k_{n-1}}, \kappa\pi\right) \times \square\left(k_1, \ldots, k_{n-1}, \frac{a}{\kappa\pi}\right)\right) \subset T^n \times \square^n\left(\frac{A_P}{\pi}\right).$$

Proof. In view of the definition of γ and since $(1 + \epsilon')A_E = A_E + \epsilon$ it is enough to show that

$$M^*\left(\Delta\left(k_1, \ldots, k_{n-1}, \frac{a}{\kappa\pi}\right)\right) \subset \Delta^n\left(\frac{A_E}{\pi}\right)$$

and that

$$M^*\left(\square\left(k_1, \ldots, k_{n-1}, \frac{a}{\kappa\pi}\right)\right) \subset \square^n\left(\frac{A_P}{\pi}\right).$$

We compute that M^* is given by the matrix

$$\begin{pmatrix} 1 & & & & \\ & 1 & & 0 & \\ & & \ddots & & \\ & 0 & & 1 & \\ \frac{1}{k_1} & \frac{1}{k_2} & \cdots & \frac{1}{k_{n-1}} & 1 \end{pmatrix}.$$

We assume first that $y \in \triangle\left(k_1, \ldots, k_{n-1}, \frac{a}{\kappa\pi}\right)$ and set $y' = M^* y$. Using the definitions of $\triangle\left(k_1, \ldots, k_{n-1}, \frac{a}{\kappa\pi}\right)$ and A_E we then find

$$y_1' + \cdots + y_n' = (k_1 + 1)\frac{y_1}{k_1} + \cdots + (k_{n-1} + 1)\frac{y_{n-1}}{k_{n-1}} + \frac{a}{\kappa\pi}\frac{y_n}{\frac{a}{\kappa\pi}}$$

$$< \max\left\{k_{n-1} + 1, \frac{a}{\kappa\pi}\right\}$$

$$= \frac{A_E}{\pi}.$$

Therefore $y' \in \triangle^n\left(\frac{A_E}{\pi}\right)$ as claimed.

We assume next that $y \in \square\left(k_1, \ldots, k_{n-1}, \frac{a}{\kappa\pi}\right)$ and set $y' = M^* y$. Using the definitions of $\square\left(k_1, \ldots, k_{n-1}, \frac{a}{\kappa\pi}\right)$ and A_P we then find

$$y' \in \square\left(k_1, \ldots, k_{n-1}, n - 1 + \frac{a}{\kappa\pi}\right) \subset \square^n\left(\frac{A_P}{\pi}\right)$$

as claimed. $\qquad\square$

4. The maps δ_E and δ_P. We define symplectic embeddings

$$\tilde{\delta}_E : \square^n(\pi) \times \triangle^n\left(\frac{A_E + \epsilon}{\pi}\right) \hookrightarrow B^{2n}(A_E + \epsilon)$$

and

$$\tilde{\delta}_P : \square^n(\pi) \times \square^n\left(\frac{A_P}{\pi}\right) \hookrightarrow C^{2n}(A_P)$$

by

$$(x_1, \ldots, x_n, y_1, \ldots, y_n) \mapsto \left(\sqrt{y_1}\cos 2x_1, \ldots, \sqrt{y_n}\cos 2x_n, \right.$$
$$\left. -\sqrt{y_1}\sin 2x_1, \ldots, -\sqrt{y_n}\sin 2x_n\right).$$

The embedding $\tilde{\delta}_E$ extends to a symplectic embedding

$$\delta_E : T^n \times \triangle^n\left(\frac{A_E + \epsilon}{\pi}\right) \hookrightarrow B^{2n}(A_E + \epsilon)$$

and the embedding $\tilde{\delta}_P$ extends to a symplectic embedding

$$\delta_P : T^n \times \square^n\left(\frac{A_P}{\pi}\right) \hookrightarrow C^{2n}(A_P).$$

This completes the construction of the symplectic embeddings involved in symplectic wrapping, and Theorem 7.1.1 is proved. $\qquad\square$

Assume that $a > 0$ and that $k_1 < \cdots < k_{n-1}$ are relatively prime numbers. As in the proof of Theorem 7.1.1 we abbreviate

$$A_E(a; k_1, \ldots, k_{n-1}) = \max\left\{(k_{n-1} + 1)\pi, \frac{a}{k_1 \cdots k_{n-1}}\right\},$$

$$A_P(a; k_1, \ldots, k_{n-1}) = \max\left\{k_{n-1}\pi, (n-1)\pi + \frac{a}{k_1 \cdots k_{n-1}}\right\}.$$

In view of Theorem 7.1.1 we are interested in the functions $w_{\text{EB}}^{2n} :]2\pi, \infty[\to \mathbb{R}$ and $w_{\text{PC}}^{2n} :]2\pi, \infty[\to \mathbb{R}$ defined by

$$w_{\text{EB}}^{2n}(a) := \inf \{ A_E (a; k_1, \ldots, k_{n-1}) \mid k_1 < \cdots < k_{n-1} \text{ are relatively prime} \},$$

$$w_{\text{PC}}^{2n}(a) := \inf \{ A_P (a; k_1, \ldots, k_{n-1}) \mid k_1 < \cdots < k_{n-1} \text{ are relatively prime} \}.$$

Notice that the infimum in the definitions of the functions w_{EB}^{2n} and w_{PC}^{2n} can be replaced by the minimum. We shall next explicitly compute the functions w_{EB}^{2n} and w_{PC}^{2n} for $n = 2$ and $n = 3$.

Corollary 7.1.4. *Assume that $a > 2\pi$.*

(i) *The ellipsoid $E(\pi, a)$ symplectically embeds into the ball $B^4 \big(w_{\text{EB}}^4(a) + \epsilon \big)$ for any $\epsilon > 0$ where*

$$w_{\text{EB}}^4(a) = \begin{cases} (k+1)\pi & \text{if } (k-1)(k+1) < \frac{a}{\pi} \leq k(k+1), \\ \frac{a}{k} & \text{if } k(k+1) < \frac{a}{\pi} \leq k(k+2). \end{cases}$$

(ii) *The polydisc $P(\pi, a)$ symplectically embeds into the cube $C^4 \big(w_{\text{PC}}^4(a) \big)$ where*

$$w_{\text{PC}}^4(a) = \begin{cases} k\pi & \text{if } (k-1)^2 < \frac{a}{\pi} \leq (k-1)k, \\ \frac{a}{k} + \pi & \text{if } (k-1)k < \frac{a}{\pi} \leq k^2. \end{cases}$$

Proof. (i) follows from Theorem 7.1.1 (i), from the identity

$$w_{\text{EB}}^4(a) = \min_{k \in \mathbb{N}} \max \left\{ (k+1)\pi, \frac{a}{k} \right\}$$

and from a straightforward computation.

(ii) follows from Theorem 7.1.1 (ii), from the identity

$$w_{\text{PC}}^4(a) = \min_{k \in \mathbb{N}} \max \left\{ k\pi, \frac{a}{k} + \pi \right\}$$

and from a straightforward computation. $\square$

Corollary 7.1.5. *Assume that $a > 2\pi$.*

(i) *The ellipsoid $E(\pi, \pi, a)$ symplectically embeds into the ball $B^6 \big(w_{\text{EB}}^6(a) + \epsilon \big)$ for any $\epsilon > 0$ where*

$$w_{\text{EB}}^6(a) = \begin{cases} (k+1)\pi & \text{if } (k-2)(k-1)(k+1) < \frac{a}{\pi} \leq (k-1)k(k+1), \\ \frac{a}{(k-1)k} & \text{if } (k-1)k(k+1) < \frac{a}{\pi} \leq (k-1)k(k+2). \end{cases}$$

(ii) *The polydisc* $P(\pi, \pi, a)$ *symplectically embeds into the cube* $C^6\left(w_{PC}^6(a)\right)$ *where*

$$w_{PC}^6(a) = \begin{cases} (k+1)\pi & \text{if } (k-1)^2 k < \frac{a}{\pi} \leq (k-1)k(k+1), \\ \frac{a}{k(k+1)} + 2\pi & \text{if } (k-1)k(k+1) < \frac{a}{\pi} \leq k^2(k+1). \end{cases}$$

Proof. (i) Fix $a > 2\pi$ and assume that $k_1 < k_2$ are relatively prime numbers. Then $k_2 - 1$ and k_2 are also relatively prime, and

$$A_E(a; k_2 - 1, k_2) \leq A_E(a; k_1, k_2).$$

It follows that

$$w_{EB}^6(a) = \min_{k \in \mathbb{N}} \max \left\{ (k+1)\pi, \frac{a}{(k-1)k} \right\}. \tag{7.1.8}$$

Assertion (i) follows from Theorem 7.1.1 (i), from the identity (7.1.8) and from a straightforward computation.

(ii) The same argument as above implies that for each $a > 2\pi$,

$$\begin{aligned} w_{PC}^6(a) &= \min_{k \in \mathbb{N}} \max \left\{ k\pi, 2\pi + \frac{a}{(k-1)k} \right\} \\ &= \min_{k \in \mathbb{N}} \max \left\{ (k+1)\pi, \frac{a}{k(k+1)} + 2\pi \right\}. \end{aligned} \tag{7.1.9}$$

Assertion (ii) follows from Theorem 7.1.1 (ii), from the identity (7.1.9) and from a straightforward computation. $\qquad\square$

For $n \geq 4$ the computation of $w_{EB}^{2n}(a)$ and $w_{PC}^{2n}(a)$ is more involved, and we do not know explicit formulae for the functions w_{EB}^{2n} and w_{PC}^{2n}. The next corollary describes the asymptotic behaviour of these functions for all $n \geq 2$.

Corollary 7.1.6. (i) *There exists a constant* C_E *depending only on n such that*

$$\frac{|E(\pi, \ldots, \pi, a)|}{\left|B^{2n}\left(w_{EB}^{2n}(a)\right)\right|} \geq 1 - C_E\, a^{-1/n} \quad \text{for all } a > 2\pi.$$

(ii) *There exists a constant* C_P *depending only on n such that*

$$\frac{|P(\pi, \ldots, \pi, a)|}{\left|C^{2n}\left(w_{PC}^{2n}(a)\right)\right|} \geq 1 - C_P\, a^{-1/n} \quad \text{for all } a > 2\pi.$$

Proof. (i) We choose $n - 1$ prime numbers $p_1 < p_2 < \cdots < p_{n-1}$ and define l to be the least common multiple of the differences

$$p_j - p_i, \quad 1 \leq i < j \leq n - 1.$$

Fix $a > 2\pi$. We set

$$m := \left\lceil \frac{1}{l}\left(\sqrt[n]{\frac{a}{\pi}} + p_{n-1}\right) \right\rceil \tag{7.1.10}$$

where $\lceil r \rceil$ denotes the minimal integer which is greater or equal to r, and

$$k_i := ml - p_{n-i}, \quad i = 1, \ldots, n-1. \tag{7.1.11}$$

We claim that the numbers $k_1 < k_2 < \cdots < k_{n-1}$ are relatively prime. Indeed, assume that

$$d \mid ml - p_i \quad \text{and} \quad d \mid ml - p_j \tag{7.1.12}$$

for some $i \neq j$. Then d divides the difference

$$(ml - p_i) - (ml - p_j) = p_j - p_i$$

and so d divides the least common multiple l. But then (7.1.12) implies that d divides both p_i and p_j, and so $d = 1$ as claimed.

Using the definitions (7.1.11) and (7.1.10) we find

$$k_1 = ml - p_{n-1} \geq \sqrt[n]{\frac{a}{\pi}}.$$

Therefore,

$$(k_{n-1} + 1) \, k_{n-1} \ldots k_1 \geq k_1^n \geq \frac{a}{\pi}.$$

We conclude that

$$(k_{n-1} + 1) \, \pi \geq \frac{a}{k_1 \cdots k_{n-1}}$$

and so

$$(k_{n-1} + 1) \, \pi = A_E \, (a; k_1, \ldots, k_{n-1}) \geq w_{\mathrm{EB}}^{2n}(a). \tag{7.1.13}$$

By definition (7.1.10) we have

$$ml \leq l \left(\frac{1}{l} \left(\sqrt[n]{\frac{a}{\pi}} + p_{n-1} \right) + 1 \right) = \sqrt[n]{\frac{a}{\pi}} + p_{n-1} + l. \tag{7.1.14}$$

Using definition (7.1.11) and the estimate (7.1.14) we find

$$k_{n-1} + 1 = ml - p_1 + 1 \leq \sqrt[n]{\frac{a}{\pi}} + p_{n-1} + l. \tag{7.1.15}$$

The estimates (7.1.13) and (7.1.15) now yield

$$\frac{\left| B^{2n} \big(w_{\mathrm{EB}}^{2n}(a) \big) \right|}{\left| E(\pi, \ldots, \pi, a) \right|} \leq \frac{(k_{n-1} + 1)^n \, \pi}{a} \leq \frac{\left(\sqrt[n]{a} + (p_{n-1} + l) \, \sqrt[n]{\pi} \right)^n}{a}. \tag{7.1.16}$$

Recall that the number $(p_{n-1} + l) \, \sqrt[n]{\pi}$ only depends on n. In view of the estimate (7.1.16) we therefore find a constant C_E depending only on n such that

$$\frac{\left| B^{2n} \big(w_{\mathrm{EB}}^{2n}(a) \big) \right|}{\left| E(\pi, \ldots, \pi, a) \right|} \leq 1 + C_E \, a^{-1/n} \quad \text{for all } a > 2\pi$$

and so

$$\frac{|E(\pi, \ldots, \pi, a)|}{\left|B^{2n}\left(w_{\mathrm{EB}}^{2n}(a)\right)\right|} \geq 1 - C_E\, a^{-1/n} \quad \text{for all } a > 2\pi.$$

Assertion (i) is proved.

(ii) can be proved in the same way as (i). $\square$

7.2 Folding versus wrapping

Recall from Chapters 4, 5 and 6 that symplectic folding yields good symplectic embedding results of $2n$-dimensional ellipsoids and polydiscs into any connected $2n$-dimensional symplectic manifold of finite volume. The reason is that symplectic folding is local in nature in the sense that each fold is achieved on a set of small volume. Symplectic wrapping described in Section 7.1, however, is global in nature and thus yields good symplectic embedding results of ellipsoids and polydiscs into special symplectic manifolds only. E.g., the best $2n$-dimensional symplectic wrapping embedding of an ellipsoid into a cube respectively of a polydisc into a ball fill less than $\frac{1}{n!}$ respectively $\frac{n!}{n^n}$ of the volume.

In this section we shall compare the embedding results for embeddings of skinny ellipsoids into balls and skinny polydiscs into cubes yielded by the two methods.

7.2.1 Embeddings $E^{2n}(\pi, \ldots, \pi, a) \hookrightarrow B^{2n}(A)$

1. The case $n = 2$. Recall from Theorem 1 that for $a \in [\pi, 2\pi]$ the ellipsoid $E(\pi, a)$ does not symplectically embed into the ball $B^4(A)$ if $A < 2\pi$. Also recall that the functions $f_{\mathrm{EB}} \colon\,]2\pi, \infty[\,\to \mathbb{R}$ and $w_{\mathrm{EB}} \equiv w_{\mathrm{EB}}^4 \colon\,]2\pi, \infty[\,\to \mathbb{R}$ constructed in 4.3.1 and defined in Corollary 7.1.4 (i) describe our results for the embedding problem $E(\pi, a) \hookrightarrow B^4(A)$ obtained by multiple symplectic folding and by symplectic wrapping.

According to Proposition 4.3.7 the difference $f_{\mathrm{EB}}(a) - \sqrt{\pi a}$ between f_{EB} and the volume condition is bounded by 2π. Computer calculations suggest that this difference is monotone increasing and converging to π as $a \to \infty$. The difference $w_{\mathrm{EB}}(a) - \sqrt{\pi a}$ between w_{EB} and the volume condition attains its local minima at $k(k + 1)\pi$, where it is equal to $m_k = (k + 1)\pi - \sqrt{k(k + 1)}\,\pi$, and it attains its local maxima at $k(k + 2)\pi$, where it is equal to $M_k = (k + 2)\pi - \sqrt{k(k + 2)}\,\pi$. The sequence (m_k) strictly decreases to $\frac{\pi}{2}$, and (M_k) strictly decreases to π. We conclude that the difference $|f_{\mathrm{EB}}(a) - w_{\mathrm{EB}}(a)|$ is bounded by 2π.

For $a > 2\pi$ small we have $f_{\mathrm{EB}}(a) < w_{\mathrm{EB}}(a)$. E.g., $f_{\mathrm{EB}}(3\pi) = 2.3801 \ldots \pi$ and $f_{\mathrm{EB}}(4\pi) = 2.6916 \ldots \pi$, and so

Fact 7.2.1. *The ellipsoid $E(\pi, 3\pi)$ symplectically embeds into $B^4(2.381\,\pi)$, and $E(\pi, 4\pi)$ symplectically embeds into $B^4(2.692\,\pi)$.*

The inequality $w_{EB}(a) < f_{EB}(a)$ happens first at $a = 5.1622\ldots\pi$. In general, the computer calculations for f_{EB} suggest that the functions w_{EB} and f_{EB} yield alternately better estimates: For all $k \in \mathbb{N}$ we seem to have $w_{EB}(a) < f_{EB}(a)$ on an interval around $k(k+1)\pi$ and $f_{EB}(a) < w_{EB}(a)$ on an interval around $k(k+2)\pi$; moreover, we seem to have

$$\lim_{k \to \infty} (f_{EB}(k(k+2)\pi) - w_{EB}(k(k+2)\pi)) = 0.$$

We checked the above statements for $k \le 5\,000$.

We extend both functions $f_{EB}(a)$ and $w_{EB}(a)$ to functions on $[\pi, \infty[$ by setting $f_{EB}(a) = a$ and $w_{EB}(a) = a$ for $a \in [\pi, 2\pi]$. The characteristic function for the embedding problem $E(\pi, a) \hookrightarrow B^4(A)$ is the function χ_{EB} on $[\pi, \infty[$ defined by

$$\chi_{EB}(a) = \inf \left\{ A \mid E(\pi, a) \text{ symplectically embeds into } B^4(A) \right\}.$$

The following proposition summarizes what we know about this function.

Proposition 7.2.2. *For $a \in [\pi, 2\pi]$ we have $\chi_{EB}(a) = a$, and on $]2\pi, \infty[$ the function χ_{EB} is bounded from below and above by*

$$\max(2\pi, \sqrt{\pi a}\,) \le \chi_{EB}(a) \le \min(w_{EB}(a), f_{EB}(a)),$$

see Figure 7.1. In particular,

$$\limsup_{\epsilon \to 0^+} \frac{\chi_{EB}(2\pi + \epsilon) - 2\pi}{\epsilon} \le \frac{3}{7}. \tag{7.2.1}$$

The function χ_{EB} is monotone increasing and hence almost everywhere differentiable. Moreover, χ_{EB} is Lipschitz continuous with Lipschitz constant at most 1; more precisely,

$$\chi_{EB}(a') - \chi_{EB}(a) \le \frac{\min(w_{EB}(a), f_{EB}(a))}{a} (a' - a) \quad \text{for all } a' \ge a \ge \pi.$$

Proof. The estimates of $\chi_{EB}(a)$ from below are provided by the second Ekeland–Hofer capacity, which yields $\chi_{EB}(a) \ge a$ for $a \in [\pi, 2\pi]$ and $\chi_{EB}(a) \ge 2\pi$ for $a > 2\pi$, and by the volume condition $|E(\pi, a)| \le |B^4(\chi_{EB}(a))|$, which translates to $\chi_{EB}(a) \ge \sqrt{\pi a}$. The estimate (7.2.1) follows from $\chi_{EB} \le f_{EB}$ and from Proposition 4.3.5. The remaining claims follow as in the proof of Proposition 4.3.11.   □

2. The case $n \ge 3$. Recall from Theorem 1 that for $a \in [\pi, 2\pi]$ the ellipsoid $E^{2n}(\pi, \ldots, \pi, a)$ does not symplectically embed into the ball $B^{2n}(A)$ if $A < a$. For $a > 2\pi$, the n'th Ekeland–Hofer capacity still implies that $E^{2n}(\pi, \ldots, \pi, a)$ does not symplectically embed into $B^{2n}(A)$ if $A < 2\pi$. This information is vacuous if $a \ge 2^n\pi$ in view of the volume condition $A \ge \sqrt[n]{\pi^{n-1}a}$.

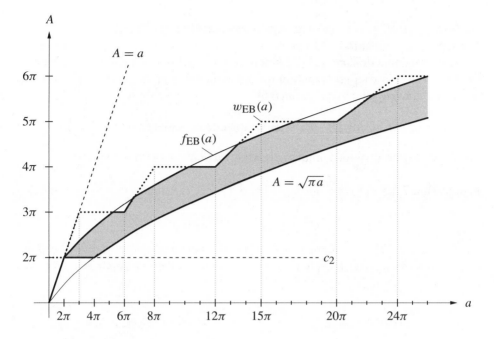

Figure 7.1. What is known about $\chi_{EB}(a)$.

If $a > 2\pi$, Theorem 3.1.1, which we proved by symplectic folding, shows that $E^{2n}(\pi, \ldots, \pi, a)$ symplectically embeds into $B^{2n}\left(\frac{a}{2} + \pi + \epsilon\right)$ for every $\epsilon > 0$. For a large, the results obtained by symplectic folding, which are Theorem 6.3.1 (i)$_E$ for $n = 3$ and the first statement in Theorem 3, are weaker than the results obtained by symplectic wrapping, which are described by the function $w_{EB}^{2n}\colon\;]2\pi, \infty[\to \mathbb{R}$ defined before Corollary 7.1.4. For $n = 3$ the difference $w_{EB}^6(a) - \sqrt[3]{\pi^2 a}$ between the function w_{EB}^6 computed in Corollary 7.1.5 (i) and the volume condition is bounded by 2π. For $n \geq 4$ it follows from Corollary 7.1.6 (i) that the difference $w_{EB}^{2n}(a) - \sqrt[n]{\pi^{n-1} a}$ between w_{EB}^{2n} and the volume condition is bounded.

7.2.2 Embeddings $P^{2n}(\pi, \ldots, \pi, a) \hookrightarrow C^{2n}(A)$

1. The case $n = 2$. Recall that the functions $f_{PC}\colon]2\pi, \infty[\to \mathbb{R}$ and $w_{PC} \equiv w_{PC}^4\colon]2\pi, \infty[\to \mathbb{R}$ defined in Proposition 4.4.4 and Corollary 7.1.4 (ii) describe our results for the embedding problem $P(\pi, a) \hookrightarrow C^4(A)$ obtained by multiple symplectic folding and by symplectic wrapping.

Comparing Proposition 4.4.4 with Corollary 7.1.4 (ii) we see that $w_{PC}(a) \leq f_{PC}(a)$ for all $a > 2\pi$, cf. Figure 7.2. Equality only holds for $a \in\;]2\pi, 4\pi[$ and $a \in [k^2\pi, (k^2 + 1)\pi]$, $k \geq 2$. The difference $w_{PC}(a) - \sqrt{\pi a}$ between w_{PC} and the volume condition attains its local minima at $k(k - 1)\pi$, $k \geq 3$, where it is equal to

$m_k = k\pi - \sqrt{k(k-1)}\pi$, and it attains its local maxima at $k^2\pi$, where it is equal to π. The sequence (m_k) strictly decreases to $\frac{\pi}{2}$.

Extend the function $w_{PC}(a)$ to a function on $[\pi, \infty[$ by setting $w_{PC}(a) = a$ for $a \in [\pi, 2\pi]$. The characteristic function for the embedding problem $P(\pi, a) \hookrightarrow C^4(A)$ is the function χ_{PC} on $[\pi, \infty[$ defined by

$$\chi_{PC}(a) = \inf \left\{ A \mid P(\pi, a) \text{ symplectically embeds into } C^4(A) \right\}.$$

The following proposition summarizes what we know about this function.

Proposition 7.2.3. *The function $\chi_{PC} \colon [\pi, \infty[\to \mathbb{R}$ is bounded from below and above by*

$$\sqrt{\pi a} \leq \chi_{PC}(a) \leq w_{PC}(a),$$

see Figure 7.2. It is monotone increasing and hence almost everywhere differentiable. Moreover, χ_{PC} is Lipschitz continuous with Lipschitz constant at most 1; more precisely,

$$\chi_{PC}(a') - \chi_{PC}(a) \leq \frac{w_{PC}(a)}{a}(a' - a) \quad \text{for all } a' \geq a \geq \pi.$$

Proof. The estimate $\sqrt{\pi a} \leq \chi_{PC}(a)$ from below is provided by the volume condition $|P(\pi, a)| \leq |C^4(\chi_{PC}(a))|$. The remaining claims follow as in the proof of Proposition 4.3.11. □

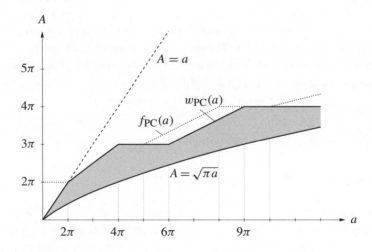

Figure 7.2. What is known about $\chi_{PC}(a)$.

2. The case $n \geq 3$. We recall that the two functions $f_{PC}^{2n} \colon \,]2\pi, \infty[\to \mathbb{R}$ and $w_{PC}^{2n} \colon \,]2\pi, \infty[\to \mathbb{R}$ defined in Proposition 5.2.1 and before Corollary 7.1.4 describe our results for the embedding problem $P^{2n}(\pi, \ldots, \pi, a) \hookrightarrow C^{2n}(A)$ obtained by multiple symplectic folding and by symplectic wrapping.

Comparing the case $n = 3$ of Proposition 5.2.1 with Corollary 7.1.5 (ii) we see that in contrast to the case $n = 2$ we have $f^6_{PC}(a) \leq w^6_{PC}(a)$ for all $a > 2\pi$. Equality only holds for $a \in \left[\left((k-1)k^2 + 2 \right) \pi, (k-1)k(k+1)\pi \right]$, $k \geq 2$. The difference $f^6_{PC}(a) - \sqrt[3]{\pi^2 a}$ between f^6_{PC} and the volume condition is bounded by

$$d_3 = \left(13 - \sqrt[3]{1586} \right)\pi \approx 1.338\pi.$$

For $n \geq 4$ the comparison of the functions f^{2n}_{PC} and w^{2n}_{PC} is more involved since we do not know an explicit formula for w^{2n}_{PC}. The difference $f^{2n}_{PC}(a) - \sqrt[n]{\pi^{n-1}a}$ between f^{2n}_{PC} and the volume condition is bounded by d_n where

$$d_4 = \left(5 - \sqrt[4]{194} \right)\pi, \quad d_5 = \left(4 - \sqrt[5]{164} \right)\pi, \quad d_n = \left(3 - \sqrt[n]{2^{n-1} + 2} \right)\pi, \quad n \geq 6.$$

The sequence (d_n), $n \geq 3$, strictly decreases to π. It follows from Corollary 7.1.6 (ii) that the difference $w^{2n}_{PC}(a) - \sqrt[n]{\pi^{n-1}a}$ is also bounded. Therefore, the difference $\left| f^{2n}_{PC} - w^{2n}_{PC} \right|$ is bounded.

We conclude this section by motivating the conjecture alluded to at the end of Section 2.3. We say that a polydisc $P^{2n}(\pi, \ldots, \pi, a)$, $a > \pi$, is *reducible* if it symplectically embeds into a cube $C^{2n}(A)$ for some $A < a$. It follows from Theorem 1.3 (ii) in [50] that a polydisc cannot be reduced by a local squeezing method. A symplectic embedding which reduces $P^{2n}(\pi, \ldots, \pi, a)$ must therefore be of global nature. In view of Proposition 5.2.1 and Corollary 7.1.4 (ii) symplectic folding and symplectic wrapping both show that $P^{2n}(\pi, \ldots, \pi, a)$ is reducible if $a > 2\pi$. However, none of the two methods can reduce $P^{2n}(\pi, \ldots, \pi, a)$ if $a \leq 2\pi$. Since we believe that only some kind of folding or wrapping can reduce a polydisc, we conjecture

Conjecture 7.2.4. *The polydisc-analogue of Theorem 1 holds true. In particular, the polydiscs $P^{2n}(\pi, \ldots, \pi, a)$ symplectically embeds into the cube $C^{2n}(A)$ for some $A < a$ if and only if $a > 2\pi$.*

Chapter 8

Proof of Theorem 4

In this chapter we prove Theorem 4 stated in Section 1.3.3 in two different ways, first by symplectic folding, and then by a symplectic lifting construction. We first state a generalization of the second statement

$$\hat{\zeta}(a) = 0 \quad \text{for all } a \in]0, \pi[\tag{8.0.1}$$

in Theorem 4, and prove a generalization of $\hat{\zeta}(\pi) = \pi$. In Section 8.2 we give a further motivation for Problem ζ coming from convex geometry and from the special behaviour of symplectic capacities on bounded convex subsets of $(\mathbb{R}^{2n}, \omega_0)$. In Section 8.3 we prove the generalization of (8.0.1) by symplectic folding, and in Section 8.4 we use symplectic lifting to prove a yet more general result which also proves the first statement in Theorem 4.

8.1   A more general statement

We consider arbitrary subsets S of $\mathbb{R}^{2n}$ which symplectically embed into the cylinder $Z^{2n}(\pi)$, and we measure the intersections $\varphi(S) \cap D_x$ by arbitrary extrinsic symplectic capacities c on $\mathbb{R}^2$ as defined in Definition C.5. We refer to Appendix C.2 for a thorough study of such capacities and only mention that any (normalized) intrinsic symplectic capacity on $\mathbb{R}^2$ as defined in Definition C.1 is a (normalized) extrinsic symplectic capacity on $\mathbb{R}^2$.

A symplectic embedding of a subset S of $\mathbb{R}^{2n}$ into another subset S' of $\mathbb{R}^{2n}$ is by definition a symplectic embedding of an open neighbourhood of S into $\mathbb{R}^{2n}$ which maps S into S'. As before we denote by $Z^{2n}(a)$ the open standard symplectic cylinder $D(a) \times \mathbb{R}^{2n-2}$, and we denote by $E_z \subset \mathbb{R}^{2n}$ the affine plane

$$E_z := \mathbb{R}^2 \times \{z\}, \quad z \in \mathbb{R}^{2n-2}.$$

Given any subset T of $Z^{2n}(\pi)$ we abbreviate

$$c\,(T \cap E_z) := c\big(\{\,(u, v) \in \mathbb{R}^2 \mid (u, v, z) \in T \cap E_z \,\}\big).$$

Assume now that the subset S of $\mathbb{R}^{2n}$ symplectically embeds into $Z^{2n}(\pi)$. We define the symplectic invariant $\xi_c(S) \in [0, \pi]$ by

$$\xi_c\,(S) := \inf_{\varphi}\, \sup_{z}\, c\,(\varphi\,(S) \cap E_z) \tag{8.1.1}$$

where φ varies over all symplectic embeddings of S into $Z^{2n}(\pi)$. The main result of this chapter is

Theorem 8.1.1. *Assume that the subset S of $\mathbb{R}^{2n}$ symplectically embeds into $Z^{2n}(a)$ for some $a < \pi$. Then $\xi_c(S) = 0$ for any extrinsic symplectic capacity c on $\mathbb{R}^2$.*

Assume now that S is a subset of $\mathbb{R}^{2n}$ which embeds into $Z^{2n}(\pi)$ by a symplectomorphism of $\mathbb{R}^{2n}$. We define the ambient symplectic invariant $\zeta_c(S) \in [0, \pi]$ by

$$\zeta_c(S) := \inf_{\varphi} \sup_z c(\varphi(S) \cap E_z)$$

where now φ varies over all symplectomorphisms of $\mathbb{R}^{2n}$ which embed S into $Z^{2n}(\pi)$. Then $\xi_c(S) \le \zeta_c(S)$.

Corollary 8.1.2. *Assume that S is a relatively compact subset of $\mathbb{R}^{2n}$ whose closure embeds into $Z^{2n}(\pi)$ by a symplectomorphism of $\mathbb{R}^{2n}$. Then $\zeta_c(S) = 0$ for any extrinsic symplectic capacity c on $\mathbb{R}^2$.*

The second statement (8.0.1) in Theorem 4 in Section 1.3.3 is a special case of Corollary 8.1.2. By Theorem 8.4.2 below, Corollary 8.1.2 can be extended to certain unbounded subsets of $\mathbb{R}^{2n}$. The following result shows that some condition on S, however, must be imposed.

Proposition 8.1.3. *For the unit circle $S^1 \subset \mathbb{R}^2$ we have $\zeta_c\big(B^{2n}(\pi)\big) \in [c(S^1), \pi]$ for any normalized extrinsic symplectic capacity c on $\mathbb{R}^2$.*

According to Corollary C.10 (i) we have $c(S^1) = 0$ for any intrinsic symplectic capacity c on $\mathbb{R}^2$, and so the statement in Proposition 8.1.3 is empty for these capacities. In fact, Theorem 8.4.2 below shows that $\zeta_c\big(B^{2n}(\pi)\big) = 0$ for intrinsic symplectic capacities. On the other hand, Proposition C.12 says that for any $a \in [0, \pi]$ there exists a normalized extrinsic symplectic capacity c on $\mathbb{R}^2$ such that $c(S^1) = a$. Examples of normalized extrinsic symplectic capacities on $\mathbb{R}^2$ with $c(S^1) = \pi$ are the first Ekeland–Hofer capacity, [20, Theorem 1], the displacement energy, [38, Theorem 1.9], and the outer cylindrical capacity $\hat{z}$, see Theorem C.8 (iii).

Proof of Corollary 8.1.2. Fix $\epsilon > 0$. We denote the closure of S by $\bar{S}$. By assumption there exists a symplectomorphism φ of $\mathbb{R}^{2n}$ such that $\varphi(\bar{S}) \subset Z^{2n}(\pi)$. Since $\bar{S}$ is compact, $\varphi(\bar{S})$ is compact in $\mathbb{R}^{2n}$. We therefore find $a \in]0, \pi[$ and $A > 0$ such that

$$\varphi(\bar{S}) \subset D(a) \times B^{2n-2}(A). \tag{8.1.2}$$

Choose $a' \in]a, \pi[$. Then

$$D(a) \times B^{2n-2}(A) \subset Z^{2n}(a'). \tag{8.1.3}$$

In view of Theorem 8.1.1 there exists a symplectic embedding

$$\psi : Z^{2n}(a') \hookrightarrow Z^{2n}(\pi) \tag{8.1.4}$$

such that

$$\sup_z c\big(\psi\big(Z^{2n}(a')\big) \cap E_z\big) < \epsilon. \tag{8.1.5}$$

Applying Proposition A.1 to the bounded starshaped domain $D(a) \times B^{2n-2}(A)$ we find a symplectomorphism Ψ of $\mathbb{R}^{2n}$ such that

$$\Psi|_{D(a) \times B^{2n-2}(A)} = \psi|_{D(a) \times B^{2n-2}(A)}. \tag{8.1.6}$$

The inclusion (8.1.2), the identity (8.1.6) and the inclusion (8.1.3) show that

$$(\Psi \circ \varphi)(\bar{S}) \subset \psi\big(Z^{2n}(a')\big). \tag{8.1.7}$$

In view of the inclusions (8.1.7) and (8.1.4) the symplectomorphism $\Psi \circ \varphi$ of $\mathbb{R}^{2n}$ embeds $\bar{S}$ into $Z^{2n}(\pi)$. Moreover, the inclusion (8.1.7), the monotonicity of the symplectic capacity c and the estimate (8.1.5) yield

$$\sup_z c\big((\Psi \circ \varphi)(\bar{S}) \cap E_z\big) \leq \sup_z c\big(\psi\big(Z^{2n}(a')\big) \cap E_z\big) < \epsilon.$$

We conclude that $\zeta_c(S) \leq \zeta_c(\bar{S}) < \epsilon$. Since $\epsilon > 0$ was arbitrary, Corollary 8.1.2 follows. $\square$

Proof of Proposition 8.1.3. We abbreviate

$$S_z^1 := \big\{(u, v, z) \in E_z \mid u^2 + v^2 = 1\big\}, \quad z \in \mathbb{R}^{2n-2}.$$

Let φ be a symplectomorphism of $\mathbb{R}^{2n}$ which embeds $B^{2n}(\pi)$ into $Z^{2n}(\pi)$. According to Lemma 1.2 in [50] there exists $z_0 \in \mathbb{R}^{2n-2}$ such that

$$S_{z_0}^1 \subset \partial\big(\varphi\big(B^{2n}(\pi)\big)\big) \cap E_{z_0}.$$

Since φ is a diffeomorphism of $\mathbb{R}^{2n}$ and $\varphi\big(B^{2n}(\pi)\big) \subset Z^{2n}(\pi)$, the boundary $\partial\big(\varphi\big(B^{2n}(\pi)\big)\big)$ of $\varphi\big(B^{2n}(\pi)\big)$ is tangent to the boundary $S^1 \times \mathbb{R}^{2n-2}$ of $Z^{2n}(\pi)$ at each point of $S_{z_0}^1$. This, the inclusion $\varphi\big(B^{2n}(\pi)\big) \subset Z^{2n}(\pi)$ and the compactness of $S_{z_0}^1$ imply that there exists an $\epsilon > 0$ such that the annulus

$$A_\epsilon := \big\{(u, v, z_0) \in E_{z_0} \mid 1 - \epsilon < |(u, v)| < 1\big\}$$

is contained in $\varphi\big(B^{2n}(\pi)\big) \cap E_{z_0}$. For each $r \in {]}1 - \epsilon, 1{[}$ we then have

$$rS_{z_0}^1 \subset A_\epsilon \subset \varphi\big(B^{2n}(\pi)\big) \cap E_z.$$

In view of the conformality and the monotonicity of the symplectic capacity c we obtain that

$$r^2 c(S^1) = c(r S^1) = c\big(r S^1_{z_0}\big) \le c(A_\epsilon) \le c\big(\varphi\big(B^{2n}(\pi)\big) \cap E_z\big)$$

for each $r \in \,]1 - \epsilon, 1[$, and so

$$c\big(\varphi\big(B^{2n}(\pi)\big) \cap E_z\big) \ge c(S^1).$$

Since this estimate holds for any symplectomorphism φ of $\mathbb{R}^{2n}$ which embeds $B^{2n}(\pi)$ into $Z^{2n}(\pi)$ we conclude that

$$\zeta_c\big(B^{2n}(\pi)\big) \ge c(S^1).$$

On the other hand, the definition of ζ_c and the normalization of c yield

$$\zeta_c\big(B^{2n}(\pi)\big) \le \sup_z \, c\big(B^{2n}(\pi) \cap E_z\big) = c\big(D(\pi)\big) = \pi.$$

The proof of Proposition 8.1.3 is complete. $\square$

Lemma 1.2 in [50] and its proof suggest that the answer to the following question is affirmative. It would be interesting to have a proof of this.

Question 8.1.4. *Is it true that $\xi_c\big(B^{2n}(\pi)\big) \ge c(S^1)$ for every extrinsic symplectic capacity c on $\mathbb{R}^2$? In particular, $\xi_{\hat{\mu}}\big(B^{2n}(\pi)\big) = \pi$?*

8.2 A further motivation for Problem ζ

Recall from Section 1.3.3 that Problem ζ was stimulated by the search for symplectic rigidity phenomena beyond the Nonsqueezing Theorem. In this section we give yet another motivation for this problem coming from convex geometry and from the special behaviour of symplectic capacities on bounded convex domains in $(\mathbb{R}^{2n}, \omega_0)$.

We denote by $\mathcal{K}^n$ the set of bounded convex domains in $\mathbb{R}^n$. Notice that $\mathcal{K}^{2n}$ is not invariant under the group $\mathcal{D}(n)$ of symplectomorphisms of $(\mathbb{R}^{2n}, \omega_0)$. This can be seen by symplectic folding or, even easier, by symplectic lifting as described in Section 8.4 below, cf. Figure 8.9. We thus consider

$$\mathcal{K}^{2n}_s = \big\{\varphi(K) \mid K \in \mathcal{K}^{2n}, \ \varphi \in \mathcal{D}(n)\big\}.$$

It seems that the symplectic geometry of sets in $\mathcal{K}^{2n}_s$ is special. One instance for this is the following observation due to Viterbo, [85].

Proposition 8.2.1. *For any two normalized intrinsic symplectic capacities c and c' on $(\mathbb{R}^{2n}, \omega_0)$ and any $K \in \mathcal{K}^{2n}_s$ we have*

$$c'(K) \le n^2 c(K).$$

We refer to Definition C.1 for the definition of an intrinsic symplectic capacity on $(\mathbb{R}^{2n}, \omega_0)$. The proof of Proposition 8.2.1 is similar to the proof of Proposition 8.2.2 below and is given in Appendix C.1. Notice that Proposition 8.2.1 does not extend to bounded starshaped domains in view of the Symplectic Hedgehog Theorem stated in Section 1.2.2.

Looking for other peculiarities of the geometry of convex domains, one can proceed as follows: Start from an inequality of Euclidean invariants on $\mathcal{K}^{2n}$, symplectify these invariants to symplectic invariants on $\mathcal{K}_s^{2n}$, and check whether the inequality survives. We refer to Section C.4 and to [76] for examples of this procedure. Here, we start from the intersection invariant σ and the projection invariant π. Fix a convex domain $K \in \mathcal{K}^n$. Given an element H of the Grassmannian $\mathcal{H}$ of hyperplanes in $\mathbb{R}^n$ through the origin, let $p_H \colon \mathbb{R}^n \to H$ be the orthogonal projection onto H and let $|p_H(K)|$ be the $(n-1)$-dimensional Lebesgue measure of $p_H(K) \subset H$. The projection invariant

$$\pi(K) = \min_{H \in \mathcal{H}} |p_H(K)|$$

of $K \in \mathcal{K}^n$ is the volume of the "smallest shadow" of K. Similarly, let $|K \cap (H + x)|$ be the $(n-1)$-dimensional Lebesgue measure of the intersection of K with a translate $H + x$ of H. The intersection invariant

$$\sigma(K) = \min_{H \in \mathcal{H}} \max_{x \in \mathbb{R}^n} |K \cap (H + x)|$$

of K is the "smallest largest intersection" of K. The following proposition was pointed out to me by Daniel Hug.

Proposition 8.2.2. *For any convex domain $K \in \mathcal{K}^n$ it holds that*

$$\sigma(K) \leq \pi(K) \leq n^{n-1}\sigma(K).$$

Proof. According to a result of John, there exists a unique open ellipsoid E of minimal volume containing K, and this ellipsoid, called *John's ellipsoid*, satisfies

$$\frac{1}{n}E \subset K \subset E, \tag{8.2.1}$$

see [28], [85] and the references therein. Since π and σ are monotone with respect to inclusion and since $\pi(E) = \sigma(E)$, we have

$$\pi(K) \leq \pi(E) = n^{n-1}\pi\left(\tfrac{1}{n}E\right) = n^{n-1}\sigma\left(\tfrac{1}{n}E\right) \leq n^{n-1}\sigma(K),$$

as claimed. $\square$

Remarks 8.2.3. 1. If K is centrally symmetric, then John's ellipsoid satisfies $\frac{1}{\sqrt{n}}E \subset K \subset E$, so that the constant n^{n-1} in Proposition 8.2.2 can be replaced by $n^{\frac{n-1}{2}}$.

2. In dimension 2, $\sigma(K) = \pi(K)$ for all $K \in \mathcal{K}^2$, see [28, Theorem 8.3.5]. This is not so in dimension $n \geq 3$: The set of K with $\sigma(K) < \pi(K)$ is open and dense in $\mathcal{K}^n$ in the Hausdorff topology, see [53]. $\diamond$

Let $H_1 = \langle x_1, \ldots, x_{n-1} \rangle \in \mathcal{H}$. Since the orthogonal group $O(n)$ acts transitively on $\mathcal{H}$,

$$\pi(K) = \min_{A \in O(n)} |p_{H_1}(A(K))|,$$

$$\sigma(K) = \min_{A \in O(n)} \max_{x \in \mathbb{R}^n} |A(K) \cap (H_1 + x)|.$$

Symplectifying these Euclidean invariants on $\mathcal{K}^{2n}$, we obtain the symplectic invariants π_s and σ_s on $\mathcal{K}_s^{2n}$ defined by

$$\pi_s(K) = \inf_{\varphi \in \mathcal{D}(n)} |p_1(\varphi(K))|,$$

$$\sigma_s(K) = \inf_{\varphi \in \mathcal{D}(n)} \sup_{z \in \mathbb{R}^{2n}} |\varphi(K) \cap E_z|,$$

where $p_1 \colon \mathbb{R}^{2n} \to E_1 := \mathbb{R}^2(x_1, y_1)$ is the projection, where again $E_z = E_1 \times \{z\}$ for $z \in \mathbb{R}^{2n-2}$, and where $|U|$ is the area of a domain $U \subset \mathbb{R}^2$. The *cylindrical capacity* of K is

$$\hat{z}(K) := \inf\{a \mid \text{there exists } \varphi \in \mathcal{D}(n) \text{ such that } \varphi(K) \subset Z^{2n}(a)\}.$$

Remarks C.17, Theorem C.18 and the Extension after Restriction Principle Proposition A.1 show that

$$\pi_s(K) = \hat{z}(K) \quad \text{for all } K \in \mathcal{K}_s^{2n}. \tag{8.2.2}$$

In particular, $\pi_s(B^{2n}(a)) = a$. Notice that $\sigma_s(B^{2n}(a)) \leq \zeta(a)$ for the function ζ considered in the introduction. If Proposition 8.2.2 would survive to $\pi_s(K) \leq c\,\sigma_s(K)$ for some $c > 0$, we would thus find the lower bound $\zeta(a) \geq \frac{1}{c}a$. However,

Proposition 8.2.4. $\sigma_s(K) = 0$ *for any* $K \in \mathcal{K}_s^{2n}$ *and* $\zeta(a)/a \to 0$ *as* $a \to 0$.

Proof. The following simple argument due to Polterovich is taken from [59].

(i) Using the monotonicity and conformality of σ_s and arguing as in the proof of Proposition 8.3.1 below we see that it suffices to prove $\sigma_s(K) = 0$ for $B = B^4(\pi)$. Choose a linear symplectomorphism M^ϵ of $\mathbb{R}^4$ mapping the symplectic plane E_1 to a symplectic plane E^ϵ so close to a Lagrangian plane that the ω_0-area of $B \cap E^\epsilon$ is less than $\epsilon \pi$. Then the ω_0-area and hence the area of $(M^\epsilon)^{-1}(B) \cap E_z$ is less than $\epsilon \pi$ for every $z \in \mathbb{R}^2$. To be more explicit, fix $\epsilon \in (0, 1)$. The linear symplectomorphism

$$M^\epsilon = \begin{bmatrix} \frac{1}{\epsilon} & 0 & 0 & 1 \\ 0 & \epsilon & 0 & 0 \\ 0 & 1 & \frac{1}{\epsilon} & 0 \\ 0 & 0 & 0 & \epsilon \end{bmatrix}$$

maps E_1 to $E^\epsilon := \{ (x_1, \epsilon x_2, x_2, 0) \mid x_1, x_2 \in \mathbb{R} \} \subset \mathbb{R}^4 = \{ (x_1, y_1, x_2, y_2) \}$. Since the ω_0-area of $B \cap (E^\epsilon + z)$ is maximal for $z = 0$, $\sigma_s(B)$ is at most the area of

$$(M^\epsilon)^{-1}(B) \cap E_1 = \left\{ (x_1, y_1) \mid \left(\tfrac{1}{\epsilon} x_1 \right)^2 + (\epsilon y_1)^2 + y_1^2 \leq 1 \right\}.$$

This set is an ellipsoid of area $\frac{\epsilon}{\sqrt{1+\epsilon^2}} \pi < \epsilon \pi$.

(ii) For fixed $\epsilon > 0$, the set $M^\epsilon\big(B^4(a)\big)$ lies in $Z^4(\pi)$ for a small enough, so that $\zeta(a) \leq \epsilon a$ for these a. In particular, $\zeta(a)/a \to 0$ as $a \to 0$. More explicitly, let $\epsilon_0^2 = (0.786\ldots)^2$ be the positive zero of $x^2 + x - 1$. A computation shows that for $\epsilon \leq \epsilon_0$ the set $M^\epsilon\big(B^4(a)\big)$ lies in $Z^4(\pi)$ provided that $a \leq \pi \epsilon^2$, so that $\zeta(a) < \sqrt{a/\pi}\, a$. $\square$

The symplectomorphisms M^ϵ used in the proof of Proposition 8.2.4, which made our first naive attempt to symplectify Proposition 8.2.2 fail, drastically increased the E_1-shadow of, say, a ball. In a second attempt we thus exclude such maps and introduce *constrained* symplectic projection and intersection invariants: For $\epsilon > 0$ the set $\mathcal{D}(K; \epsilon)$ of symplectomorphisms of $\mathbb{R}^{2n}$ mapping K into $Z^{2n}(\hat{z}(K) + \epsilon)$ is non-empty, and we set

$$\pi_s^z(K) = \lim_{\epsilon \to 0} \inf_{\varphi \in \mathcal{D}(K;\epsilon)} |p_1(\varphi(K))|,$$

$$\sigma_s^z(K) = \lim_{\epsilon \to 0} \inf_{\varphi \in \mathcal{D}(K;\epsilon)} \sup_{z \in \mathbb{R}^{2n}} |\varphi(K) \cap E_z|.$$

Clearly, $\pi_s(K) \leq \pi_s^z(K) \leq \hat{z}(K)$, so that the identity (8.2.2) implies

$$\pi_s^z(K) = \pi_s(K) = \hat{z}(K) \quad \text{for all } K \in \mathcal{K}_s^{2n}.$$

In particular, $\pi_s^z\big(B^{2n}(a)\big) = a$. Notice that $\sigma_s^z\big(B^{2n}(\pi)\big) = \lim_{a \nearrow \pi} \zeta(a)$. An inequality of the form $\pi_s^z\big(B^{2n}(\pi)\big) \leq c\, \sigma_s^z\big(B^{2n}(\pi)\big)$ for some $c > 0$ would thus be equivalent to the asymptotics $\lim_{a \nearrow \pi} \zeta(a) \geq \frac{1}{c} \pi$. Corollary 8.1.2 shows that there are no such inequalities, so that our second attempt of symplectifying Proposition 8.2.2 also fails.

8.3   Proof by symplectic folding

In this section we prove Theorem 8.1.1 by symplectic folding. We fix an extrinsic symplectic capacity c on $\mathbb{R}^2$. We assume without loss of generality that c is normalized.

Step 1. Reduction to a 4-dimensional problem

Proposition 8.3.1. *Assume that*

$$\xi_c\big(Z^4(a)\big) = 0 \quad \text{for all } a \in \,]0, \pi[.$$

Then Theorem 8.1.1 holds true.

Proof. Let S be a subset of $\mathbb{R}^{2n}$ for which there exist $a < \pi$ and a symplectic embedding $\varphi \colon S \hookrightarrow Z^{2n}(a)$. We fix $\epsilon > 0$. By assumption we find a symplectic embedding $\psi \colon Z^4(a) \hookrightarrow Z^4(\pi)$ such that

$$\sup_{z \in \mathbb{R}^2} c\big(\psi\big(Z^4(a)\big) \cap E_z\big) < \epsilon. \tag{8.3.1}$$

The composition

$$\rho \colon S \xrightarrow{\varphi} Z^{2n}(a) = Z^4(a) \times \mathbb{R}^{2n-4} \xrightarrow{\psi \times \mathrm{id}_{2n-4}} Z^4(\pi) \times \mathbb{R}^{2n-4} = Z^{2n}(\pi)$$

symplectically embeds S into $Z^{2n}(\pi)$. Moreover, the monotonicity of the symplectic capacity c and the estimate (8.3.1) yield

$$\begin{aligned}
\sup_{z \in \mathbb{R}^{2n-2}} c\big(\rho(S) \cap E_z\big) &= \sup_{z \in \mathbb{R}^{2n-2}} c\big((\psi \times \mathrm{id}_{2n-4})\,(\varphi(S)) \cap E_z\big) \\[2mm]
&\leq \sup_{z \in \mathbb{R}^{2n-2}} c\big((\psi \times \mathrm{id}_{2n-4})\big(Z^{2n}(a)\big) \cap E_z\big) \\[2mm]
&= \sup_{z \in \mathbb{R}^2} c\big(\psi\big(Z^4(a)\big) \cap E_z\big) \\[2mm]
&< \epsilon.
\end{aligned}$$

Since $\epsilon > 0$ was arbitrary, we conclude that $\xi_c(S) = 0$, as claimed. $\qquad\square$

Step 2. Reformulation of the 4-dimensional problem. We shall use coordinates u, v, x, y on $(\mathbb{R}^4, du \wedge dv + dx \wedge dy)$. Fix $a < \pi$ and $\epsilon > 0$. We may assume $a > \pi/2$ and $\epsilon < (\pi - a)/9$. Denote by $R \subset \{ (u, v) \mid 0 < u < \pi, \, 0 < v < 1 \}$ the convex hull of the image of the map γ drawn in Figure 8.4 below. The domain R is a rectangle with smooth corners. We set

$$\mathcal{A} := \{ (u, v, x, y) \mid \epsilon < u, \, 0 < v < 1 - \epsilon, \, 0 < x < 1, \, 0 < y < a \},$$

$$\mathcal{Z} := R \times \mathbb{R}^2.$$

Proposition 8.3.2. *Assume there exists a symplectic embedding* $\Phi \colon \mathcal{A} \hookrightarrow \mathcal{Z}$ *such that*

$$\sup_{z \in \mathbb{R}^2} c\,(\Phi(\mathcal{A}) \cap E_z) \leq 2\epsilon. \tag{8.3.2}$$

Then there exists a symplectic embedding $\Psi \colon Z^4(a) \hookrightarrow Z^4(\pi)$ *such that*

$$\sup_{z \in \mathbb{R}^2} c\big(\Psi\big(Z^4(a)\big) \cap E_z\big) \leq 2\epsilon. \tag{8.3.3}$$

Proof. We start with

Lemma 8.3.3.

  (i) *There exists a symplectomorphism*

$$\alpha \colon \mathbb{R}^2 \to \{\, (u, v) \in \mathbb{R}^2 \mid \epsilon < u, \, 0 < v < 1 - \epsilon \,\}.$$

  (ii) *There exists a symplectomorphism*

$$\sigma \colon D(a) \to \{\, (x, y) \in \mathbb{R}^2 \mid 0 < x < 1, \, 0 < y < a \,\}.$$

  (iii) *There exists a symplectomorphism ω of $\mathbb{R}^2$ such that $\omega(R) \subset D(\pi)$.*

Proof. (i) follows from the proof of Lemma 3.1.5 or from Proposition B.6 (ii). Here we give an explicit construction. Choose orientation preserving diffeomorphisms $g \colon \mathbb{R} \to \,]0, \infty[$ and $h \colon \mathbb{R} \to \,]0, 1 - \epsilon[$, and denote by g' and h' their derivatives. Then the maps

$$\alpha_1 \colon \mathbb{R}^2 \to \,]0, \infty[\, \times \mathbb{R}, \qquad\qquad (u, v) \mapsto \left(g(u), \frac{v}{g'(u)} \right),$$

$$\alpha_2 \colon \,]0, \infty[\, \times \mathbb{R} \to \,]\epsilon, \infty[\, \times \,]0, 1 - \epsilon[, \quad (u, v) \mapsto \left(\frac{u}{h'(v)} + \epsilon, \, h(v) \right)$$

are symplectomorphisms. The map $\alpha := \alpha_2 \circ \alpha_1$ is therefore as desired.

  (ii) follows from Lemma 3.1.5 or from Proposition B.6 (i).

  (iii) Choose $r \in \mathbb{R}$ so large that $R \subset D(\pi r^2)$, and choose a diffeomorphism χ of $D(2\pi r^2)$ which is a translation near the origin, maps the boundary of $D(\text{area}(R))$ to the boundary of R, and is the identity near the boundary of $D(2\pi r^2)$. By Lemma 3.1.5 and its proof, we may assume that χ is a symplectomorphism. Extend χ to the symplectomorphism of $\mathbb{R}^2$ which is the identity outside $D(2\pi r^2)$, and let ω be the inverse of this extension. Then $\omega(R) = D(\text{area}(R)) \subset D(\pi)$, as desired.   □

    Let $\Phi \colon \mathcal{A} \hookrightarrow \mathcal{Z}$ be a symplectic embedding as assumed in the proposition, let α, σ and ω be symplectomorphisms as guaranteed by Lemma 8.3.3, and denote the linear symplectomorphism $(u, v, x, y) \mapsto (x, y, u, v)$ of $\mathbb{R}^4$ by τ. Then the composition

$$\Psi \colon Z^4(a) \xrightarrow{(\alpha \times \sigma) \circ \tau} \mathcal{A} \xrightarrow{\Phi} \mathcal{Z} \xrightarrow{\omega \times \text{id}} Z^4(\pi)$$

is a symplectic embedding. Moreover, the identity $(\alpha \times \sigma)\left(\tau \left(Z^4(a) \right) \right) = \mathcal{A}$, the monotonicity of the symplectic capacity c on $\mathbb{R}^2$ and the assumed estimate (8.3.2) imply that

$$\sup_{z \in \mathbb{R}^2} c\left(\Psi\left(Z^4(a) \right) \cap E_z \right) = \sup_{z \in \mathbb{R}^2} c\left((\omega \times \text{id}) \left(\Phi(\mathcal{A}) \right) \cap E_z \right)$$

$$= \sup_{z \in \mathbb{R}^2} c\left(\Phi(\mathcal{A}) \cap E_z \right)$$

$$\leq 2\epsilon.$$

This completes the proof of Proposition 8.3.2.   □

Step 3. Construction of the embedding Φ. We are going to construct a symplectic embedding $\Phi\colon \mathcal{A} \hookrightarrow \mathcal{Z}$ satisfying the estimate (8.3.2) by a variant of multiple symplectic folding. Instead of folding alternatingly on the right and on the left, we will always fold on the right. If we neglect all quantities which can be chosen arbitrarily small, the image $\Phi(\mathcal{A}) \subset \mathcal{Z}$ will look as in Figure 8.1.

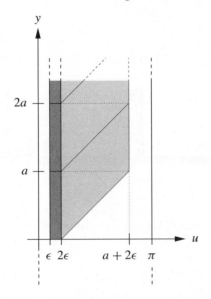

Figure 8.1. The embedding $\Phi\colon \mathcal{A} \hookrightarrow \mathcal{Z}$ for $a = \frac{3\pi}{5}$.

The symplectic embedding Φ will be the composition of the three maps $\beta \times \mathrm{id}$, φ and $\gamma \times \mathrm{id}$. Here, the smooth area preserving embedding

$$\beta\colon \{\, (u,v) \mid \epsilon < u,\ 0 < v < 1 - \epsilon \,\} \hookrightarrow \mathbb{R}^2$$

is similar to the map constructed in Step 1 of Section 4.2. Indeed, set $\delta = \frac{\epsilon}{2a}$. The map β restricts to the identity on $\{\epsilon < u \le 2\epsilon\}$, and it maps $\{2\epsilon < u \le 3\epsilon\}$ to the black region in $\{2\epsilon < u \le a + 9\epsilon\}$ drawn in Figure 8.2. Moreover,

$$\beta(u,v) = \beta(u - i2\epsilon, v) + (i(a + 8\epsilon), 0)$$

for $u \in \,]\epsilon + i2\epsilon,\ \epsilon + (i+1)2\epsilon]$ and $i = 1, 2, 3, \ldots$, see Figure 8.2.

The "lifting" map φ is similar to the map constructed in Step 3 of Section 4.2. Indeed, choose a cut off function $f_0\colon \mathbb{R} \to [0, 1 - \epsilon - \delta]$ with support $[3\epsilon, a + 8\epsilon]$ such that

$$\hat{a} := \int_{\mathbb{R}} f_0(s)\, ds > a.$$

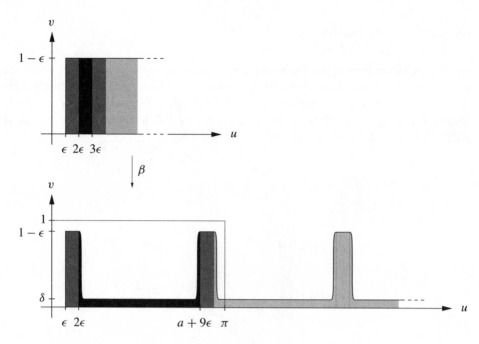

Figure 8.2. The left part of the image of β.

Set $f_i(s) = f_0(s - i(a + 8\epsilon))$, $i = 1, 2, 3, \ldots$, and $f(s) = \sum_{i \geq 0} f_i(s)$. The symplectomorphism $\varphi \colon \mathbb{R}^4 \to \mathbb{R}^4$ is defined by

$$\varphi(u, v, x, y) = \left(u, v + f(u)x, x, y + \int_0^u f(s)\, ds \right). \qquad (8.3.4)$$

The image $\varphi\left((\beta \times \mathrm{id})(\mathcal{A}) \right) \subset \mathbb{R}^4$ is illustrated in Figure 8.3. Denote the projection $(u, v, x, y) \mapsto (u, v)$ by p. The left part of the set $p\left(\varphi\left((\beta \times \mathrm{id})(\mathcal{A}) \right) \right)$ is equal to the upper domain in Figure 8.4.

We finally spiral the set $\varphi\left((\beta \times \mathrm{id})(\mathcal{A}) \right)$ into $\mathcal{Z}$. The smooth area preserving local embedding $\gamma \colon p\left(\varphi\left((\beta \times \mathrm{id})(\mathcal{A}) \right) \right) \to R$, which is explained by Figure 8.4, restricts to the identity on $\{ \epsilon < u \leq a + 8\epsilon \}$, and it maps the black region in $\{ a + 8\epsilon < u \leq a + 9\epsilon \}$ to the black region in its image. Moreover,

$$\gamma(u, v) = \gamma(u - i(a + 8\epsilon), v)$$

for $u \in\,]\epsilon + i(a + 8\epsilon),\ \epsilon + (i + 1)(a + 8\epsilon)]$ and $i = 1, 2, 3, \ldots$, see Figure 8.4. The map γ is constructed in the same way as the map in Step 4 of Section 3.2. Since γ is an embedding on each part

$$\{ (u, v) \in p\left(\varphi\left((\beta \times \mathrm{id})(\mathcal{A}) \right) \right) \mid \epsilon + i(a + 8\epsilon) < u \leq \epsilon + (i + 1)(a + 8\epsilon) \},$$

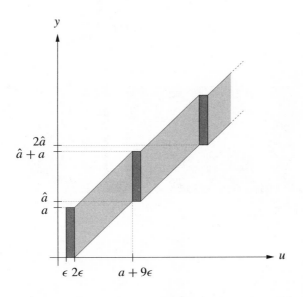

Figure 8.3. The left part of the image $\varphi\left((\beta \times \mathrm{id})(\mathcal{A})\right)$.

$i = 0, 1, 2, \ldots$, of its domain, and since $\hat{a} > a$, the map

$$\gamma \times \mathrm{id} \colon \varphi\left((\beta \times \mathrm{id})(\mathcal{A})\right) \to \mathbb{R}^4$$

is a symplectic embedding, cf. Figure 8.3 and Figure 8.5. We conclude that the composition

$$\Phi := (\gamma \times \mathrm{id}) \circ \varphi \circ (\beta \times \mathrm{id})$$

is a symplectic embedding of $\mathcal{A}$ into $\mathcal{Z}$.

Step 4. Verification of the estimate (8.3.2) in Proposition 8.3.2. We have to show that for any point $(x_0, y_0) \in \mathbb{R}^2$ the estimate

$$c\left(\Phi(\mathcal{A}) \cap E_{(x_0, y_0)}\right) \leq 2\epsilon \tag{8.3.5}$$

holds true. To this end, we may assume that $x_0 \in]0, 1[$ since otherwise the intersection $\Phi(\mathcal{A}) \cap E_{(x_0, y_0)}$ is empty. Moreover, by construction of the embedding Φ we have

$$\Phi(u + 2\epsilon, v, x, y) = \Phi(u, v, x, y) + (0, 0, 0, \hat{a})$$

for all $(u, v, x, y) \in \mathcal{A}$. It follows that for each $y_0 \in \mathbb{R}$ there exists $i \in \mathbb{Z}$ such that $y_0 + i\hat{a} \in]\hat{a}, 2\hat{a}]$ and

$$p\left(\Phi(\mathcal{A}) \cap E_{(x_0, y_0)}\right) \subset p\left(\Phi(\mathcal{A}) \cap E_{(x_0, y_0 + i\hat{a})}\right),$$

cf. Figure 8.5. We may therefore assume that $(x_0, y_0) \in]0, 1[\times]\hat{a}, 2\hat{a}]$. In order to verify the estimate (8.3.5) we distinguish two cases.

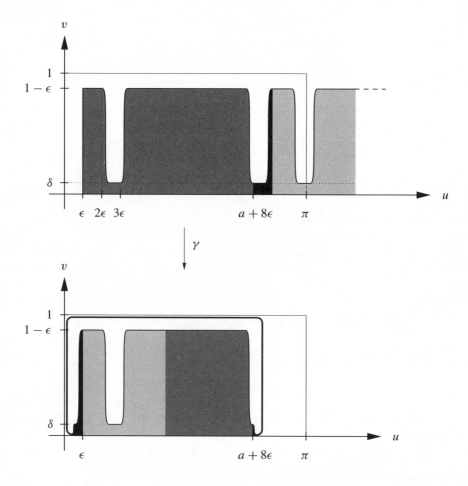

Figure 8.4. The folding map γ.

Case A. Assume first $y_0 \in]\hat{a}, \hat{a} + a[$. Then $E_{(x_0, y_0)}$ intersects the "floor"

$$F_2 := \{ \Phi(u, v, x, y) \mid 3\epsilon < u \leq 4\epsilon \}$$
$$= \{ (u, v, x, y) \mid \epsilon < u \leq 2\epsilon, \, 0 < v < 1 - \epsilon, \, 0 < x < 1, \, \hat{a} < y < 2\hat{a} \}$$

and the two "stairs"

$$S_i := \{ \Phi(u, v, x, y) \mid 2i\epsilon < u \leq (2i + 1)\epsilon \}, \quad i = 1, 2,$$

only, cf. Figure 8.5. The set $F := F_2 \cap E_{(x_0, y_0)}$ has area $\epsilon(1 - \epsilon)$. The set $S_1 \cap E_{(x_0, y_0)}$ consists of the black thickened "arc" A drawn in Figure 8.6 and of the branch $B_1 = B_1(x_0, y_0)$, which for $(x_0, y_0) = \left(\frac{1}{2}, \hat{a} + \frac{a}{2} \right)$ looks like in Figure 8.6. Indeed, since the preimage of S_1 under Φ is contained in $\{0 < y < a\}$, we read off from definition

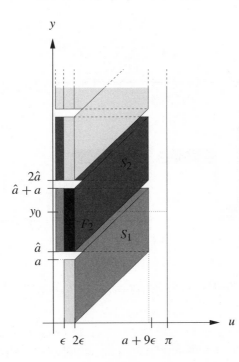

Figure 8.5. The floor F_2 and the stairs S_1 and S_2.

(8.3.4) that

$$B_1(x_0, y_0) = \{ (u, v + f(u)x_0, x_0, y_0) \mid u_1 < u < a + 8\epsilon, \, 0 < v < \delta \} \qquad (8.3.6)$$

where $u_1 = u_1(y_0)$ is defined through the equation

$$\int_{3\epsilon}^{u_1} f(s)\, ds = y_0 - a. \qquad (8.3.7)$$

Similarly, the branch $B_2 = B_2(x_0, y_0) = S_2 \cap E_{(x_0, y_0)}$ looks for $(x_0, y_0) = \left(\frac{1}{2}, \hat{a} + \frac{a}{2}\right)$ as in Figure 8.6. Indeed, since the preimage of S_2 under Φ is contained in the set $\{\hat{a} < y < \hat{a} + a\}$, we read off from (8.3.4) that

$$B_2(x_0, y_0) \cap \{u > 3\epsilon\} = \{ (u, v + f(u)x_0, x_0, y_0) \mid 3\epsilon < u < u_2, \, 0 < v < \delta \} \qquad (8.3.8)$$

where $u_2 = u_2(y_0)$ is defined through the equation

$$\int_{3\epsilon}^{u_2} f(s)\, ds = y_0 - \hat{a}. \qquad (8.3.9)$$

Subtracting (8.3.9) from (8.3.7) and using that $f(s) < 1$ for all $s \in \mathbb{R}$ we obtain

$$0 < \hat{a} - a = \int_{u_2}^{u_1} f(s)\, ds < u_1 - u_2. \qquad (8.3.10)$$

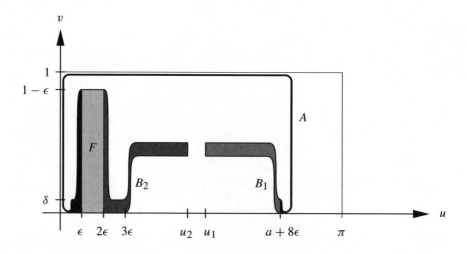

Figure 8.6. The set $\Phi(\mathcal{A}) \cap E_{(x_0,y_0)}$ for $(x_0, y_0) = \left(\frac{1}{2}, \hat{a} + \frac{a}{2}\right)$.

Let μ be the area of the set $\Phi(\mathcal{A}) \cap E_{(x_0,y_0)}$. In order to estimate the capacity of $\Phi(\mathcal{A}) \cap E_{(x_0,y_0)}$ we next embed $\Phi(\mathcal{A}) \cap E_{(x_0,y_0)}$ into a set whose capacity is known.

Lemma 8.3.4. *There exists a symplectomorphism ϕ of $\mathbb{R}^2$ such that*

$$\phi\left(p\left(\Phi(\mathcal{A}) \cap E_{(x_0,y_0)}\right)\right) \subset D\left(\mu + \epsilon^2\right).$$

Proof.   The set $p\left(\Phi(\mathcal{A}) \cap E_{(x_0,y_0)}\right) \subset \mathbb{R}^2$ is diffeomorphic to an open disc, and in view of the estimate (8.3.10) its boundary is a piecewise smooth embedded closed curve in $\mathbb{R}^2$, cf. Figure 8.6. We therefore find a simply connected domain $U \subset \mathbb{R}^2$ such that the boundary of U is a smoothly embedded closed curve in $\mathbb{R}^2$, and such that $p\left(\Phi(\mathcal{A}) \cap E_{(x_0,y_0)}\right) \subset U$ and area$(U) = \mu + \epsilon^2$. Applying the argument given in the proof of Lemma 8.3.3 (iii) to U, we find a symplectomorphism ϕ of $\mathbb{R}^2$ such that $\phi(U) = D(\mu + \epsilon^2)$. Then $\phi\left(p\left(\Phi(\mathcal{A}) \cap E_{(x_0,y_0)}\right)\right) \subset \phi(U) = D(\mu + \epsilon^2)$, as desired.   □

In view of the monotonicity and the conformality of the normalized symplectic capacity c on $\mathbb{R}^2$, Lemma 8.3.4 implies

$$\begin{aligned}
c\left(\Phi(\mathcal{A}) \cap E_{(x_0,y_0)}\right) &= c\left(\phi\left(p\left(\Phi(\mathcal{A}) \cap E_{(x_0,y_0)}\right)\right)\right) \\
&\leq c\left(D(\mu + \epsilon^2)\right) \qquad\qquad (8.3.11) \\
&= \left(\mu + \epsilon^2\right).
\end{aligned}$$

It remains to estimate the number $\mu = \text{area}\left(\Phi(\mathcal{A}) \cap E_{(x_0,y_0)}\right)$. We abbreviate $Q = \{2\epsilon < u \leq 3\epsilon, \ 0 < v < 1 - \epsilon\}$. Using the definitions of β and γ and the identities

(8.3.6) and (8.3.8) we can estimate

$$\text{area}(B_2 \cup B_1 \cup A) = \text{area}\left(p(B_2) \cup p(B_1) \cup \gamma^{-1}(p(A))\right)$$
$$< \text{area}\left(\{\,(u, v + f(u)x_0) \mid (u, v) \in \beta(Q)\,\}\right)$$
$$= \text{area}\left(\{\,(u, v) \mid (u, v) \in \beta(Q)\,\}\right)$$
$$= \text{area}\left(\beta(Q)\right) = \text{area}(Q) = \epsilon(1 - \epsilon).$$

Therefore,

$$\mu = \text{area}\left(\Phi(\mathcal{A}) \cap E_{(x_0, y_0)}\right) = \text{area}(F) + \text{area}(B_2 \cup B_1 \cup A)$$
$$< \epsilon(1 - \epsilon) + \epsilon(1 - \epsilon).$$

In view of (8.3.11) we conclude that $c\left(\Phi(\mathcal{A}) \cap E_{(x_0, y_0)}\right) < 2\epsilon$.

Case B. Assume now $y_0 \in [\hat{a} + a, 2\hat{a}]$. Then $E_{(x_0, y_0)}$ intersects the stairs S_2 only, and

$$\Phi(\mathcal{A}) \cap E_{(x_0, y_0)} = \{\,(u, v + f(u)x_0, x_0, y_0) \mid u_l < u < u_r, \ 0 < v < \delta\,\}$$

where $u_l = u_l(y_0)$ and $u_r = u_r(y_0)$ lie in $[3\epsilon, a + 8\epsilon]$ and are defined through the equations

$$\int_{3\epsilon}^{u_l} f(s)\, ds = y_0 - (\hat{a} + a) \quad \text{and} \quad \int_{3\epsilon}^{u_r} f(s)\, ds = y_0 - \hat{a}.$$

The set $p\left(\Phi(\mathcal{A}) \cap E_{(x_0, y_0)}\right) \subset \mathbb{R}^2$ is diffeomorphic to an open disc, and its boundary is piecewise smooth. Moreover,

$$\text{area}\left(\Phi(\mathcal{A}) \cap E_{(x_0, y_0)}\right) = (u_r - u_l)\,\delta < (a + 5\epsilon)\,\frac{\epsilon}{2a} < \epsilon.$$

Arguing as above, we find $c\left(\Phi(\mathcal{A}) \cap E_{(x_0, y_0)}\right) < 2\epsilon$. This finishes the verification of the estimate (8.3.5) and hence of the estimate (8.3.2) in Proposition 8.3.2.

Step 5. End of the proof of Theorem 8.1.1. We have constructed a symplectic embedding $\Phi \colon \mathcal{A} \hookrightarrow \mathcal{Z}$ satisfying the estimate (8.3.2). In view of Proposition 8.3.2 there exists a symplectic embedding $\Psi \colon Z^4(a) \hookrightarrow Z^4(\pi)$ satisfying the estimate (8.3.3). Since $a < \pi$ and $\epsilon > 0$ were arbitrary, we conclude that $\xi_c(Z^4(a)) = 0$ for all $a \in \,]0, \pi[$. Proposition 8.3.1 now implies that Theorem 8.1.1 holds true. $\qquad\square$

8.4 Proof by symplectic lifting

In this section we give another proof of Theorem 8.1.1 which relies on a symplectic lifting construction. In order to explain the idea of the construction, we assume that $S = B^4(a) \subset Z^4(\pi)$. We slice $B^4(a)$ by planes $u = \text{const}$ and then lift a large part of the interior of the i'th slice by i into the y-direction, cf. Figures 8.8 and 8.9 below. Symplectic lifting is even more elementary than symplectic folding and in the problem at hand leads to stronger results.

We consider again a (normalized) extrinsic symplectic capacity c on $\mathbb{R}^2$. If c satisfies the stronger monotonicity axiom

A1. Monotonicity: $c(S) \le c(T)$ if S symplectically embeds into T,

then c is called a (normalized) *intrinsic* symplectic capacity on $\mathbb{R}^2$. Examples of normalized intrinsic symplectic capacities on $\mathbb{R}^2$ are the outer Lebesgue measure $\bar{\mu}$, the Gromov width and the Hofer–Zehnder capacity, see Appendix C.1. For examples of extrinsic symplectic capacities on $\mathbb{R}^2$ which are not intrinsic we refer to Proposition C.7.

Definition 8.4.1. A subset S of $Z^{2n}(\pi)$ is *partially bounded* if at least one of the coordinate functions $x_2, \ldots, x_n, y_2, \ldots, y_n$ is bounded on S.

The following theorem was proved in [74].

Theorem 8.4.2. *Consider a subset S of $Z^{2n}(\pi)$ and an extrinsic symplectic capacity c on $\mathbb{R}^2$.*

(i) $\xi_c(S) = 0$ *if c is intrinsic.*

(ii) $\xi_c(S) = 0$ *if $S \subset Z^{2n}(a)$ for some $a < \pi$.*

If S is partially bounded, then (i) *and* (ii) *hold with $\xi_c(S)$ replaced by $\zeta_c(S)$.*

Notice that Theorem 8.4.2 (ii) is Theorem 8.1.1, and that Theorem 4 in Section 1.3.3 is a special case of the last assertion in Theorem 8.4.2. We do not know $\zeta_c(Z^{2n}(a))$ for any symplectic capacity c on $\mathbb{R}^2$ and any $a \in]0, \pi[$. We also do not know anything about $\zeta_c(Z^{2n}(\pi))$ if $c(S^1) = 0$.

Proof of Theorem 8.4.2. Consider a bounded subset T of $\mathbb{R}^2$. The simply connected hull $\hat{T}$ of T is the union of its closure $\overline{T}$ and the bounded components of $\mathbb{R}^2 \setminus \overline{T}$. We denote by μ the Lebesgue measure on $\mathbb{R}^2$, and we abbreviate $\hat{\mu}(T) = \mu(\hat{T})$. Notice that the outer Lebesgue measure $\bar{\mu}$ is a normalized intrinsic symplectic capacity on $\mathbb{R}^2$ and that $\hat{\mu}$ is a normalized extrinsic symplectic capacity on $\mathbb{R}^2$ which is not intrinsic. Since T is bounded, $c(T) \le \bar{\mu}(T)$ for every normalized intrinsic symplectic capacity c on $\mathbb{R}^2$ and $c(T) \le \hat{\mu}(T)$ for every normalized extrinsic symplectic capacity c on $\mathbb{R}^2$, see Theorem C.8. We can thus assume that $c = \bar{\mu}$ in (i) and $c = \hat{\mu}$ in (ii).

As in the previous proof of Theorem 8.1.1, the main ingredient in the proof of Theorem 8.4.2 is a special embedding result in dimension 4. We shall again use coordinates $z = (u, v, x, y)$ on $(\mathbb{R}^4, du \wedge dv + dx \wedge dy)$, denote by $E_{(x,y)} \subset \mathbb{R}^4$ the affine plane $E_{(x,y)} = \mathbb{R}^2 \times \{(x, y)\}$, and for $S \subset \mathbb{R}^4$ abbreviate

$$\bar{\mu}\big(S \cap E_{(x,y)}\big) = \bar{\mu}\big(p\big(S \cap E_{(x,y)}\big)\big), \qquad \hat{\mu}\big(S \cap E_{(x,y)}\big) = \hat{\mu}\big(p\big(S \cap E_{(x,y)}\big)\big).$$

Fix an integer $k \geq 2$. We set

$$\epsilon = \frac{\pi}{k}, \quad \delta = \frac{\epsilon}{4k},$$

and we define closed rectangles P, P' and Q in $\mathbb{R}^2(u, v)$ by

$$P = [0, \pi] \times [0, 1],$$
$$P' = [\delta, \pi - \delta] \times [\delta, 1 - \delta],$$
$$Q = [3\delta, \pi - 3\delta] \times [3\delta, 1 - 3\delta].$$

The support of a map $\varphi \colon \mathbb{R}^4 \to \mathbb{R}^4$ is defined by

$$\operatorname{supp} \varphi = \overline{\{z \in \mathbb{R}^4 \mid \varphi(z) \neq z\}}.$$

Proposition 8.4.3. *There exists a symplectomorphism φ of $\mathbb{R}^4$ with $\operatorname{supp} \varphi \subset P' \times \mathbb{R}^2$ and such that for each $(x, y) \in \mathbb{R}^2$,*

$$\mu\left(\varphi\left(P' \times \mathbb{R} \times [0, 1]\right) \cap E_{(x,y)}\right) \leq 2\epsilon, \tag{8.4.1}$$
$$\hat{\mu}\left(\varphi\left(Q \times \mathbb{R} \times [0, 1]\right) \cap E_{(x,y)}\right) \leq 2\epsilon. \tag{8.4.2}$$

Proof. We define closed rectangles R, R' and R'' in $\mathbb{R}^2(u, v)$ by

$$R = [0, \epsilon] \times [0, 1],$$
$$R' = [\delta, \epsilon - \delta] \times [\delta, 1 - \delta],$$
$$R'' = [2\delta, \epsilon - 2\delta] \times [2\delta, 1 - 2\delta],$$

and we define closed rectangular annuli A and A' in $\mathbb{R}^2(u, v)$ by

$$A = \overline{R \setminus R'}, \quad A' = \overline{R' \setminus R''}.$$

Then $R = A \cup A' \cup R''$, cf. Figure 8.7.

We choose smooth cut off functions $f_1, f_2 \colon \mathbb{R} \to [0, 1]$ such that

$$f_1(t) = \begin{cases} 0, & t \notin [\delta, \epsilon - \delta], \\ 1, & t \in [2\delta, \epsilon - 2\delta], \end{cases}$$

$$f_2(t) = \begin{cases} 0, & t \notin [\delta, 1 - \delta], \\ 1, & t \in [2\delta, 1 - 2\delta], \end{cases}$$

and we define the smooth function $H \colon \mathbb{R}^4 \to \mathbb{R}$ by

$$H(u, v, x, y) = -f_1(u) f_2(v)(1 + \epsilon)x.$$

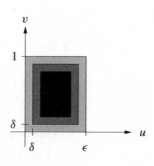

Figure 8.7. The decomposition $R = A \cup A' \cup R''$.

The Hamiltonian vector field X_H of H defined by

$$\omega_0\,(X_H(z), \cdot) = dH(z), \quad z \in \mathbb{R}^{2n}, \tag{8.4.3}$$

is given by

$$X_H(u, v, x, y) = (1 + \epsilon) \begin{pmatrix} -f_1(u)\,f_2'(v)x \\ f_1'(u)\,f_2(v)x \\ 0 \\ f_1(u)\,f_2(v) \end{pmatrix}. \tag{8.4.4}$$

The time-1-map ϕ_H is a lifting map with the following properties.

(P1) supp $\phi_H \subset R' \times \mathbb{R}^2$,

(P2) ϕ_H fixes $A \times \mathbb{R}^2$,

(P3) ϕ_H embeds $A' \times \mathbb{R}^2$ into $A' \times \mathbb{R}^2$,

(P4) ϕ_H translates $R'' \times \mathbb{R}^2$ by $(1 + \epsilon)1_y$,

where we abbreviated $1_y = (0, 0, 0, 1)$.

For each subset T of $\mathbb{R}^2(u, v)$ and each $i \in \{1, \ldots, k\}$ we define the translate T_i of T by

$$T_i = \{(u + (i - 1)\epsilon, v) \mid (u, v) \in T\}.$$

With this notation we have

$$P = \bigcup_{i=1}^{k} R_i = \bigcup_{i=1}^{k} A_i \cup A_i' \cup R_i'',$$

cf. Figure 8.8.

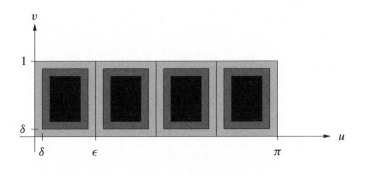

Figure 8.8. The decomposition $P = \bigcup_{i=1}^{k} R_i = \bigcup_{i=1}^{k} A_i \cup A_i' \cup R_i''$ for $k = 4$.

Abbreviate $H_i(u, v, x, y) = iH(u - (i-1)\epsilon, v, x, y)$. We define the smooth function $\widetilde{H} \colon \mathbb{R}^4 \to \mathbb{R}$ by

$$\widetilde{H}(z) = \sum_{i=1}^{k} H_i(z)$$

and we define the symplectomorphism φ of $\mathbb{R}^4$ by $\varphi = \phi_{\widetilde{H}}$. In view of the identity (8.4.4) we see that φ is of the form

$$\varphi(u, v, x, y) = (u', v', x, y'), \qquad (8.4.5)$$

and in view of the Properties (P1)–(P4) we find

$(\widetilde{P1})$ $\operatorname{supp} \varphi \subset P' \times \mathbb{R}^2$,

$(\widetilde{P2})$ φ fixes $\bigcup_{i=1}^{k} A_i \times \mathbb{R}^2$,

$(\widetilde{P3})$ φ embeds $A_i' \times \mathbb{R}^2$ into $A_i' \times \mathbb{R}^2$, $i = 1, \ldots, k$,

$(\widetilde{P4})$ φ translates $R_i'' \times \mathbb{R}^2$ by $i(1 + \epsilon)1_y$, $i = 1, \ldots, k$.

Verification of the estimates (8.4.1) and (8.4.2). Fix $(x, y) \in \mathbb{R}^2$. We abbreviate

$$\mathcal{P}' = p\big(\varphi\big(P' \times \mathbb{R} \times [0, 1]\big) \cap E_{(x,y)}\big),$$
$$\mathcal{Q} = p\big(\varphi\big(Q \times \mathbb{R} \times [0, 1]\big) \cap E_{(x,y)}\big).$$

Lemma 8.4.4. *We have $\mu(\mathcal{P}') \leq 2\epsilon$.*

Proof. Using the definitions $\epsilon = \frac{\pi}{k}$ and $\delta = \frac{\epsilon}{4k}$ we estimate

$$\mu\big(A_i \cup A_i'\big) = \epsilon - (\epsilon - 4\delta)(1 - 4\delta) \leq \frac{\epsilon}{k}, \quad i = 1, \ldots, k. \qquad (8.4.6)$$

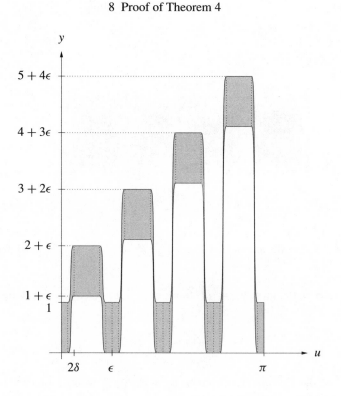

Figure 8.9. The intersection of $\varphi\,(P \times \mathbb{R} \times [0, 1])$ with a plane $\{(u, v, x, y) \mid v, x \text{ constant}\}$ for $v \in [2\delta, 1 - 2\delta]$.

Case A: $y \in [i^*(1 + \epsilon), i^*(1 + \epsilon) + 1]$. According to Properties $(\widetilde{\text{P2}})$–$(\widetilde{\text{P4}})$ we have $\mathcal{P}' \cap R_i'' = \emptyset$ if $i \neq i^*$, and so

$$\mathcal{P}' \subset R_{i*} \cup \bigcup_{i=1}^{k} A_i \cup A_i'.$$

Together with the estimate (8.4.6) we therefore find

$$\mu(\mathcal{P}') \leq \epsilon + k\frac{\epsilon}{k} = 2\epsilon. \tag{8.4.7}$$

Case B: $y \notin \bigcup_{i=1}^{k}[i(1 + \epsilon), i(1 + \epsilon) + 1]$. According to Properties $(\widetilde{\text{P2}})$–$(\widetilde{\text{P4}})$ we have $\mathcal{P}' \cap R_i'' = \emptyset$ for all i, and so

$$\mathcal{P}' \subset \bigcup_{i=1}^{k} A_i \cup A_i'.$$

Therefore,

$$\mu(\mathcal{P}') \leq \epsilon. \tag{8.4.8}$$

The estimates (8.4.7) and (8.4.8) yield that $\mu(\mathcal{P}') \leq 2\epsilon$. $\qquad\square$

Lemma 8.4.5. *We have $\hat{\mu}(\mathcal{Q}) \leq 2\epsilon$.*

Proof. In view of the special form (8.4.5) of the map φ we have

$$\mathcal{Q} = p\big(\varphi\,(Q \times \{x\} \times [0, 1]) \cap E_{(x,y)}\big).$$

For $i = 1, \dots, k$ we abbreviate the intersections

$$\mathscr{A}_i = Q \cap A_i, \quad \mathscr{A}_i' = Q \cap A_i', \quad \mathscr{R}_i'' = Q \cap R_i''. \tag{8.4.9}$$

Each of the sets $\mathscr{A}_i$ and $\mathscr{A}_i'$ consists of one closed rectangle if $i \in \{1, k\}$ and of two closed rectangles if $i \in \{2, \dots, k-1\}$, cf. Figure 8.10. The crucial observation in the proof is that for each i the simply connected hull of the part

$$p\big(\varphi\big(\mathscr{A}_i' \times \{x\} \times [0, 1]\big) \cap E_{(x,y)}\big)$$

of $\mathcal{Q}$ is a simply connected subset of A_i'. Indeed, according to property $\widetilde{(P3)}$ the closed and simply connected set $\varphi\big(\mathscr{A}_i' \times \{x\} \times [0, 1]\big)$ is contained in $A_i' \times \{x\} \times \mathbb{R}$, and so the simply connected hull of $\varphi\big(\mathscr{A}_i' \times \{x\} \times [0, 1]\big) \cap E_{(x,y)}$ is a simply connected subset of $A_i' \times \{(x, y)\}$.

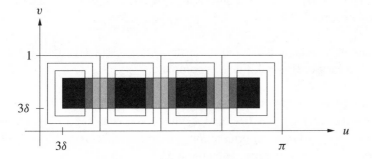

Figure 8.10. The subsets $\mathscr{A}_i$, $\mathscr{A}_i'$ and $\mathscr{R}_i''$ of Q, $i = 1, \dots, 4$.

Denote by $\hat{\mathcal{Q}}$ the simply connected hull of $\mathcal{Q}$.

Case A: $y \in [0, 1]$. According to Properties $\widetilde{(P2)}$–$\widetilde{(P4)}$ we have $\mathcal{Q} \cap A_i = \mathscr{A}_i$ and $\mathcal{Q} \cap R_i'' = \emptyset$ for all i. In view of the above observation we conclude that

$$\hat{\mathcal{Q}} \subset \bigcup_{i=1}^{k} A_i \cup A_i'.$$

Together with the estimate (8.4.6) we therefore find

$$\mu(\hat{\mathcal{Q}}) \le k\frac{\epsilon}{k} = \epsilon. \tag{8.4.10}$$

Case B: $y \in [i^*(1+\epsilon), i^*(1+\epsilon)+1]$. According to Properties $(\widetilde{P2})$–$(\widetilde{P4})$ we have $\mathcal{Q} \cap A_i = \emptyset$ for all i and $\mathcal{Q} \cap R_i'' = \emptyset$ if $i \ne i^*$. In view of the above observation we conclude that

$$\hat{\mathcal{Q}} \subset R_{i^*} \cup \bigcup_{i=1}^{k} A_i'.$$

Therefore,

$$\mu(\hat{\mathcal{Q}}) \le \epsilon + \epsilon = 2\epsilon. \tag{8.4.11}$$

Case C: $y \notin [0,1] \cup \bigcup_{i=1}^{k}[i(1+\epsilon), i(1+\epsilon)+1]$. According to Properties $(\widetilde{P2})$–$(\widetilde{P4})$ we have $\mathcal{Q} \cap A_i = \mathcal{Q} \cap R_i'' = \emptyset$ for all i. In view of the above observation we conclude that

$$\hat{\mathcal{Q}} \subset \bigcup_{i=1}^{k} A_i'.$$

Therefore,

$$\mu(\hat{\mathcal{Q}}) \le \epsilon. \tag{8.4.12}$$

The estimates (8.4.10), (8.4.11) and (8.4.12) yield that $\hat{\mu}(\mathcal{Q}) = \mu(\hat{\mathcal{Q}}) \le 2\epsilon$. This completes the proof of Lemma 8.4.5. $\square$

In view of Lemmata 8.4.4 and 8.4.5 the estimates (8.4.1) and (8.4.2) hold true. The proof of Proposition 8.4.3 is thus complete. $\square$

End of the proof of Theorem 8.4.2 (i). Fix $k \ge 2$ and set $\epsilon = \frac{\pi}{k}$. Arguing as in the proof of Lemma 8.3.3 (iii) we find a symplectomorphism α of $\mathbb{R}^2(u, v)$ such that $P' \subset \alpha\left(B^2(\pi)\right)$. Choose an orientation preserving diffeomorphism $f \colon \mathbb{R} \to \;]0, 1[$ and denote by f' its derivative. Then the map

$$\beta \colon \mathbb{R}^2 \to \mathbb{R} \times \;]0, 1[, \quad (x, y) \mapsto \left(\frac{x}{f'(y)}, f(y)\right) \tag{8.4.13}$$

is a symplectomorphism. We define the symplectic embedding $\Phi \colon \mathbb{R}^{2n} \hookrightarrow \mathbb{R}^{2n}$ by

$$\Phi = \left((\alpha^{-1} \times \mathrm{id}) \circ \varphi \circ (\alpha \times \beta)\right) \times \mathrm{id}_{2n-4}$$

where φ is the map guaranteed by Proposition 8.4.3. Since

$$\mathrm{supp}\, \varphi \subset P' \times \mathbb{R}^2 \subset \alpha\left(B^2(\pi)\right) \times \mathbb{R}^2 \tag{8.4.14}$$

we have $\Phi\big(Z^{2n}(\pi)\big) \subset Z^{2n}(\pi)$. Recall that $D_z = B^2(\pi) \times \{z\}$, $z \in \mathbb{R}^{2n-2}$. For each subset S of $Z^{2n}(\pi)$ and each point $z = (x, y, z_3, \ldots, z_n) \in \mathbb{R}^{2n-2}$ we have

$$
\begin{aligned}
\Phi(S) \cap D_z &\subset \Phi\big(Z^{2n}(\pi)\big) \cap D_z \\
&= \big((\alpha^{-1} \times \mathrm{id}) \circ \varphi \circ (\alpha \times \beta)\big)\big(Z^4(\pi)\big) \cap D_{(x,y)} \\
&\subset \big((\alpha^{-1} \times \mathrm{id}) \circ \varphi\big)\big(\alpha\big(B^2(\pi)\big) \times \mathbb{R} \times [0, 1]\big) \cap E_{(x,y)}.
\end{aligned}
$$

Using this, the facts that $\bar{\mu}$ is monotone and α^{-1} preserves μ, the inclusions (8.4.14) and the estimates (8.4.1) and (8.4.6) we can estimate

$$
\begin{aligned}
\bar{\mu}\big(\Phi(S) \cap D_z\big) &\leq \mu\big(\varphi\big(\alpha\big(B^2(\pi)\big) \times \mathbb{R} \times [0, 1]\big) \cap E_{(x,y)}\big) \\
&= \mu\big(\varphi\big(P' \times \mathbb{R} \times [0, 1]\big) \cap E_{(x,y)}\big) + \mu\big(\alpha\big(B^2(\pi)\big) \setminus P'\big) \\
&\leq 3\epsilon.
\end{aligned}
$$

Since this holds true for all $z \in \mathbb{R}^{2n-2}$ and since $k \geq 2$ was arbitrary, we conclude that $\xi_{\bar{\mu}}(S) = 0$.

Assume now that $S \subset Z^{2n}(\pi)$ is partially bounded. There exists $i \in \{2, \ldots, n\}$ and $b > 0$ such that $x_i(S) \subset \,] - b, b[$ or $y_i(S) \subset \,] - b, b[$. We can assume without loss of generality that $i = 2$. If $x(S) \subset \,] - b, b[$, we define the symplectomorphism σ of $\mathbb{R}^2(x, y)$ by $\sigma(x, y) = (-y, x)$, and we let σ be the identity mapping otherwise. Define the symplectomorphism τ of $\mathbb{R}^2(x, y)$ by

$$
\tau(x, y) = \left(2bx, \frac{1}{2b}y + \frac{1}{2}\right).
$$

The composition $\mathrm{id}_2 \times (\tau \circ \sigma) \times \mathrm{id}_{2n-4}$ maps S into $B^2(\pi) \times \mathbb{R} \times]0, 1[\times \mathbb{R}^{2n-4}$. Replacing the symplectomorphism β in (8.4.13) by the symplectomorphism $\tau \circ \sigma$ of $\mathbb{R}^2$ and proceeding as above, we find that $\zeta_{\bar{\mu}}(S) = 0$.

End of the proof of Theorem 8.4.2 (ii). Choose $a < \pi$ so large that $S \subset Z^{2n}(a)$. We choose $k \geq 2$ so large that $a < \mu(Q)$. Arguing again as in the proof of Lemma 8.3.3 (iii) we then find a symplectomorphism α of $\mathbb{R}^2(u, v)$ such that

$$
\alpha\big(B^2(a)\big) \subset Q \quad \text{and} \quad \alpha\big(B^2(\pi)\big) \supset P',
$$

cf. Figure 8.11. Choose a symplectomorphism $\beta \colon \mathbb{R}^2 \to \mathbb{R} \times]0, 1[$ as above and define the symplectic embedding $\Phi \colon \mathbb{R}^{2n} \hookrightarrow \mathbb{R}^{2n}$ by

$$
\Phi = \big((\alpha^{-1} \times \mathrm{id}) \circ \varphi \circ (\alpha \times \beta)\big) \times \mathrm{id}_{2n-4}.
$$

Since $\operatorname{supp} \varphi \subset P' \times \mathbb{R}^2 \subset \alpha(B^2(\pi)) \times \mathbb{R}^2$ we have $\Phi(Z^{2n}(a)) \subset Z^{2n}(\pi)$. For each $z = (x, y, z_3, \ldots, z_n) \in \mathbb{R}^{2n-2}$ we have

$$
\begin{aligned}
\Phi(S) \cap D_z &\subset \Phi\big(Z^{2n}(a)\big) \cap D_z \\
&= \big((\alpha^{-1} \times \mathrm{id}) \circ \varphi \circ (\alpha \times \beta)\big)\big(Z^4(a)\big) \cap D_{(x,y)} \\
&\subset \big((\alpha^{-1} \times \mathrm{id}) \circ \varphi\big)\big(Q \times \mathbb{R} \times [0, 1]\big) \cap E_{(x,y)}.
\end{aligned}
$$

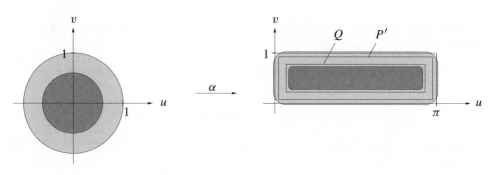

Figure 8.11. The symplectomorphism α.

Using this, the facts that $\hat{\mu}$ is monotone and α^{-1} preserves $\hat{\mu}$ and the estimate (8.4.2) we can estimate

$$\hat{\mu}\left(\Phi(S) \cap D_z\right) \leq \hat{\mu}\left(\varphi\left(Q \times \mathbb{R} \times [0,1]\right) \cap E_{(x,y)}\right)$$
$$\leq 2\epsilon.$$

Since this holds true for all $z \in \mathbb{R}^{2n-2}$ and since we can choose k as large as we like, we conclude that $\xi_{\hat{\mu}}(S) = 0$.

If $S \subset Z^{2n}(a)$ is partially bounded, we replace β by a symplectomorphism $\tau \circ \sigma$ as above and find that $\zeta_{\hat{\mu}}(S) = 0$. The proof of Theorem 8.4.2 is complete. $\square$

Addendum: Symplectic lifting via Hamiltonian deformations Hofer-close to the identity.   The symplectic embeddings in the definitions of $\xi_c(S)$ and $\zeta_c(S)$ were not further specified. Following a suggestion of Polterovich, we next ask whether the vanishing phenomenon described by Theorem 8.4.2 persists if we restrict ourselves to symplectic embeddings which are close to the identity mapping in a symplectically relevant sense. We denote by $\mathcal{H}(2n)$ the set of smooth and bounded functions $H \colon \mathbb{R}^{2n} \to \mathbb{R}$ whose support is contained in $Z^{2n}(\pi)$ and whose Hamiltonian vector field X_H defined by (8.4.3) generates a flow on $\mathbb{R}^{2n}$. The time-1-map of this flow is then again denoted by ϕ_H. Moreover, we abbreviate

$$\|H\| = \sup_{z \in \mathbb{R}^{2n}} H(z) - \inf_{z \in \mathbb{R}^{2n}} H(z). \qquad (8.4.15)$$

For each subset S of $Z^{2n}(\pi)$ and each extrinsic symplectic capacity c on $\mathbb{R}^2$ we define

$$\eta_c(S) = \inf_H \left\{ \sup_z c\left(\phi_H(S) \cap E_z\right) + \|H\| \right\}$$

where H varies over $\mathcal{H}(2n)$. Clearly, $\xi_c(S) \leq \zeta_c(S) \leq \eta_c(S)$. The following result improves the last statement in Theorem 8.4.2.

Theorem 8.4.6. *Consider a partially bounded subset S of $Z^{2n}(\pi)$ and an extrinsic symplectic capacity c on $\mathbb{R}^2$.*

(i) $\eta_c(S) = 0$ *if c is intrinsic.*

(ii) $\eta_c(S) = 0$ *if $S \subset Z^{2n}(a)$ for some $a < \pi$.*

Of course, $\eta_c(Z^{2n}(\pi)) = \pi$ for every normalized extrinsic symplectic capacity c on $\mathbb{R}^2$.

Question 8.4.7. *Is it true that $\eta_{\bar{\mu}}(Z^{2n}(a)) = 0$ for all $a \in]0, \pi[$?*

Theorem 8.4.6 can be proved by refining the lifting construction used in the proof of Theorem 8.4.2. The proof is much trickier, however, since one needs to control the behaviour of the lifting maps on those parts of their compact supports which are not translated in the y-direction. We refer to [74] for the proof.

In order to see Theorem 8.4.6 in its right perspective we consider the set $\mathcal{H}(Z^{2n}(\pi))$ of smooth functions $H = H(t, z) \colon [0, 1] \times \mathbb{R}^{2n} \to \mathbb{R}$ whose support is contained in $[0, 1] \times Z^{2n}(\pi)$ and whose Hamiltonian flow ϕ_H^t exists for all $t \in [0, 1]$, set $\phi_H = \phi_H^1$, and for $H \in \mathcal{H}(Z^{2n}(\pi))$ abbreviate

$$\|H\| = \int_0^1 \left(\sup_{z \in \mathbb{R}^{2n}} H(t, z) - \inf_{z \in \mathbb{R}^{2n}} H(t, z) \right) dt.$$

Notice that $\mathcal{H}(2n) \subset \mathcal{H}(Z^{2n}(\pi))$ and that this definition extends definition (8.4.15). Abbreviating
$$\text{Ham}(Z^{2n}(\pi)) = \{\phi_H \mid H \in \mathcal{H}(Z^{2n}(\pi))\}$$
we define the energy $E(\phi)$ of $\phi \in \text{Ham}(Z^{2n}(\pi))$ by

$$E(\phi) = \inf \{\|H\| \mid \phi = \phi_H \text{ for some } H \in \mathcal{H}(Z^{2n}(\pi))\}.$$

In the framework of Hofer geometry the energy of a Hamiltonian diffeomorphism is its distance from the identity mapping, see [39], [50], [70]. Notice that

$$\eta_c(S) \geq \inf_\phi \left\{ \sup_z c(\phi(S) \cap E_z) + E(\phi) \right\}$$

where ϕ varies over $\text{Ham}(Z^{2n}(\pi))$. Theorem 8.4.6 therefore says that the vanishing phenomenon described by Theorem 8.4.2 persists if we restrict ourselves to Hamiltonian diffeomorphism of $Z^{2n}(\pi)$ whose Hofer distance to the identity mapping is arbitrarily small.

Chapter 9

Packing symplectic manifolds by hand

This last chapter is devoted to explicit symplectic packings of some symplectic manifolds by equal balls. After recalling the symplectic packing problem, we give several motivations for it, and then collect the known packing numbers of interest to us. In Section 9.3 we describe a simple construction of explicit maximal packings of the 4-ball and $\mathbb{CP}^2$ and of ruled symplectic 4-manifolds by few equal balls, and in Section 9.4 we briefly look at packings in higher dimensions.

We consider again a connected $2n$-dimensional symplectic manifold (M, ω) of finite volume $\mathrm{Vol}(M, \omega) = \frac{1}{n!} \int_M \omega^n$, and as before we abbreviate the Lebesgue measure of an open subset U of $\mathbb{R}^{2n}$ by $|U|$. In Chapters 2 to 6 we studied the numbers

$$p_a^E(M, \omega) = \sup_\lambda \frac{|\lambda E(\pi, \dots, \pi, a)|}{\mathrm{Vol}(M, \omega)}$$

where the supremum is taken over all those λ for which the ellipsoid $\lambda E^{2n}(\pi, \dots, \pi, a)$ symplectically embeds into (M, ω), and we proved in Chapter 6 that

$$\lim_{a \to \infty} p_a^E(M, \omega) = 1. \tag{9.0.1}$$

Notice that the invariant $p_\pi^E(M, \omega)$ measures the maximal ball which symplectically embeds into (M, ω). Instead of "stretching the ball to ellipsoids", we shall now increase the number of balls and study for each $k \in \mathbb{N}$ the k'th *symplectic packing number*

$$p_k(M, \omega) = \sup_a \frac{k \left| B^{2n}(a) \right|}{\mathrm{Vol}(M, \omega)}$$

where the supremum is taken over all those a for which the disjoint union $\coprod_{i=1}^k B^{2n}(a)$ of k equal balls symplectically embeds into (M, ω). The problem of understanding the numbers $p_k(M, \omega) \in \,]0, 1]$ is called the *symplectic packing problem*, a problem much studied in recent years. If $p_k(M, \omega) < 1$, one says that there is a *packing obstruction*, and if $p_k(M, \omega) = 1$, one says that (M, ω) admits a *full packing* by k balls. The first examples of packing obstructions were found by Gromov, [31], and many further packing obstructions and also some exact values of p_k were obtained by McDuff and Polterovich in [61]. Finally, Biran showed in [7], [8] that

$$P(M, \omega) := \inf \{k_0 \in \mathbb{N} \mid p_k(M, \omega) = 1 \quad \text{for all } k \geq k_0\} < \infty \tag{9.0.2}$$

for an interesting class of closed symplectic 4-manifolds containing sphere bundles over a surface and for all closed symplectic 4-manifolds with $[\omega] \in H^2(M; \mathbb{Q})$.

According to Lemma 6.2.2 and [61, Remark 1.5.G],

$$\lim_{k \to \infty} p_k(M, \omega) = 1 \qquad (9.0.3)$$

for every connected symplectic manifold (M, ω) of finite volume. By (9.0.1), the asymptotics of p_a^E and p_k are thus the same. The analogue of Biran's result (9.0.2) for p_a^E is not known for any connected symplectic manifold of finite volume.

Question 9.0.8. *For which connected symplectic manifolds (M, ω) of finite volume does there exist a_0 such that $p_a^E(M, \omega) = 1$ for all $a \geq a_0$?*

Besides sporadic results on the first packing number p_1 and on packing numbers for ellipsoids in [81], [54], all known computations of packing numbers are contained in [61],[7], [8]. We refer to Biran's excellent survey [10] for the methods used, and only mention that in [61], [7], [8] the problem of symplectically embedding k equal balls into (M, ω) is first reformulated as the problem of deforming a symplectic form on the k-fold blow-up of (M, ω) along a certain family of cohomology classes, and that this problem is then solved using tools from classical algebraic geometry, Seiberg–Witten–Taubes theory, and Donaldson's symplectic submanifold theorem, respectively. As a consequence, the symplectic packings found are not explicit. For some of the symplectic manifolds considered in [61], [7], [8] and some values of k, explicit maximal symplectic packings were constructed by Karshon [41], Traynor [81], Kruglikov [45], and Maley, Mastrangeli and Traynor [54]. In this final chapter we describe a very simple and explicit construction realizing the packing numbers $p_k(M, \omega)$ for those symplectic 4-manifolds (M, ω) and numbers k considered in [41], [81], [45], [54] as well as for some other closed symplectic 4-manifolds and small values of k. In the range of k for which these constructions fail to give maximal packings, they give a feeling that the balls in the packings from [61], [7], [8] must be "wild". We shall also construct an explicit full packing of $B^{2n}(a)$ by l^n equal balls for each $l \in \mathbb{N}$ in a most simple way.

As in the previous chapters, balls and ellipsoids will always be endowed with the standard symplectic form ω_0. Since the packing numbers $p_k(B^{2n}(a), \omega_0)$ do not depend on a, we shall usually pack the unit ball $B^{2n} := (B^{2n}(\pi), \omega_0)$.

9.1 Motivations for the symplectic packing problem

1. Higher Gromov widths. The Gromov width

$$w_G(M, \omega) := \sup\{a \mid B^{2n}(a) \text{ symplectically embeds into } (M, \omega)\}$$

of a symplectic manifold (M, ω) measures the size of a largest round symplectic chart of (M, ω). It is the smallest normalized symplectic capacity as defined in [39, p. 51], and we refer to [7], [8], [12], [27], [31], [40], [42], [48], [49], [56], [60], [61], [64], [77] and to Theorem C.8 (i) for results on the Gromov width and to Section 9.3 below for explicit symplectic embeddings realizing $w_G(M, \omega)$ or estimating it from below. If (M, ω) has finite volume, the first packing number $p_1(M, \omega)$ is equivalent to the Gromov width,

$$p_1(M, \omega) \operatorname{Vol}(M, \omega) = \frac{1}{n!} \big(w_G(M, \omega)\big)^n.$$

Similarly, the higher packing numbers $p_k(M, \omega)$, $k \geq 2$, are equivalent to the *higher Gromov widths*

$$w_G^k(M, \omega) := \sup \Big\{ a \mid \coprod_{i=1}^k B^{2n}(a) \text{ symplectically embeds into } (M, \omega) \Big\},$$

which form a distinguished sequence of embedding capacities as considered in [16].

2. "Superrecurrence for symplectomorphisms" via packing obstructions? In view of the Poincaré recurrence theorem, for which we refer to [35], volume preserving mappings have strong recurrence properties. The solution of the Arnold conjecture for the torus by Conley and Zehnder, which we stated in Section 1.2.1, demonstrated that Hamiltonian diffeomorphisms have yet stronger recurrence properties. As was pointed out to me by Leonid Polterovich, the original motivation for Gromov to study the packing numbers p_k was his search for recurrence properties of arbitrary symplectomorphisms which are stronger than those of volume preserving mappings.

We explain the relation between "superrecurrence for symplectomorphisms" and symplectic packing obstructions by means of an example. Let B and B' be the open balls in $\mathbb{R}^{2n}$ centred at the origin of volumes $2^n - \frac{1}{2}$ and 1, respectively. For every compactly supported volume preserving diffeomorphism φ of B set

$$R(\varphi) = \min \big\{ m \in \mathbb{N} \mid \varphi^m(B') \cap B' = \emptyset \big\}.$$

Of course, $R(\varphi) \leq 2^n - 2$, and using Moser's deformation argument, for which we refer to [39, p. 11], it is easy to construct a φ with $R(\varphi) = 2^n - 2$. The packing obstruction $p_2(B) = \frac{1}{2^{n-1}}$ proved by Gromov in [31] shows, however, that $R(\varphi) = 1$ if φ is symplectic.

This motivation for symplectic packings lost some of its appeal by the work of McDuff–Polterovich and Biran. Indeed, in dynamics one usually asks for recurrence into *small* neighbourhoods of a point. To establish recurrence of small balls we would need packing obstructions for *large* k. In view of (9.0.3), these obstructions asymptotically always vanish, and in view of (9.0.2), they completely vanish for many symplectic 4-manifolds.

3. Between Euclidean and volume preserving

Volume preserving packings. Consider a connected n-dimensional manifold M endowed with a volume form Ω such that the volume $\mathrm{Vol}(M, \Omega) = \int_M \Omega$ is finite, and denote the Lebesgue measure of an open subset U of $\mathbb{R}^n$ by $|U|$. We write $B^n(A)$ for the open ball of radius $\sqrt{A/\pi}$ in $\mathbb{R}^n$. For $k \in \mathbb{N}$ we set

$$v_k(M, \Omega) = \sup \left\{ \frac{k\,|B^n(A)|}{\mathrm{Vol}(M, \Omega)} \right\}$$

where the supremum is taken over all A for which there exists a volume preserving embedding $\coprod_{i=1}^k B^n(A) \hookrightarrow (M, \Omega)$. Moser's deformation method readily implies that $v_k(M, \Omega) = 1$ for all $k \in \mathbb{N}$. Proposition 2 of Section 1.3.2 shows more: For any partition $M = \coprod_{i=1}^k M_i$ of M into subsets M_i such that $\mathrm{Int}\, M_i$ is connected and $\mathrm{Vol}\,(\mathrm{Int}\, M_i, \Omega) = \frac{1}{k} \mathrm{Vol}(M, \Omega)$ for all i there exists a volume preserving embedding $\coprod_{i=1}^k B^n(A) \hookrightarrow \coprod_{i=1}^k \mathrm{Int}\, M_i$ with $|B^n(A)| = \frac{1}{k} \mathrm{Vol}(M, \Omega)$. If the volume form Ω comes from a symplectic form ω, the sequence $(1 - p_k(M, \omega))_{k \in \mathbb{N}}$ is a measure for how far the symplectic geometry of (M, ω) is from the volume geometry of (M, Ω).

Euclidean packings. Given a bounded domain U in $\mathbb{R}^n$, define its k'th *Euclidean packing number* as

$$\Delta_k(U) = \sup \left\{ \frac{k\,|B^n(a)|}{|U|} \right\}$$

where the supremum is taken over all a for which k disjoint translates of $B^n(a)$ fit into U. Then

$$\Delta_k(U) \le p_k(U) \le v_k(U) = 1 \quad \text{for all } k \in \mathbb{N},$$

and it is interesting to understand "on which side" $p_k(U)$ lies. To fix the ideas, we assume that U is the unit ball $B^n := B^n(\pi)$ in $\mathbb{R}^n$. The precise values of $\Delta_k(B^n)$ are known only for small k: If $1 \le k \le n + 1$, the smallest ball containing k balls of radius 1 has radius $1 + \sqrt{2 - 2/k}$, and the centres of the balls are arranged as vertices of a regular $(k - 1)$-dimensional simplex inscribed in the ball and concentric with it. Moreover, if $n + 2 \le k \le 2n$, the smallest ball B containing k balls of radius 1 has radius $1 + \sqrt{2}$, and the packing configuration of $2n$ balls in B is unique up to isometry, the centres being the midpoints of the faces of an n-dimensional Euclidean cube whose edges have length $2\sqrt{2}$. In particular,

$$\Delta_k\left(B^n\right) = \begin{cases} \dfrac{k}{\left(1 + \sqrt{2 - \frac{2}{k}}\right)^n} & \text{if } 1 \le k \le n + 1, \\[3mm] \dfrac{k}{\left(1 + \sqrt{2}\right)^n} & \text{if } n + 2 \le k \le 2n. \end{cases} \tag{9.1.1}$$

While for $1 \le k \le n + 1$ these numbers were known to Rankin in 1955, for $n + 2 \le k \le 2n$ they were obtained only recently by W. Kuperberg, [46]. An obvious upper bound for $\Delta_k(B^n)$ is

$$\Delta_k\left(B^n\right) \le \frac{k}{2^n} \quad \text{for all } k \ge 2. \tag{9.1.2}$$

Given a bounded domain U in $\mathbb{R}^n$, let $\text{conv}(U)$ be the convex hull of U. For each $k \geq 1$ we set

$$\text{conv}_k(B^n) = \sup \frac{k\,|B^n|}{|\text{conv}(U)|} \qquad (9.1.3)$$

where the supremum is taken over all configurations U of k disjoint translates of B^n in $\mathbb{R}^n$. Since B^n is convex, $\Delta_k(B^n) \leq \text{conv}_k(B^n)$ for all $k \in \mathbb{N}$. Let $S_k^n = \text{conv}(U)$ be the sausage obtained by choosing

$$U = \coprod_{i=0}^{k-1}(B^n + i\boldsymbol{u}) \qquad (9.1.4)$$

where $\boldsymbol{u}$ is a unit vector in $\mathbb{R}^n$. With $\kappa_n := |B^n|$ we then have $|S_k^n| = \kappa_n + 2(k-1)\kappa_{n-1}$. The sausage conjecture of L. Fejes Tóth from 1975 states that equality in (9.1.3) is attained exactly for U as in (9.1.4), and this conjecture was proved by Betke and Henk, [5], for $n \geq 42$. Therefore,

$$\Delta_k(B^n) \leq \text{conv}_k(B^n) = \frac{k\kappa_n}{\kappa_n + 2(k-1)\kappa_{n-1}} < \frac{k}{k-1}\sqrt{\frac{\pi}{2}}\sqrt{\frac{1}{n+1}} \quad \text{if } n \geq 42. \qquad (9.1.5)$$

For arbitrary n, an older result of Gritzmann, [30], states that

$$\Delta_k(B^n) \leq \text{conv}_k(B^n) < \left(2 + \sqrt{3}\right)\sqrt{\frac{\pi}{2}}\sqrt{\frac{1}{n}}.$$

In order to get an idea of the values $\Delta_k(B^n)$ for large k we notice that the limit

$$\Delta^n := \lim_{k \to \infty} \Delta_k(B^n)$$

exists and is equal to the highest density of a packing of $\mathbb{R}^n$, see Section 2.1 of Chapter 3.3 in [34]. The highest density of a packing of $\mathbb{R}^2$ is

$$\Delta^2 = \frac{\pi}{\sqrt{12}} = 0.9069\ldots$$

as in the familiar hexagonal lattice packing in which each disk touches 6 others (Thue, 1910). The highest density of a packing of $\mathbb{R}^3$ is

$$\Delta^3 = \frac{\pi}{\sqrt{18}} = 0.74048\ldots$$

as in the face centred cubic lattice packing which is usually found in fruit stands and in which each ball touches 12 other balls. This was conjectured by Keppler in 1611, and Gauss proved in 1831 that no lattice packing has a higher density. The Keppler conjecture was settled only recently by Hales, see [33] and the references therein. For

$4 \leq n \leq 36$, the currently best upper bound for Δ^n was given recently by Cohn and Elkies in [17]. E.g.,

$$\frac{\pi^2}{16} = 0.61685 \leq \Delta^4 \leq 0.647742.$$

Here, the lower bound is the density of the packing associated with the "checkerboard lattice" consisting of all vectors $(a, b, c, d) \in \mathbb{Z}^4$ with $a + b + c + d \in 2\mathbb{Z}$, and it is known that this is the highest possible density for a 4-dimensional lattice packing. A result of Blichfeldt from 1929 states that

$$\Delta^n \leq (n + 2)2^{-(n+2)/2}, \tag{9.1.6}$$

and the best known lower and upper bounds for Δ^n of asymptotic nature are

$$cn2^{-n} \leq \Delta^n \leq 2^{-(0.599+o(1))n} \quad \text{as } n \to \infty$$

for any constant $c < \log 2$, see Section 2 of Chapter 3.3 in [34].

We refer to [19], to Sections 3.3 and 3.4 of [34], and to [88] for more information on Euclidean packings, its long history and its many relations and applications to other branches of mathematics (such as discrete geometry, group theory, number theory and crystallography) and to problems in physics, chemistry, engineering and computer science.

The symplectic packing numbers $p_k(B^4)$ are listed in Table 9.1 below. For $n \geq 3$, the results known about $p_k(B^{2n})$ are

$$p_k(B^{2n}) = \frac{k}{2^n} \quad \text{for } 2 \leq k \leq 2^n, \tag{9.1.7}$$

$$p_{l^n}(B^{2n}) = 1 \quad \text{for all } l \in \mathbb{N}, \tag{9.1.8}$$

see [61, Corollary 1.5.C and 1.6.B] and Section 9.4.1 below. The identities (9.1.8) yield another proof of (9.0.3) for the ball,

$$\lim_{k \to \infty} p_k(B^{2n}) = 1 \quad \text{for all } n. \tag{9.1.9}$$

Of course, $\Delta_k(B^2) < p_k(B^2) = v_k(B^2) = 1$ for all $k \geq 2$. Comparing (9.1.1) or (9.1.2) for $n = 4$ with the values $p_k(B^4)$ listed in Table 9.1 we see that

$$\Delta_k(B^4) < p_k(B^4) \quad \text{for all } k \geq 2.$$

Moreover, (9.1.2) and (9.1.7) show that

$$\Delta_k(B^{2n}) \leq \frac{1}{2^n} p_k(B^{2n}) \quad \text{for } 2 \leq k \leq 2^n \text{ and all } n \in \mathbb{N}.$$

Inequality (9.1.5) and (9.1.8) yield an explicit $k(2n)$ such that

$$\Delta_k(B^{2n}) < p_k(B^{2n}) \quad \text{for all } k \geq k(2n) \text{ and } 2n \geq 42.$$

It is conceivable that $\Delta_k(B^{2n}) < p_k(B^{2n})$ for all $k \geq 2$ and $n \in \mathbb{N}$, but we do not know the answer to

Question 9.1.1. *Is it true that* $\Delta_{28}(B^6) < p_{28}(B^6)$?

Finally, comparing (9.1.9) with (9.1.6) we see that $p_k(B^{2n})$ is much larger than $\Delta_k(B^{2n})$ for sufficiently large k and large n.

4. Relations to algebraic geometry. A symplectic packing of (M, ω) by k equal balls corresponds to a symplectic blow-up of (M, ω) at k points with equal weights. Via this correspondence, the symplectic packing problem is intimately related to old problems in algebraic geometry: The symplectic packing problem for the complex projective plane $\mathbb{CP}^2$ (completely solved by Biran in [7]) is related to an old (and still open) conjecture of Nagata on the minimal degree of an irreducible algebraic curve in $\mathbb{CP}^2$ passing through $N \geq 9$ points with given multiplicities, see [9], [10], [61], [87] for details. Moreover, the symplectic packing problem is closely related to the problem of computing Seshadri constants of ample line bundles, which are a measure of their local positivity, see [9], [10], [12], [52].

9.2   The packing numbers of the 4-ball and $\mathbb{CP}^2$ and of ruled symplectic 4-manifolds

In this section we collect the known packing numbers of interest to us.

9.2.1   The packing numbers of the 4-ball and $\mathbb{CP}^2$.

Let ω_{SF} be the unique U(3)-invariant Kähler form on $\mathbb{CP}^2$ whose integral over $\mathbb{CP}^1$ equals 1. According to a result of Taubes, [80], every symplectic form on $\mathbb{CP}^2$ is diffeomorphic to $a\,\omega_{\text{SF}}$ for some $a > 0$. In view of the symplectomorphism

$$\left(B^4(\pi), \omega_0\right) \to \left(\mathbb{CP}^2 \setminus \mathbb{CP}^1, \pi\,\omega_{\text{SF}}\right), \quad z = (z_1, z_2) \mapsto \left[z_1 : z_2 : \sqrt{1 - |z|^2}\right] \tag{9.2.1}$$

further discussed in [62, Example 7.14] we have $p_k\left(B^4\right) \leq p_k\left(\mathbb{CP}^2\right)$ for all k. It is shown in [61, Remark 2.1.E] that in fact

$$p_k(B^4) = p_k(\mathbb{CP}^2) \quad \text{for all } k. \tag{9.2.2}$$

A complete list of these packing numbers was obtained in [7] (see Table 9.1).

Table 9.1. $p_k(B^4) = p_k(\mathbb{CP}^2)$

k	1	2	3	4	5	6	7	8	≥ 9
p_k	1	$\frac{1}{2}$	$\frac{3}{4}$	1	$\frac{20}{25}$	$\frac{24}{25}$	$\frac{63}{64}$	$\frac{288}{289}$	1

Explicit maximal packings were found by Karshon [41] for $k \leq 3$ and by Traynor [81] for $k \leq 6$ and $k = l^2$ ($l \in \mathbb{N}$). We will give even simpler maximal packings for these values of k in 9.3.1.

9.2.2 The packing numbers of ruled symplectic 4-manifolds.

Denote by Σ_g the closed orientable surface of genus g. There are exactly two orientable S^2-bundles with base Σ_g, namely the trivial bundle $\pi : \Sigma_g \times S^2 \to \Sigma_g$ and the nontrivial bundle $\pi : \Sigma_g \ltimes S^2 \to \Sigma_g$, see [62, Lemma 6.9]. Such a manifold M is called a *ruled surface*. A symplectic form ω on a ruled surface is called *compatible* with the given ruling π if it restricts on each fibre to a symplectic form. Such a symplectic manifold is then called a *ruled symplectic 4-manifold*. It is known that every symplectic structure on a ruled surface is diffeomorphic to a form compatible with the given ruling π via a diffeomorphism which acts trivially on homology, and that two cohomologous symplectic forms compatible with the same ruling are isotopic [51]. A symplectic form ω on a ruled surface M is thus determined up to diffeomorphism by the class $[\omega] \in H^2(M; \mathbb{R})$. In order to describe the set of cohomology classes realized by (compatible) forms on M we fix an orientation of Σ_g and an orientation of the fibres of the given ruled surface M. These orientations determine an orientation of M in a natural way, see below. We say that a compatible symplectic form ω is *admissible* if its restriction to each fibre induces the given orientation and if ω induces the natural orientation on M. Notice that every symplectic form on M is diffeomorphic to an admissible form for a suitable choice of orientations of Σ_g and the fibres.

Consider first the trivial bundle $\Sigma_g \times S^2$, and let $\{B = [\Sigma_g \times \text{pt}], F = [\text{pt} \times S^2]\}$ be a basis of $H^2(M; \mathbb{Z})$. Here and henceforth we identify homology and cohomology via Poincaré duality. The natural orientation of $\Sigma_g \times S^2$ is such that $B \cdot F = 1$. A cohomology class $C = bB + aF$ can be represented by an admissible form if and only if $C \cdot F > 0$ and $C \cdot C > 0$, i.e.,

$$a > 0 \quad \text{and} \quad b > 0,$$

standard representatives being split forms. We write $\Sigma_g(a) \times S^2(b)$ for this ruled symplectic 4-manifold.

In case of the nontrivial bundle $\Sigma_g \ltimes S^2$ a basis of $H^2(\Sigma_g \ltimes S^2; \mathbb{Z})$ is given by $\{A, F\}$, where A is the class of a section with selfintersection number -1 and F is the fibre class. The homology classes of sections of $\Sigma_g \ltimes S^2$ of self-intersection number k are $A_k = A + \frac{k+1}{2}F$ with k odd. The natural orientation of $\Sigma_g \ltimes S^2$ is such that $A_k \cdot F = A \cdot F = 1$ for all k. Set $B = A + F/2$. Then $\{B, F\}$ is a basis of $H^2(\Sigma_g \ltimes S^2; \mathbb{R})$ with $B \cdot B = F \cdot F = 0$ and $B \cdot F = 1$. As for the trivial bundle, the necessary condition for a cohomology class $bB + aF$ to be representable by an admissible form is $a > 0$ and $b > 0$. It turns out that this condition is sufficient only if $g \geq 1$: A cohomology class $bB + aF$ can be represented by an admissible form if and only if

$$a > b/2 > 0 \qquad \text{if } g = 0,$$
$$a > 0 \text{ and } b > 0 \quad \text{if } g \geq 1,$$

see [62, Theorem 6.11]. We write $(\Sigma_g \ltimes S^2, \omega_{ab})$ for this ruled symplectic 4-manifold. A "standard Kähler form" in the class $[\omega_{ab}]$ is explicitly constructed in [57, Section 3] and [62, Exercise 6.14]. When constructing our explicit symplectic packings, it will always be clear which symplectic form in $[\omega_{ab}]$ is chosen.

We begin with the trivial sphere bundle over the sphere.

Proposition 9.2.1. *Assume that $a \geq b$. Abbreviate $p_k = p_k(S^2(a) \times S^2(b))$, and denote by $\lceil x \rceil$ the minimal integer which is greater than or equal to x. Then*

$$p_k = \frac{k}{2}\frac{b}{a} \qquad \text{if} \quad \left\lceil \frac{k}{2} \right\rceil \frac{b}{a} \leq 1.$$

Moreover,

$$p_1 = \frac{b}{2a}, \quad p_2 = \frac{b}{a}, \quad p_3 = \frac{3}{2ab}\left\{b, \frac{a+b}{3}\right\}^2 \qquad \text{on} \quad \left]0, \frac{1}{2}, 1\right],$$

$$p_4 = \frac{4}{3}p_3, \quad p_5 = \frac{5}{2ab}\left\{b, \frac{a+2b}{5}\right\}^2 \qquad \text{on} \quad \left]0, \frac{1}{3}, 1\right],$$

$$p_6 = \frac{3}{ab}\left\{b, \frac{a+2b}{5}, \frac{2a+2b}{7}\right\}^2 \qquad \text{on} \quad \left]0, \frac{1}{3}, \frac{3}{4}, 1\right],$$

$$p_7 = \frac{7}{2ab}\left\{b, \frac{a+3b}{7}, \frac{3a+4b}{13}, \frac{4a+4b}{15}\right\}^2 \qquad \text{on} \quad \left]0, \frac{1}{4}, \frac{8}{11}, \frac{7}{8}, 1\right].$$

In particular, for $k \leq 7$ we have $p_k(S^2(a) \times S^2(b)) = 1$ exactly for $(k = 2, \frac{b}{a} = 1)$, $(k = 4, \frac{b}{a} = \frac{1}{2})$, $(k = 6, \frac{b}{a} = \frac{1}{3})$, $(k = 6, \frac{b}{a} = \frac{3}{4})$ and $(k = 7, \frac{b}{a} = \frac{7}{8})$.

We explain our notation by an example:

$$p_3 = \tfrac{3}{2ab}b^2 \text{ if } 0 < \tfrac{b}{a} \leq \tfrac{1}{2} \quad \text{and} \quad p_3 = \tfrac{3}{2ab}\left(\tfrac{a+b}{3}\right)^2 \text{ if } \tfrac{1}{2} \leq \tfrac{b}{a} \leq 1.$$

In 9.3.2 we will construct explicit maximal packings of $S^2(a) \times S^2(b)$ for all k with $\lceil \frac{k}{2} \rceil \frac{b}{a} \leq 1$, for $k \leq 6$ and $0 < b \leq a$ arbitrary, and for $k = 7$ and $0 < \frac{b}{a} \leq \frac{3}{5}$, as well as explicit full packings for $k = 2ml^2$ if $a = mb$ $(l, m \in \mathbb{N})$. These explicit packings will give to the above quantities a transparent geometric meaning.

The following corollary slightly refines Corollary 5.B of [7].

Corollary 9.2.2. *We have* $\max\left(2\frac{a}{b}, 8\right) \leq P(S^2(a) \times S^2(b)) \leq 8\frac{a}{b}$ *except possibly for* $\frac{b}{a} = \frac{7}{8}$, *in which case* $P(S^2(a) \times S^2(b)) \in \{7, 8, 9\}$. *For* $S^2(1) \times S^2(1)$ *we thus have*

Table 9.2. $p_k(S^2(1) \times S^2(1))$

k	1	2	3	4	5	6	7	≥ 8
p_k	$\frac{1}{2}$	1	$\frac{2}{3}$	$\frac{8}{9}$	$\frac{9}{10}$	$\frac{48}{49}$	$\frac{224}{225}$	1

Proposition 9.2.3. *Assume that* $a > \frac{b}{2} > 0$. *Abbreviate* $p_k = p_k(S^2 \ltimes S^2, \omega_{ab})$, *and set* $\langle k \rangle = k$ *if* k *is odd and* $\langle k \rangle = k + 1$ *if* k *is even. Then*

$$p_k = \frac{k}{2}\frac{b}{a} \qquad if \quad \frac{\langle k \rangle}{2}\frac{b}{a} \leq 1.$$

Moreover,

$$p_1 = \frac{b}{2a}, \quad p_2 = \frac{1}{ab}\left\{b, \frac{2a+b}{4}\right\}^2 \qquad on \quad \left]0, \frac{2}{3}, 2\right[,$$

$$p_3 = \frac{3}{2}p_2, \quad p_4 = \frac{2}{ab}\left\{b, \frac{2a+3b}{8}\right\}^2 \qquad on \quad \left]0, \frac{2}{5}, 2\right[,$$

$$p_5 = \frac{5}{2ab}\left\{b, \frac{2a+3b}{8}, \frac{2a+b}{5}\right\}^2 \qquad on \quad \left]0, \frac{2}{5}, \frac{6}{7}, 2\right[,$$

$$p_6 = \frac{3}{ab}\left\{b, \frac{2a+5b}{12}, \frac{2a+2b}{7}, \frac{2a+b}{5}\right\}^2 \qquad on \quad \left]0, \frac{2}{7}, \frac{10}{11}, \frac{4}{3}, 2\right[,$$

$$p_7 = \frac{7}{2ab}\left\{b, \frac{2a+5b}{12}, \frac{6a+9b}{28}, \frac{4a+4b}{15}, \frac{4a+3b}{13}, \frac{6a+3b}{16}\right\}^2$$

$$on \quad \left]0, \frac{2}{7}, \frac{1}{2}, \frac{22}{23}, \frac{8}{7}, \frac{14}{9}, 2\right[.$$

In particular, for $k \leq 7$ *we have* $p_k(S^2 \ltimes S^2, \omega_{ab}) = 1$ *exactly for* $(k = 3, \frac{b}{a} = \frac{2}{3})$, $(k = 5, \frac{b}{a} = \frac{2}{5})$, $(k = 6, \frac{b}{a} = \frac{4}{3})$ $(k = 7, \frac{b}{a} = \frac{2}{7})$ $(k = 7, \frac{b}{a} = \frac{8}{7})$ *and* $(k = 7, \frac{b}{a} = \frac{14}{9})$.

In 9.3.2 we will construct explicit maximal packings of $(S^2 \ltimes S^2, \omega_{ab})$ for all k with $\frac{\langle k \rangle}{2}\frac{b}{a} \leq 1$, for $k \leq 5$ and $0 < \frac{b}{2} < a$ arbitrary, and for $k = 6$ and $\frac{b}{a} \in]0, \frac{2}{3}] \cup [\frac{4}{3}, 2[$. Moreover, given ω_{ab} with $\frac{b}{a} = \frac{2l}{2m-l}$ for some $l, m \in \mathbb{N}$ with $m > l$, we will construct explicit full packings of $(S^2 \ltimes S^2, \omega_{ab})$ by $l(2m - l)$ balls.

Corollary 9.2.4. *We have*

$$
\max\left(2\frac{a}{b}, 8\right) \le P(S^2 \ltimes S^2, \omega_{ab}) \le
\begin{cases}
\dfrac{8a}{b} & \text{if } b \le a \\[2mm]
\dfrac{8ab}{(2a-b)^2} & \text{if } b \ge a
\end{cases}
$$

except possibly for $\frac{b}{a} \in \left\{\frac{2}{7}, \frac{8}{7}, \frac{14}{9}\right\}$, *in which case the lower bound for* $P(S^2 \ltimes S^2, \omega_{ab})$ *is 7. For* $(S^2 \ltimes S^2, \omega_{11})$ *we thus have*

Table 9.3. $p_k(S^2 \ltimes S^2, \omega_{11})$

k	1	2	3	4	5	6	7	≥ 8
p_k	$\frac{1}{2}$	$\frac{9}{16}$	$\frac{27}{32}$	$\frac{25}{32}$	$\frac{9}{10}$	$\frac{48}{49}$	$\frac{14}{15}$	1

Proposition 9.2.5. *Let* $g \ge 1$ *and let* $a > 0$ *and* $b > 0$. *Then*

$$
p_k(\Sigma_g(a) \times S^2(b)) = p_k(\Sigma_g \ltimes S^2, \omega_{ab}) = \min\left\{1, \frac{k\,b}{2\,a}\right\}.
$$

In particular, $P(\Sigma_g(a) \times S^2(b)) = P(\Sigma_g \ltimes S^2, \omega_{ab}) = \left\lceil \frac{2a}{b} \right\rceil$.

In 9.3.2 we will construct explicit maximal packings of $\Sigma_g(a) \times S^2(b)$ and of $(\Sigma_g \ltimes S^2, \omega_{ab})$ for all k with $\left\lceil \frac{k}{2} \right\rceil \frac{b}{a} \le 1$ and explicit full packings for $k = 2ml^2$ if $a = mb$ or $b = ma$ $(l, m \in \mathbb{N})$.

For the proofs of Propositions 9.2.1, 9.2.3 and 9.2.5 and Corollaries 9.2.2 and 9.2.4 we refer to [7] and [75].

9.3   Explicit maximal packings in four dimensions

In this section we realize most of the packing numbers listed in the previous section by explicit symplectic packings. Sometimes, we shall give two different maximal packings. It is known that for the 4-ball and $\mathbb{CP}^2$ and for ruled symplectic 4-manifolds, any two packings by k closed balls of equal size are symplectically isotopic, see [6], [58].

As we have seen in Lemmata 3.1.8 and 5.3.1, for each $\epsilon > 0$ there exist explicit symplectic embeddings $B^4(a) \hookrightarrow \triangle^2(a + \epsilon) \times \square^2(1)$ and $\triangle^2(a) \times \square^2(1) \hookrightarrow B^4(a + \epsilon)$. We shall omit $\epsilon > 0$ and think of $B^4(a)$ as $\triangle^2(a) \times \square^2(1)$. Variations of the map α in Figure 3.3 yield different shapes. We recall that α is represented by (α) in Figure 9.1. Let $\bar{\alpha}$ be its mirror represented by $(\bar{\alpha})$, and let β, γ and δ be represented

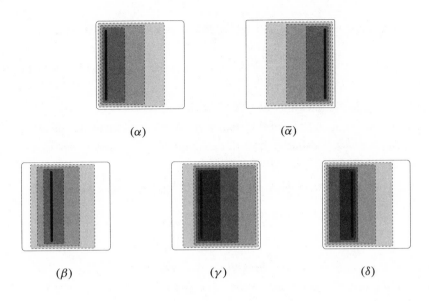

Figure 9.1. The images of the disc.

by (β), (γ) and (δ), respectively. Applying α to both the z_1- and the z_2-plane yields a shape whose x_1-x_2-shadow is (arbitrarily close to) the simplex $[\alpha\alpha]$ in Figure 9.2, applying $\bar{\alpha}$ to the z_1-plane and α to the z_2-plane yields $[\bar{\alpha}\alpha]$, and other combinations yield various other shapes.

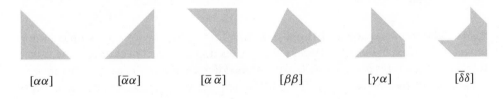

Figure 9.2. The x_1-x_2-shadows.

The symplectic 4-manifolds (M, ω) we shall consider contain a domain of equal volume which is explicitly symplectomorphic to $U \times \square^2(1) \subset \mathbb{R}^2(x) \times \mathbb{R}^2(y)$. In order to construct an explicit symplectic packing of (M, ω) by k equal balls it thus suffices to insert k disjoint x_1-x_2-shadows of equal width as in Figure 9.2 into U.

Remark 9.3.1. Besides of being explicit, the 4-dimensional symplectic packings constructed in [41], [81] and in this section have yet another advantage over the packings found in [61], [7], [8]: The symplectic packings of (M, ω) by k balls obtained from the method in [61], [7], [8] are maximal in the following sense. For every $\epsilon > 0$ there exists a symplectic embedding $\varphi_\epsilon\colon \coprod_{i=1}^{k} B^{2n}(a) \hookrightarrow (M, \omega)$ such that

$$\frac{\text{Vol}\,(\text{Im}\,\varphi_\epsilon, \omega)}{\text{Vol}\,(M, \omega)} \geq p_k(M, \omega) - \epsilon. \tag{9.3.1}$$

Karshon's symplectic packings of $(\mathbb{CP}^2, \omega_{\text{SF}})$ by 2 and 3 balls $B^4(\frac{\pi}{2})$ given by the map (9.2.1) and compositions of this map with coordinate permutations fill *exactly* $\frac{1}{2}$ and $\frac{3}{4}$ of $(\mathbb{CP}^2, \omega_{\text{SF}})$. Similarly, the 4-dimensional packings in [81] and in this section are maximal in the following sense:

There exists a symplectic embedding $\varphi\colon \coprod_{i=1}^{k} B^4(a) \hookrightarrow (M, \omega)$ such that

$$\frac{\text{Vol}\,(\text{Im}\,\varphi, \omega)}{\text{Vol}\,(M, \omega)} = p_k(M, \omega). \tag{9.3.2}$$

Moreover, φ is explicit in the following sense: The image $\coprod_{i=1}^{k} \varphi\left(B^4(a)\right)$ of φ is explicit, and given $a' < a$ one can construct φ such that its restriction to $\coprod_{i=1}^{k} B^4(a')$ is given pointwise.

Indeed, choose a sequence $a' < a_j \nearrow a$. The packings in [81] and our packings $\varphi(a_j)\colon \coprod_{i=1}^{k} B^4(a_j) \hookrightarrow (M, \omega)$ can be chosen such that

$$\text{Im}\,\varphi(a_j) \subset \text{Im}\,\varphi(a_{j+1}) \quad \text{for all } j.$$

The claim now follows from a result of McDuff, [55], stating that two symplectic embeddings of a closed ball into a larger ball are isotopic via a symplectic isotopy of the larger ball. $\diamond$

9.3.1 Maximal packings of the 4-ball and $\mathbb{CP}^2$.

In view of the symplectomorphism (9.2.1) and the identity (9.2.2) we only need to construct packings of the 4-ball. It follows from Table 9.1 that any k of the embeddings in Figure 9.3(a) yield a maximal packing of B^4 by k balls, $k = 2, 3, 4$, and that any k of the embeddings in Figure 9.3(b) yield a maximal packing by $k = 5, 6$ balls. Figure 9.3(c) shows a full packing by 9 balls.

Explicit maximal packings of B^4 by $k \leq 6$ balls were first constructed by Traynor in [81]. As the symplectic wrapping embeddings described in Section 7.1, her symplectic embeddings of a 4-ball into a larger 4-ball are of the form $\delta_E \circ \gamma \circ \beta \circ \alpha_E$. This time, $\beta = \text{diag}\left[\pi, \pi, \frac{1}{\pi}, \frac{1}{\pi}\right]$, and $\gamma\colon \mathbb{R}^2(x) \times \mathbb{R}^2(y) \to T^2 \times \mathbb{R}^2(y)$ is of the form $\gamma = (\text{id}\circ\tau) \circ \left((p \circ M) \times M^*\right)$ where τ is a translation of $\mathbb{R}^2(y)$ and $M \in \text{SL}(2; \mathbb{R})$. For $k \leq 5$, the matrix M can be chosen in $\text{SL}(2; \mathbb{Z})$, but for $k = 6$ also elements $M \in \text{SL}(2; \mathbb{R}) \setminus \text{SL}(2; \mathbb{Z})$ for which $p \circ M\colon \square^2(\pi) \to T^2$ is injective must be considered.

We cannot realize the packing numbers $p_7\left(B^4\right) = \frac{63}{64}$ and $p_8\left(B^4\right) = \frac{288}{289}$ by our packing method. This method as well as Traynor's method and their combination only fill $\frac{7}{9}$ and $\frac{8}{9}$ of the 4-ball by 7 and 8 equal balls, respectively.

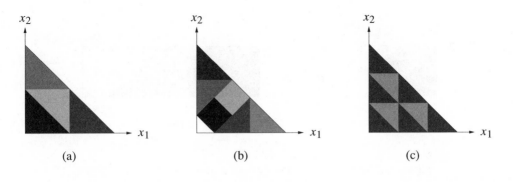

Figure 9.3. Maximal packings of B^4 for $k \leq 6$ and $k = l^2$.

Question 9.3.2. *Is there an explicit embedding of 7 or 8 equal balls into the 4-ball filling more than $\frac{7}{9}$ and $\frac{8}{9}$ of the volume?*

9.3.2 Maximal packings of ruled symplectic 4-manifolds.

Given a ruled symplectic 4-manifold (M, ω_{ab}), let $c_k(a, b)$ be the supremum of those A for which $\coprod_{i=1}^{k} B^{2n}(A)$ symplectically embeds into (M, ω_{ab}), so that

$$p_k(M, \omega_{ab}) = \frac{k \, c_k^2(a, b)}{2 \, \text{Vol}(M, \omega_{ab})}. \tag{9.3.3}$$

We shall write c instead of $c_k(a, b)$ if (M, ω_{ab}) and k are clear from the context.

Maximal packings of $S^2(a) \times S^2(b)$. As in Proposition 9.2.1 we assume that $a \geq b$. Represent the symplectic structure of $S^2(a) \times S^2(b)$ by a split form. Using Lemma 3.1.5 we symplectically identify $S^2(a) \setminus \text{pt}$ with $]0, a[\times]0, 1[$ and $S^2(b) \setminus \text{pt}$ with $]0, b[\times]0, 1[$. Then

$$\square(a, b) \times \square^2(1) = S^2(a) \times S^2(b) \setminus \{S^2(a) \times \text{pt} \cup \text{pt} \times S^2(b)\}.$$

Besides for $k \in \{6, 7\}$, we will construct the explicit maximal packings promised after Proposition 9.2.1 by constructing packings of $\square(a, b) \times \square^2(1)$ which realize the packing numbers of $S^2(a) \times S^2(b)$ computed in Proposition 9.2.1 and hence are maximal. (It is, in fact, known that *all* packing numbers of $\square(a, b) \times \square^2(1)$ and $S^2(a) \times S^2(b)$ agree, see [61, Remark 2.1.E]).

To construct explicit maximal packings for all k with $\lceil \frac{k}{2} \rceil \frac{b}{a} \leq 1$ is a trivial matter. Figure 9.4 shows a maximal packing by 1 and 2 respectively 5 and 6 balls.

Let now $k = 3, 4$ and $\frac{b}{a} \geq \frac{1}{2}$. Figure 9.5 shows maximal packings of $S^2(a) \times S^2(b)$ by k balls for $\frac{b}{a} = \frac{1}{2}$, $\frac{b}{a} = \frac{3}{4}$ and $\frac{b}{a} = 1$. For $\frac{b}{a} > \frac{1}{2}$ the (x_1, x_2)-coordinates of the

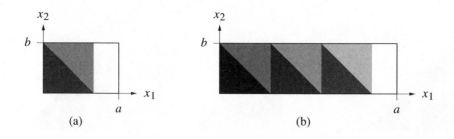

Figure 9.4. Maximal packings of $S^2(a) \times S^2(b)$ by k balls, $\lceil \frac{k}{2} \rceil \frac{b}{a} \leq 1$.

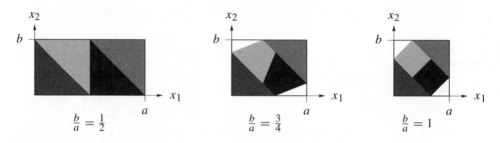

Figure 9.5. Maximal packings of $S^2(a) \times S^2(b)$ by 3 and 4 balls, $\frac{b}{a} \geq \frac{1}{2}$.

vertices of the "upper left ball" are

$$(0, c), \quad (a - c, b), \quad (c, c), \quad (a - c, b - c),$$

where $c = \frac{a+b}{3}$. As in most of the subsequent figures of this chapter, the three pictures in Figure 9.5 should be seen as moments of a movie starting at $\frac{b}{a} = \frac{1}{2}$ and ending at $\frac{b}{a} = 1$. Each ball in this movie moves in a smooth way.

Next, let $k = 5$ and $\frac{b}{a} \geq \frac{1}{3}$. In order to construct a smooth family of maximal packings of $S^2(a) \times S^2(b)$ by 5 balls, we think of the maximal packing for $\frac{b}{a} = \frac{1}{3}$ rather as in Figure 9.6 than as in Figure 9.4(a). The x_1-width of all balls is $\frac{a+2b}{5}$, and the "upper left ball" has 5 vertices for $\frac{1}{3} < \frac{b}{a} \leq \frac{3}{4}$ and 7 vertices for $\frac{b}{a} > \frac{3}{4}$.

For $k \in \{6, 7\}$, we cannot realize the packing numbers $p_k(S^2(a) \times S^2(b))$ by directly packing rectangles as for $k \leq 4$. We shall instead construct certain maximal packings of $\mathbb{CP}^2$ which correspond to maximal packings of $S^2(a) \times S^2(b)$. As noticed in [7], the correspondence between symplectic packings and the symplectic blow-up operation implies

Lemma 9.3.3. *Packing $S^2(a) \times S^2(b)$ by k equal balls $\coprod_{i=1}^{k} B^4(c)$ corresponds to packing $(\mathbb{CP}^2, (a+b-c)\omega_{SF})$ by the $k+1$ balls $B^4(a-c) \coprod B^4(b-c) \coprod_{i=1}^{k-1} B^4(c)$.*

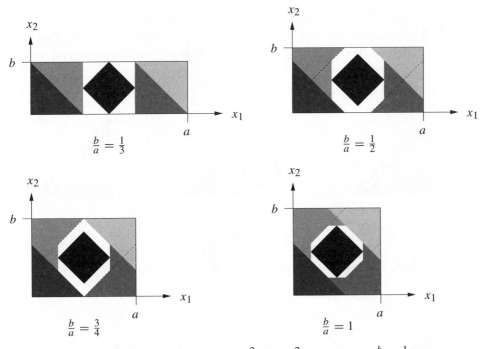

Figure 9.6. Maximal packings of $S^2(a) \times S^2(b)$ by 5 balls, $\frac{b}{a} \geq \frac{1}{3}$.

In order to make this correspondence plausible, we choose $\frac{b}{a} = \frac{2}{3}$ and $c = c_6(a, b) = \frac{a+2b}{5}$, and we think of $(\mathbb{CP}^2, (a + b - c)\omega_{SF})$ as the simplex of width $a + b - c$ and of $S^2(a) \times S^2(b)$ as the rectangle of width a and length b. As Figure 9.7

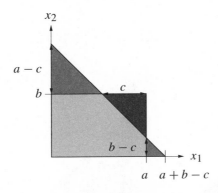

Figure 9.7. $(\mathbb{CP}^2, (a + b - c)\omega_{SF}) \setminus B^4(a - c) \coprod B^4(b - c) = S^2(a) \times S^2(b) \setminus B^4(c)$.

illustrates, the space obtained by removing a ball $B^4(c)$ from $S^2(a) \times S^2(b)$ coincides with the space obtained by removing the balls $B^4(a - c) \coprod B^4(b - c)$ from $(\mathbb{CP}^2, (a + b - c)\omega_{SF})$.

Figures 9.8, 9.9 and 9.10 describe explicit packings of $(\mathbb{CP}^2, (a + b - c)\omega_{\mathrm{SF}})$ by balls $B^4(a-c) \coprod B^4(b-c) \coprod_{i=1}^{k-1} B^4(c)$ for $k \in \{6, 7\}$ and c as in Proposition 9.2.1. The lower left triangle represents $B^4(a - c)$ and the black "ball" represents $B^4(b - c)$.

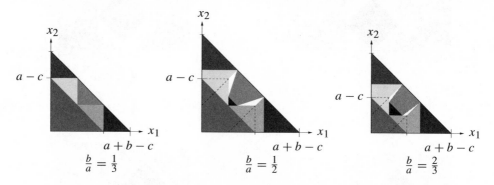

Figure 9.8. Maximal packings of $S^2(a) \times S^2(b)$ by 6 balls, $\frac{1}{3} \leq \frac{b}{a} \leq \frac{3}{4}$.

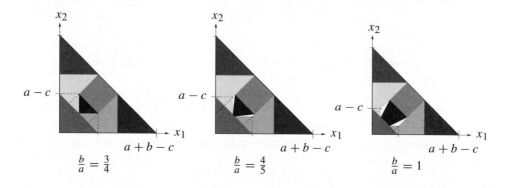

Figure 9.9. Maximal packings of $S^2(a) \times S^2(b)$ by 6 balls, $\frac{3}{4} \leq \frac{b}{a} \leq 1$.

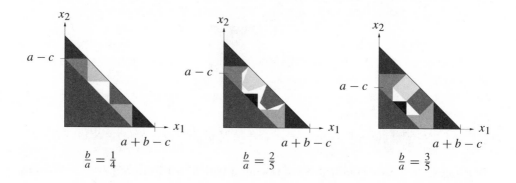

Figure 9.10. Maximal packings of $S^2(a) \times S^2(b)$ by 7 balls, $\frac{1}{4} \leq \frac{b}{a} \leq \frac{3}{5}$.

From these packings one obtains explicit packings of $S^2(a) \times S^2(b)$ as follows: First symplectically blow up $(\mathbb{CP}^2, (a + b - c)\omega_{SF})$ twice by removing the balls $B^4(a - c)$ and $B^4(b - c)$ and collapsing the remaining boundary spheres to exceptional spheres in homology classes D_1 and D_2. The resulting manifold, which is symplectomorphic to $S^2(a) \times S^2(b)$ blown up at one point with weight c, still contains the $k - 1$ explicitly embedded balls $B^4(c)$, and according to [7, Theorem 4.1.A] the exceptional sphere in class $L - D_1 - D_2$ can be symplectically blown down with weight c to yield the k'th ball $B^4(c)$ in $S^2(a) \times S^2(b)$. Here and in the sequel, $L = [\mathbb{CP}^1]$ denotes the class of a line in $\mathbb{CP}^2$.

Finally, the construction of full packings of $S^2(mb) \times S^2(b)$ by $2ml^2$ balls for $l, m \in \mathbb{N}$ is also straightforward. Figure 9.11 shows such a packing for $l = m = 2$.

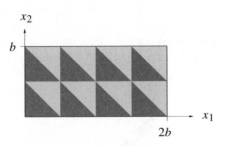

Figure 9.11. A full packing of $S^2(2b) \times S^2(b)$ by 16 balls.

Maximal packings of $(S^2 \ltimes S^2, \omega_{ab})$. In order to describe our maximal packings of $(S^2 \ltimes S^2, \omega_{ab})$, it will be convenient to work with the parameters $\alpha = a - \frac{b}{2}, \beta = b$, so that $\alpha > 0, \beta > 0$ and $\omega_{ab} = \beta A + (\alpha + \beta)F$. According to [62, Example 7.4], the manifold $S^2 \ltimes S^2$ is diffeomorphic to the complex blow-up $\widetilde{N}_1$ of $\mathbb{CP}^2$ at one point via a diffeomorphism under which the class L of a line corresponds to $A + F$ and the class D_1 of the exceptional divisor corresponds to A. We can therefore view $(S^2 \ltimes S^2, \omega_{ab})$ as $\widetilde{N}_1$ endowed with the symplectic form in class $(\alpha + \beta)L - \alpha D_1$ obtained by symplectically blowing up $(\mathbb{CP}^2, (\alpha + \beta)\omega_{SF})$ with weight α. Since symplectically blowing up with weight α corresponds to removing a ball $B^4(\alpha)$ and collapsing the remaining boundary sphere to an exceptional sphere in class D_1, we can think of this symplectic manifold as the truncated simplex obtained by removing the simplex of width α from the simplex of width $\alpha + \beta$.

Denote by $\lfloor x \rfloor$ the integer part of $x \geq 0$. In the parameters α and β, the packings promised after Proposition 9.2.3 are explicit maximal packings of $(S^2 \ltimes S^2, \omega_{ab})$ for all k with $\lfloor \frac{k}{2} \rfloor \frac{\beta}{\alpha} \leq 1$, for $k \leq 5$ and $\alpha, \beta > 0$ arbitrary, and for $k = 6$ and $\frac{\beta}{\alpha} \in]0, 1] \cup [4, \infty[$. Moreover, given ω_{ab} with $\frac{\beta}{\alpha} = \frac{l}{m-l}$ for some $l, m \in \mathbb{N}$ with $m > l$, we will construct explicit full packings of $(S^2 \ltimes S^2, \omega_{ab})$ by $l(2m - l)$ balls.

Set $c_k = c_k(a, b) = c_k(S^2 \ltimes S^2, \omega_{ab})$. Replacing a by $\alpha + \beta/2$ and b by β in the list in Proposition 9.2.3 and using $\beta(2\alpha + \beta) = 2\,\mathrm{Vol}(S^2 \ltimes S^2, \omega_{ab})$ and (9.3.3), we find that

$$c_1 = \beta, \quad c_2 = c_3 = \left\{ \beta, \frac{\alpha + \beta}{2} \right\} \quad \text{on }]0, 1, \infty[,$$

$$c_4 = \left\{ \beta, \frac{\alpha + 2\beta}{4} \right\} \quad \text{on } \left]0, \frac{1}{2}, \infty\right[,$$

$$c_5 = \left\{ \beta, \frac{\alpha + 2\beta}{4}, \frac{2\alpha + 2\beta}{5} \right\} \quad \text{on } \left]0, \frac{1}{2}, \frac{3}{2}, \infty\right[,$$

$$c_6 = \left\{ \beta, \frac{\alpha + 3\beta}{6}, \frac{2\alpha + 3\beta}{7}, \frac{2\alpha + 2\beta}{5} \right\} \quad \text{on } \left]0, \frac{1}{3}, \frac{5}{3}, 4, \infty\right[.$$

To construct packings with $p_k = k\frac{\beta}{2\alpha+\beta}$ for all k with $\lfloor \frac{k}{2} \rfloor \frac{\beta}{\alpha} \leq 1$ is very easy. Figure 9.12(a) shows a maximal packing by 1 ball, and Figures 9.12(b1) and (b2) show maximal packings by 4 and 5 balls for $\frac{\beta}{\alpha} = \frac{1}{2}$ and $\frac{\beta}{\alpha} < \frac{1}{2}$, respectively.

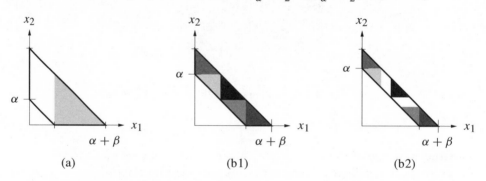

    (a)   (b1)   (b2)

Figure 9.12. Maximal packings of $(S^2 \ltimes S^2, \omega_{ab})$ by k balls, $\lfloor \frac{k}{2} \rfloor \frac{\beta}{\alpha} \leq 1$.

Figure 9.13 shows maximal packings for $k = 2, 3$ and $\frac{\beta}{\alpha} \geq 1$.

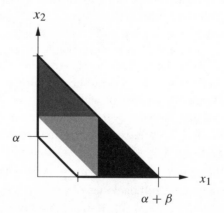

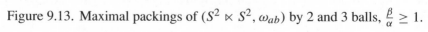

Figure 9.13. Maximal packings of $(S^2 \ltimes S^2, \omega_{ab})$ by 2 and 3 balls, $\frac{\beta}{\alpha} \geq 1$.

Also our maximal packings by 4 balls are easy to understand (Figure 9.14 and Figure 9.15(a)): $2\,c_4 = \beta + \frac{\alpha}{2}$ just means that the two middle gray balls touch each other. As long as $\frac{\beta}{\alpha} \leq \frac{3}{2}$, there is enough room for a fifth (black) ball between these two balls. If $\frac{\beta}{\alpha} > \frac{3}{2}$, there is enough space for a fifth ball if and only if the capacity c of the balls satisfies $2c + \frac{c}{2} \leq \alpha + \beta$; hence $c_5 = \frac{2\alpha+2\beta}{5}$ (Figures 9.15(b1) and (b2)).

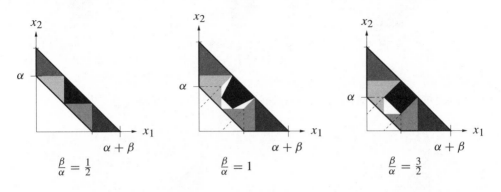

Figure 9.14. Maximal packings of $(S^2 \ltimes S^2, \omega_{ab})$ by 4 and 5 balls, $\frac{1}{2} \leq \frac{\beta}{\alpha} \leq \frac{3}{2}$.

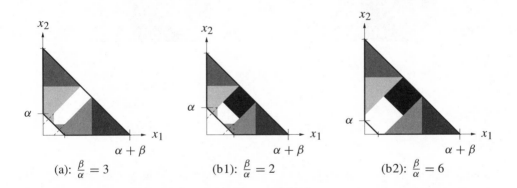

Figure 9.15. Maximal packings of $(S^2 \ltimes S^2, \omega_{ab})$ by 4 and 5 balls, $\frac{\beta}{\alpha} \geq \frac{3}{2}$.

Let now $k = 6$. Figure 9.16 shows maximal packings for $\frac{1}{3} \leq \frac{\beta}{\alpha} \leq 1$. For $\frac{\beta}{\alpha} > \frac{1}{3}$ the vertices of the "lower middle ball" are

$$(\alpha+\beta-2c_6, c_6), \quad \left(\frac{\alpha+\beta}{2}, \frac{\alpha+\beta}{2}\right), \quad (\alpha+\beta-c_6, c_6), \quad \left(\frac{\alpha+\beta}{2}, \frac{\alpha+\beta}{2} - c_6\right).$$

Maximal packings for $\frac{\beta}{\alpha} \geq 4$ are illustrated in Figure 9.17.

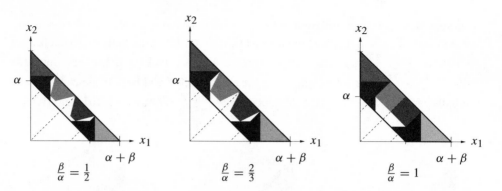

Figure 9.16. Maximal packings of $(S^2 \ltimes S^2, \omega_{ab})$ by 6 balls, $\frac{1}{3} \leq \frac{\beta}{\alpha} \leq 1$.

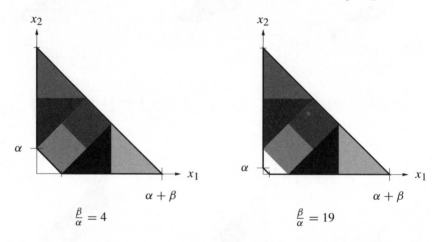

Figure 9.17. Maximal packings of $(S^2 \ltimes S^2, \omega_{ab})$ by 6 balls, $\frac{\beta}{\alpha} \geq 4$.

Remark 9.3.4. It is not a coincidence that we were not able to construct maximal packings of $(S^2 \ltimes S^2, \omega_{ab})$ by 6 balls for all ratios $\frac{\beta}{\alpha} > 0$. Indeed, a maximal packing of $(S^2 \ltimes S^2, \omega_{ab})$ by 6 equal balls for $\frac{\beta}{\alpha} = \frac{5}{3}$ corresponds to a maximal packing of the 4-ball by 7 equal balls. $\diamond$

Finally, suppose that $\frac{\beta}{\alpha} = \frac{l}{m-l}$ for some $l, m \in \mathbb{N}$ with $m > l$. We can then fill $(S^2 \ltimes S^2, \omega_{ab})$ by $l(2m - l)$ balls by decomposing $S^2 \ltimes S^2$ into l shells and filling the i-th shell with $2m + 1 - 2i$ balls (see Figure 9.18, where $l = 2$ and $m = 4$).

Maximal packings of $\Sigma_g(a) \times S^2(b)$ and $(\Sigma_g \ltimes S^2, \omega_{ab})$ for $g \geq 1$. Fix $a > 0$ and $b > 0$. We represent the symplectic structure of the product $\Sigma_g(a) \times S^2(b)$ by a split form. Removing a wedge of $2g$ loops from $\Sigma(a)$ and a point from $S^2(b)$ we see that $\Sigma_g(a) \times S^2(b)$ contains $\square(a, b) \times \square^2(1)$. The explicit construction of the "standard

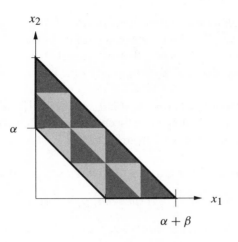

Figure 9.18. A full packing of $(S^2 \ltimes S^2, \omega_{ab})$, $\frac{\beta}{\alpha} = 1$, by 12 balls.

Kähler form" in class $[\omega_{ab}]$ given in [57, Section 3] and [62, Exercise 6.14] shows that also $(\Sigma_g \ltimes S^2, \omega_{ab})$ endowed with this standard form contains $\square(a, b) \times \square^2(1)$. The explicit maximal packings promised after Proposition 9.2.5 can thus be constructed as for $S^2(a) \times S^2(b)$, see Figures 9.4 and 9.11.

9.3.3 Explicit packings of $\Sigma_g(a) \times \Sigma_h(b)$ for $g, h \geq 1$. We consider 4-manifolds of the form $\Sigma_g \times \Sigma_h$ with $g, h \geq 1$. The space of symplectic structures on such manifolds in not understood, but no symplectic structure different from $\Sigma_g(a) \times \Sigma_h(b)$ for some $a > 0, b > 0$ is known. For $\Sigma_g(a) \times \Sigma_h(b)$, no obstructions to full packings are known. Recall from (9.0.2) that for $\frac{a}{b} \in \mathbb{Q}$,

$$P\left(\Sigma_g(a) \times \Sigma_h(b)\right) := \inf \left\{k_0 \in \mathbb{N} \mid p_k\left(\Sigma_g(a) \times \Sigma_h(b)\right) = 1 \text{ for all } k \geq k_0\right\}$$

is finite. In fact, Biran showed in Corollary 1.B and Section 5 of [8] that

$$P\left(T^2(1) \times T^2(1)\right) \leq 2 \tag{9.3.4}$$

and that

$$P\left(\Sigma_g(a) \times \Sigma_h(b)\right) \leq \begin{cases} 8ab & \text{if } a, b \in \mathbb{N}, \\ 2ab & \text{if } a, b \in \mathbb{N} \setminus \{1\}. \end{cases} \tag{9.3.5}$$

If $\frac{a}{b} \notin \mathbb{Q}$ or if $1 \leq k < P\left(\Sigma_g(a) \times \Sigma_h(b)\right)$, there is not much known about $p_k\left(\Sigma_g(a) \times \Sigma_h(b)\right)$: We can assume without loss of generality that $a \geq b$. Since the symplectic packing numbers of $S^2(a) \times S^2(b)$ and $\square(a, b) \times \square^2(1)$ agree, and since $\square(a, b) \times \square^2(1)$ symplectically embeds into $\Sigma_g(a) \times \Sigma_h(b)$,

$$p_k\left(S^2(a) \times S^2(b)\right) \leq p_k\left(\Sigma_g(a) \times \Sigma_h(b)\right) \quad \text{for all } k \in \mathbb{N}, \tag{9.3.6}$$

and Figures 9.4, 9.5, 9.6 and 9.11 describe some explicit packings of $\Sigma_g(a) \times \Sigma_h(b)$. A comparison of Corollary 9.2.2 with the estimates (9.3.4) and (9.3.5) and with Proposition 9.3.5 below shows, however, that in general the inequalities (9.3.6) are not equalities and that for $\Sigma_g(a) \times \Sigma_h(b)$ not all of the packings in Figures 9.4, 9.5 and 9.6 are maximal.

Elaborating an idea of Polterovich, [62, Exercise 12.4], Jiang constructed in [40, Corollary 3.3 and 3.4] explicit symplectic embeddings of one ball which improve the estimate $\frac{b}{2a} \leq p_1\big(\Sigma_g(a) \times \Sigma_h(b)\big)$ from (9.3.6).

Proposition 9.3.5 (Jiang). *Let $\Sigma(a)$ be any closed surface of area $a \geq 1$.*

(i) *There exists a constant $C > 0$ such that $p_1\big(\Sigma(a) \times T^2(1)\big) \geq C$.*

(ii) *If $h \geq 2$, there exists a constant $C = C(h) > 0$ depending only on h such that $w_G\big(\Sigma(a) \times \Sigma_h(1)\big) \geq C \log a$. In other words,*

$$p_1\big(\Sigma(a) \times \Sigma_h(1)\big) \geq \frac{(C \log a)^2}{2a}.$$

Notice that for $\Sigma = S^2$ Biran's result $p_1\big(S^2(a) \times \Sigma_h(1)\big) = \min\big(1, \frac{a}{2}\big)$ stated in Proposition 9.2.5 is much stronger. We shall use Jiang's embedding method to prove the following quantitative version of Proposition 9.3.5 (i).

Proposition 9.3.6. *If $a \geq 1$,*

$$p_1\big(\Sigma(a) \times T^2(1)\big) \geq \frac{\max\{a + 1 - \sqrt{2a+1}, 2\}}{4a}.$$

In particular, the constant C in Proposition 9.3.5 (i) *can be chosen to be $C = 1/8$.*

Proof. Set $R(a) = \{(x, y) \in \mathbb{R}^2 \mid 0 < x < 1, \; 0 < y < a\}$, and consider the linear symplectic map

$$\varphi \colon (R(a) \times R(a), dx_1 \wedge dy_1 + dx_2 \wedge dy_2) \to (\mathbb{R}^2 \times \mathbb{R}^2, dx_1 \wedge dy_1 + dx_2 \wedge dy_2),$$
$$(x_1, y_1, x_2, y_2) \mapsto (x_1 + y_2, y_1, -y_2, y_1 + x_2).$$

Let $\mathrm{pr} \colon \mathbb{R}^2 \to T^2 = \mathbb{R}/\mathbb{Z} \times \mathbb{R}/\mathbb{Z}$ be the projection onto the standard symplectic torus. Then $(\mathrm{id}_2 \times \mathrm{pr}) \circ \varphi \colon R(a) \times R(a) \to \mathbb{R}^2 \times T^2$ is a symplectic embedding. Indeed, given (x_1, y_1, x_2, y_2) and (x_1', y_1', x_2', y_2') with

$$x_1 + y_2 = x_1' + y_2' \tag{9.3.7}$$
$$y_1 = y_1' \tag{9.3.8}$$
$$-y_2 \equiv -y_2' \mod \mathbb{Z} \tag{9.3.9}$$
$$y_1 + x_2 \equiv y_1' + x_2' \mod \mathbb{Z} \tag{9.3.10}$$

equations (9.3.8) and (9.3.10) imply $x_2 \equiv x_2' \mod \mathbb{Z}$, whence $x_2 = x_2'$. Moreover, (9.3.9) and (9.3.7) show that $y_2 - y_2' = x_1' - x_1 \equiv 0 \mod \mathbb{Z}$, and so $x_1 = x_1'$ and $y_2 = y_2'$. Next observe that

$$\big((\mathrm{id}_2 \times \mathrm{pr}) \circ \varphi\big)\big(R(a) \times R(a)\big) \subset \,] - a, 0[\times] - a - 1, a + 1[\times T^2.$$

Thus $R(a) \times R(a)$ symplectically embeds into $\Sigma(2a(a+1)) \times T^2(1)$, and since $B^4(a)$ symplectically embeds into $R(a) \times R(a)$ and $B^4(1)$ symplectically embeds into $\Sigma(a) \times T^2(1)$ for any $a \geq 1$, Proposition 9.3.6 follows. $\qquad\square$

Remark 9.3.7. Gromov's Nonsqueezing Theorem stated in Section 1.2 was generalized by Lalonde and McDuff in [48] to *all* $(2n-2)$-dimensional symplectic manifolds (V, ω).

General Nonsqueezing Theorem. *If the ball $B^{2n}(a)$ symplectically embeds into $(V \times B^2(A), \omega \oplus \omega_0)$, then $A \geq a$.*

Gromov's original proof of his Nonsqueezing Theorem combined with the existence of Gromov–Witten invariants for arbitrary closed symplectic manifolds shows the Nonsqueezing Theorem

$$w_G\big(V \times S^2, \omega_V \oplus \omega_{S^2}\big) \leq \int_{S^2} \omega_{S^2} \tag{9.3.11}$$

for any closed symplectic manifold (V, ω), see [63, Section 9.3] and [60, Proposition 1.18]. This Nonsqueezing Theorem can be generalized by either viewing S^2 as a fibre or as the base of a symplectic fibration.

(i) A *ruled symplectic manifold* (M, ω) is a 2-sphere bundle $S^2 \to M \to V$ over a closed manifold V endowed with a symplectic form ω which restricts to an area form on each fibre. Examples are ruled symplectic 4-manifolds and products $(V \times S^2, \omega_V \oplus \omega_{S^2})$. The proof of (9.3.11) extends to this situation and shows the Nonsqueezing Theorem

$$w_G\,(M, \omega) \leq \int_{S^2} \omega, \tag{9.3.12}$$

see again [63, Section 9.3]. Recall that for the spaces $S^2(a) \times S^2(b)$ we assume $a \geq b$. Using $w_G = c_1$ and (9.3.3), we read off from Propositions 9.2.1, 9.2.3 and 9.2.5 that for ruled symplectic 4-manifolds,

$$w_G(S^2(a) \times S^2(b)) = w_G(S^2 \ltimes S^2, \omega_{ab}) = b, \tag{9.3.13}$$

$$w_G(\Sigma_g(a) \times S^2(b)) = w_G(\Sigma_g \ltimes S^2, \omega_{ab}) = \min\{\sqrt{2ab}, b\} \quad \text{if } g \geq 1. \tag{9.3.14}$$

Since $\int_{S^2} \omega_{ab} = b$ in all cases, we see that the upper bound for the Gromov width predicted by the Nonsqueezing Theorem (9.3.12) and the volume condition is sharp for ruled symplectic 4-manifolds. Explicit maximal embeddings were given for $g = 0$ in

Figures 9.4 (a) and 9.12 (a) and for $g \geq 1$ and $a \geq b$ in Figure 9.4 (a), but no explicit maximal embedding is known for $g \geq 1$ and $a < b$.

(ii) A locally trivial fibration $V \to M \to S^2$ of a closed connected symplectic manifold (M, ω) is called a *symplectic fibration over* S^2 if all fibres $(V, \omega|_V)$ are symplectic and mutually symplectomorphic. The *area* of such a fibration is naturally defined as

$$\text{area}(M, \omega) = \frac{\text{Vol}(M, \omega)}{\text{Vol}(V, \omega|_V)} = \frac{\int_M \omega^n}{n \int_V (\omega|_V)^{n-1}}.$$

In [60, Proposition 1.20 (i)] McDuff proved the Nonsqueezing Theorem

$$w_G (M, \omega) \leq \text{area} (M, \omega)$$

provided that $[\omega|_V]$ vanishes on $\pi_2(V)$. The symplectic fibrations $(S^2 \ltimes S^2, \omega_{ab})$, for which $w_G(S^2 \ltimes S^2, \omega_{ab}) = b$ in view of (9.3.13) and $\text{area}(S^2 \ltimes S^2, \omega_{ab}) = a$, show that this assumption cannot be omitted.

(iii) The Nonsqueezing Theorem (9.3.11) does not remain valid if the factor S^2 is replaced by a closed surface with positive genus. This was first noticed by Lalonde, [47], and follows from (9.3.14) as well as from Propositions 9.3.5 and 9.3.6. $\Diamond$

9.3.4   Maximal packings of 4-dimensional ellipsoids.

We finally construct some explicit maximal packings of 4-dimensional ellipsoids $E(\pi, a)$ with $a \geq \pi$.

Proposition 9.3.8. (i) *For each $k \in \mathbb{N}$ the ellipsoid $E(\pi, k\pi)$ admits an explicit full symplectic packing by k balls.*

(ii) *$p_1 (E(\pi, a)) = \frac{\pi}{a}$ and $p_2 (E(\pi, a)) = \min \left(\frac{2\pi}{a}, \frac{a}{2\pi}\right)$, and these packing numbers can be realized by explicit symplectic packings.*

The statement (i) was proved in [81, Theorem 6.3 (2)], and (ii) was proved in [54, Corolary 3.11]. Their embeddings are different from ours.

Proof of Proposition 9.3.8. (i) Recall from Lemma 5.3.1 that we can think of $B^4(\pi)$ as $\triangle^2(\pi) \times \square^2(1)$ and of $E(\pi, k\pi)$ as $\triangle(\pi, k\pi) \times \square^2(1)$. The linear symplectic map $(x_1, x_2, y_1, y_2) \mapsto \left(x_1, kx_2, y_1, \frac{1}{k}y_2\right)$ maps $\triangle^2(\pi) \times \square^2(1)$ to $\triangle (\pi, k\pi) \times \square \left(1, \frac{1}{k}\right)$, and it is clear how to insert k copies of this set into $\triangle (\pi, k\pi) \times \square^2(1)$.

(ii) The estimates $p_1 (E(\pi, a)) \leq \frac{\pi}{a}$ and $p_2 (E(\pi, a)) \leq \frac{2\pi}{a}$ follow from the inclusion $E(\pi, a) \subset Z^4(\pi)$ and from Gromov's Nonsqueezing Theorem, and the estimate $p_2 (E(\pi, a)) \leq \frac{a}{2\pi}$ follows from $E(\pi, a) \subset B^4(a)$ and from Gromov's result $p_2 \left(B^4(a)\right) \leq \frac{1}{2}$ stated in (9.1.7). The inclusion $B^4(\pi) \subset E(\pi, a)$ shows that $p_1 (E(\pi, a)) = \frac{\pi}{a}$, and explicit symplectic packings of $E(\pi, a)$ by two balls realizing $p_2 (E(\pi, a)) = \min \left(\frac{2\pi}{a}, \frac{a}{2\pi}\right)$ can be constructed as in the proof of (i). $\square$

9.4 Maximal packings in higher dimensions

In dimensions $2n \geq 6$, only few maximal symplectic packings by equal balls are known.

1. Balls and $(\mathbb{CP}^n, \omega_{SF})$. As in dimension 4 we denote by ω_{SF} the unique $U(n+1)$-invariant Kähler form on $\mathbb{CP}^n$ whose integral over $\mathbb{CP}^1$ equals 1. The embedding (9.2.1) generalizes to all dimensions, and

$$p_k\left(B^{2n}\right) = p_k(\mathbb{CP}^n, \omega_{SF}) \quad \text{for all } k,$$

see [61, Remark 2.1.E]. Recall from (9.1.7) and (9.1.8) that

$$p_k\left(B^{2n}\right) = \frac{k}{2^n} \quad \text{for } 2 \leq k \leq 2^n,$$

$$p_{l^n}\left(B^{2n}\right) = 1 \quad \text{for all } l \in \mathbb{N}.$$

An explicit maximal packing of $(\mathbb{CP}^n, \omega_{SF})$ by $k \leq n+1$ balls was found by Karshon in [41], and explicit full packings of B^{2n} by l^n balls for each $l \in \mathbb{N}$ were given by Traynor in [81]. Taking $l = 2$, any k balls of such a packing yield a maximal packing by k balls. The following different construction of an explicit full packing of B^{2n} by l^n equal balls is mentioned in [81, Remark 5.13]. Recall from Lemma 5.3.1 that we can think of $B^{2n}(\pi)$ as $\triangle^n(\pi) \times \square^n(1)$ and of $B^{2n}\left(\frac{\pi}{l}\right)$ as $\triangle^n\left(\frac{\pi}{l}\right) \times \square^n(1)$. The matrix $\mathrm{diag}\left[l, \ldots, l, \frac{1}{l}, \ldots, \frac{1}{l}\right] \in \mathrm{Sp}(n; \mathbb{R})$ maps $\triangle^n\left(\frac{\pi}{l}\right) \times \square^n(1)$ to $\triangle^n(\pi) \times \square^n\left(\frac{1}{l}\right)$. It is clear how to insert l^n copies of $\triangle^n(\pi) \times \square^n\left(\frac{1}{l}\right)$ into $\triangle^n(\pi) \times \square^n(1)$.

2. Products of balls, complex projective spaces and surfaces. Set $n = \sum_{i=1}^{d} n_i$ and let $a_1, \ldots, a_d \in \pi\mathbb{N}$. According to [61, Theorem 1.5.A], the product

$$\left(\mathbb{CP}^{n_1} \times \cdots \times \mathbb{CP}^{n_d}, a_1\,\omega_{SF} \oplus \cdots \oplus a_d\,\omega_{SF}\right)$$

admits a full symplectic packing by $\frac{n!}{n_1! \cdots n_d!} a_1^{n_1} \cdots a_d^{n_d}$ equal $2n$-dimensional balls. These full packings can be constructed in an explicit way. Indeed, explicit full packings of $B^{2n_i}(a_i)$ by $a_i^{n_i}$ equal balls as in 1. above can be used to construct explicit full packings of

$$B^{2n_1}(a_1) \times \cdots \times B^{2n_d}(a_d)$$

by $\frac{n!}{n_1! \cdots n_d!} a_1^{n_1} \cdots a_d^{n_d}$ balls, see [45, Section 3.2]. In particular, there are explicit full packings of the polydisc $P(a_1, \ldots, a_n)$ and of the products of surfaces $\Sigma_{g_1}(a_1) \times \cdots \times \Sigma_{g_n}(a_n)$ with $a_i \in \pi\mathbb{N}$ by $n!\, a_1 \cdots a_n$ equal balls, see also [81, Section 4.1], [54, Theorem 4.1], and Figure 9.11 above for the case $n = 2$. An explicit packing construction in [54, Theorem 1.21] yields the lower bounds

$$p_7\left(C^6(\pi)\right) \geq \frac{224}{375} \quad \text{and} \quad p_8\left(C^6(\pi)\right) \geq \frac{9}{16}.$$

The technique in the proof of Proposition 9.3.6 can be used to generalize Proposition 9.3.5 (i): For any closed surface Σ endowed with an area form σ and any constant symplectic form ω on the $2n$-dimensional torus T^{2n}, there exists a constant $C > 0$ such that $p_1(\Sigma \times T^{2n}, a\sigma \oplus \omega) \geq C$ for all $a \geq 1$, see [40, Theorem 3.1].

3. Ellipsoids. Generalizing the result of Proposition 9.3.8 (ii), the packing numbers $p_1(E(a_1, \ldots, a_n)) = \frac{a_1^n}{a_1 \cdots a_n}$ and $p_2 (E(a_1, \ldots, a_n)) = \frac{2}{a_1 \cdots a_n} \min \left(a_1^n, \left(\frac{a_n}{2}\right)^n\right)$ of a $2n$-dimensional ellipsoid were computed and realized by explicit symplectic packings in [54, Corollary 3.11].

Remark 9.4.1. Karshon's explicit packing of $(\mathbb{CP}^n, \omega_{SF})$ by $k \leq n + 1$ balls is maximal in the sense of (9.3.1). Since in dimensions ≥ 6 it is not yet known whether the space of symplectic embeddings of a closed ball into a larger ball is connected, all other explicit (and non-explicit) maximal symplectic packings known in dimensions ≥ 6 are maximal only in the sense of (9.3.2). $\diamond$

We conclude this chapter with addressing two widely open problems. As before, we consider connected symplectic manifolds of finite volume.

Question 9.4.2. *Which connected symplectic manifolds (M, ω) of finite volume satisfy $p_k(M, \omega) = 1$ for all $k \geq 1$?*

Examples are 2-dimensional manifolds, $(\mathbb{CP}^2, \omega_{SF})$ symplectically blown up at $N \geq 9$ points with weights close enough to $1/\sqrt{N}$ and $S^2(1) \times S^2(1)$ symplectically blown up at $N \geq 8$ points with weights close enough to $1/\sqrt{N}$ (see [7, Section 5]), the ruled symplectic 4-manifolds $\Sigma_g(a) \times S^2(b)$ and $(\Sigma_g \ltimes S^2, \omega_{ab})$ with $g \geq 1$ and $b \geq 2a$ and their symplectic blow-ups (see Proposition 9.2.5 and [7, Theorem 6.A]), as well as certain closed symplectic 4-manifolds described in [7, Theorem 2.F] and their symplectic blow-ups.

A related problem is

Question 9.4.3. *Which connected symplectic manifolds (M, ω) of finite volume satisfy $p_1(M, \omega) = 1$?*

Examples different from the above ones are the ball B^{2n} and $(\mathbb{CP}^n, \omega_{SF})$, and, more generally, the complement $(\mathbb{CP}^n \setminus \Gamma, \omega_{SF})$ of a closed complex submanifold Γ of $\mathbb{CP}^n$ (see [61, Corollary 1.5.B]).

Appendix

A   The Extension after Restriction Principle

In this appendix we give a proof of the well-known Extension after Restriction Principle for symplectic embeddings of bounded starshaped domains. This principle was used to derive the Symplectic Hedgehog Theorem from Sikorav's Nonsqueezing Theorem for the torus T^n and in the proof of Theorem 1 to bring Ekeland–Hofer capacities into play. We shall also discuss an extension of this principle to symplectic embeddings of unbounded starshaped domains, which was proved in [71].

A domain U in $\mathbb{R}^{2n}$ is called *starshaped* if U contains a point p such that for every point $z \in U$ the straight line between p and z is contained in U. A Hamiltonian diffeomorphism of $\mathbb{R}^{2n}$ is called *compactly supported* if it is generated by a compactly supported Hamiltonian function $H : \mathbb{R} \times \mathbb{R}^{2n} \to \mathbb{R}$.

Proposition A.1. *Assume that $\varphi \colon U \hookrightarrow \mathbb{R}^{2n}$ is a symplectic embedding of a bounded starshaped domain $U \subset \mathbb{R}^{2n}$. Then for any subset $A \subset U$ whose closure in $\mathbb{R}^{2n}$ is contained in U there exists a compactly supported Hamiltonian diffeomorphism Φ_A of $\mathbb{R}^{2n}$ such that $\Phi_A|_A = \varphi|_A$.*

Proof. We closely follow [20]. We first prove that there exists a symplectic isotopy $\phi \colon \mathbb{R} \times U \to \mathbb{R}^{2n}$ such that $\phi^0 = \mathrm{id}_U$ and $\phi^1 = \varphi$, and then cut off a Hamiltonian function generating this isotopy.

Step 1. Alexander's trick. We can assume that 0 is a star point of U and that $\varphi(0) = 0$ because translations of $\mathbb{R}^{2n}$ are symplectically isotopic to the identity. We can also assume that $d\varphi(0) = \mathrm{id}$ because the linear symplectic group $\mathrm{Sp}(n; \mathbb{R})$ retracts onto $\mathrm{U}(n)$ (see [62, Section 2.2]) and is hence connected. Since U is starshaped with respect to 0 we can use "Alexander's trick" and define a continuous path $\varphi^t, t \in \mathbb{R}$, of symplectic embeddings $\varphi^t \colon U \hookrightarrow \mathbb{R}^{2n}$ by setting

$$\varphi^t(z) := \begin{cases} z & \text{if } t \leq 0, \\ \frac{1}{t}\varphi(tz) & \text{if } t \geq 0. \end{cases} \tag{A.1}$$

The path φ^t is smooth except possibly at $t = 0$. In order to smoothen φ^t, we define the smooth function $\eta \colon \mathbb{R} \to \mathbb{R}$ by

$$\eta(t) := \begin{cases} 0 & \text{if } t \leq 0, \\ e^2 e^{-2/t} & \text{if } t \geq 0, \end{cases} \tag{A.2}$$

where e denotes the Euler number, and for $t \in \mathbb{R}$ and $z \in U$ we set

$$\phi^t(z) := \varphi^{\eta(t)}(z). \tag{A.3}$$

Then ϕ^t, $t \in \mathbb{R}$, is a smooth path of symplectic embeddings $\phi^t : U \hookrightarrow \mathbb{R}^{2n}$ such that $\phi^0 = \mathrm{id}_U$ and $\phi^1 = \varphi$.

Step 2. Cutting off the isotopy. Since U is starshaped, it is contractible, and so the same holds true for all the open sets $\phi^t(U)$, $t \in \mathbb{R}$. We therefore find a smooth time-dependent Hamiltonian function

$$H : \bigcup_{t \in \mathbb{R}} \{t\} \times \phi^t(U) \to \mathbb{R} \tag{A.4}$$

generating the path ϕ^t, i.e., ϕ^t is the solution of the Hamiltonian system

$$\left. \begin{aligned} \tfrac{d}{dt}\phi^t(z) &= J\nabla H\left(t, \phi^t(z)\right), & z \in U, \ t \in \mathbb{R}, \\ \phi^0(z) &= z, & z \in U. \end{aligned} \right\} \tag{A.5}$$

Here, J denotes the standard complex structure defined by

$$\omega_0(z, w) = \langle Jz, w \rangle, \quad z, w \in \mathbb{R}^{2n}.$$

Fix now a subset A of U whose closure $\bar{A}$ in $\mathbb{R}^{2n}$ is contained in U. Since U is bounded, the set $\bar{A}$ is compact, and so the set

$$K = \bigcup_{t \in [0,1]} \{t\} \times \phi^t(\bar{A}) \subset \mathbb{R} \times \mathbb{R}^{2n}$$

is also compact and hence bounded. We therefore find a bounded neighbourhood V of K which is open in $\mathbb{R} \times \mathbb{R}^{2n}$ and is contained in the set $\bigcup_{t \in \mathbb{R}}\{t\} \times \phi^t(U)$. By Whitney's Theorem, there exists a smooth function f on $\mathbb{R} \times \mathbb{R}^{2n}$ which is equal to 1 on K and vanishes outside V. Since V is bounded, the function $fH : \mathbb{R} \times \mathbb{R}^{2n} \to \mathbb{R}$ has compact support, and so the Hamiltonian system associated with fH can be solved for all $t \in \mathbb{R}$. We define Φ_A to be the resulting time-1-map. Then Φ_A is a globally defined symplectomorphism of $\mathbb{R}^{2n}$ with compact support, and since $f \equiv 1$ on K we have $\Phi_A|_A = \phi^1|_A = \varphi|_A$. The proof of Proposition A.1 is thus complete. $\square$

Addendum: An extension of the Extension after Restriction Principle to unbounded starshaped domains. In Proposition A.1 we assumed that U is bounded. In this addendum we discuss how essential this assumption is.

Definition A.2. Consider a symplectic embedding $\varphi : U \hookrightarrow \mathbb{R}^{2n}$ of a starshaped domain $U \subset \mathbb{R}^{2n}$. We say that the pair (U, φ) has the *extension property* if for each subset $A \subset U$ whose closure in $\mathbb{R}^{2n}$ is contained in U there exists a Hamiltonian diffeomorphism Φ_A of $\mathbb{R}^{2n}$ such that $\Phi_A|_A = \varphi|_A$.

By Proposition A.1 the pair (U, φ) has the extension property whenever U is bounded. As the following example shows, this is not true for all symplectic embeddings of starshaped domains.

Example A.3. We let $U \subset \mathbb{R}^2$ be the strip $]1, \infty[\times] - 1, 1[$. The symplectic embedding $\varphi \colon U \hookrightarrow \mathbb{R}^2$ defined by $\varphi(x, y) = (1/x, -x^2 y)$ maps $(k, 0)$ to $(1/k, 0)$, $k = 2, 3, \ldots$, and so there does not exist any subset A of U containing the set $\{(k, 0) \mid k = 2, 3, \ldots\}$ for which $\varphi|_A$ extends to a diffeomorphism of $\mathbb{R}^2$. $\diamond$

Observe that if (U, φ) has the extension property, then φ is proper in the sense that each subset $A \subset U$ whose closure in $\mathbb{R}^{2n}$ is contained in U and whose image $\varphi(A)$ is bounded is bounded. The map φ in Example A.3 is not proper in this sense. However, the map φ in the following example is proper in this sense, and still (U, φ) does not have the extension property.

Example A.4. Let $U \subset \mathbb{R}^2$ be the strip $\mathbb{R} \times] - 1, 0[$, and let

$$A = \left\{ (x, y) \in U \mid \left| y + \tfrac{1}{2} \right| \leq f(x) \right\}$$

where $f \colon \mathbb{R} \to]0, \tfrac{1}{2}[$ is a smooth function such that

$$\int_{\mathbb{R}} \left(\tfrac{1}{2} - f(x) \right) dx < \infty, \tag{A.6}$$

cf. Figure A.1. Using the method used in Step 4 of Section 3.2 we find a symplectic embedding $\varphi \colon U \hookrightarrow \mathbb{R}^2$ such that

$$\varphi(x, y) = (x, y) \text{ if } x \geq 1 \quad \text{and} \quad \varphi(x, y) = (-x, -y) \text{ if } x \leq -1,$$

cf. Figure A.1. In view of the estimate (A.6) the component C of $\mathbb{R}^2 \setminus \varphi(A)$ which contains the point $(1, 0)$ has finite volume. Any symplectomorphism Φ_A of $\mathbb{R}^2$ such that $\Phi_A|_A = \varphi|_A$ would map the "upper" component of $\mathbb{R}^2 \setminus A$, which has infinite volume, to C. This is impossible. $\diamond$

Example A.4 shows that the assumption (A.7) on φ in Theorem A.6 below cannot be omitted. For technical reasons in the proof of Theorem A.6 we shall also impose a mild convexity condition on the starshaped domain U. The length of a smooth curve $\gamma \colon [0, 1] \to \mathbb{R}^{2n}$ is defined by

$$\text{length}(\gamma) := \int_0^1 \left| \gamma'(s) \right| ds.$$

On any domain $U \subset \mathbb{R}^{2n}$ we define a distance function $d_U \colon U \times U \to \mathbb{R}$ by

$$d_U(z, z') := \inf \{ \text{length}(\gamma) \}$$

where the infimum is taken over all smooth curves $\gamma \colon [0, 1] \to U$ with $\gamma(0) = z$ and $\gamma(1) = z'$. Then $\left| z - z' \right| \leq d_U(z, z')$ for all $z, z' \in U$.

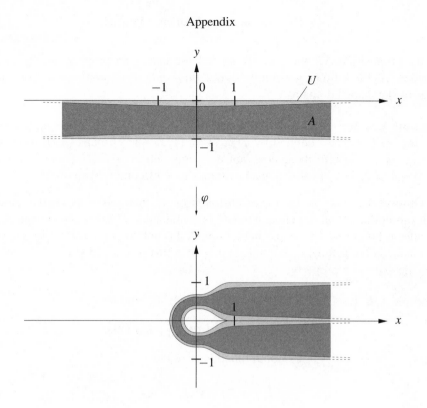

Figure A.1. A pair (U, φ) which does not have the extension property.

Definition A.5. We say that a domain $U \subset \mathbb{R}^{2n}$ is a *Lipschitz domain* if there exists a constant $\lambda > 0$ such that

$$d_U(z, z') \leq \lambda |z - z'| \quad \text{for all } z, z' \in U.$$

Each *convex* domain $U \subset \mathbb{R}^{2n}$ is a Lipschitz domain with Lipschitz constant $\lambda = 1$. Figure A.2 shows a starshaped domain in $\mathbb{R}^2$ with smooth boundary which is not a Lipschitz domain.

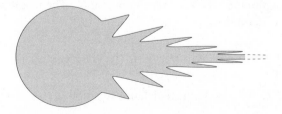

Figure A.2. A starshaped domain with smooth boundary which is not a Lipschitz domain.

Theorem A.6. *Assume that* $\varphi \colon U \hookrightarrow \mathbb{R}^{2n}$ *is a symplectic embedding of a starshaped Lipschitz domain* $U \subset \mathbb{R}^{2n}$ *such that there exists a constant* $L > 0$ *satisfying*

$$\left|\varphi(z) - \varphi(z')\right| \geq L \left|z - z'\right| \quad \text{for all } z, z' \in U. \tag{A.7}$$

Then the pair (U, φ) *has the extension property.*

Outline of the proof. One first easily verifies that one can assume that 0 is a star point of U and that $\varphi(0) = 0$ and $d\varphi(0) = \mathrm{id}$. Then apply Alexander's trick as in Step 1 of the proof of Proposition A.1 and consider a function H as in (A.4) satisfying (A.5).

Fix a subset A of U whose closure in $\mathbb{R}^{2n}$ is contained in U. If the set U is not bounded, then A does not need to be relatively compact, and so there might be no cut off fH of H whose Hamiltonian flow exists for all $t \in [0, 1]$. One therefore needs to extend the Hamiltonian H more carefully. One first verifies that assumption (A.7) on φ implies that ∇H is linearly bounded. Since it is not clear how to extend a linearly bounded gradient field to a linearly bounded gradient field, one then passes to the function

$$G(t, w) = \frac{H(t, w)}{g\left(|w|\right)}$$

where $g\left(|w|\right) = |w|$ for $|w|$ large. The assumption that U is a Lipschitz domain implies that G is Lipschitz continuous in w and can hence be extended to a continuous function $\hat{G}$ on $[0, 1] \times \mathbb{R}^{2n}$ which is Lipschitz continuous in w. After smoothing $\hat{G}$ in w to $\widetilde{G}$ one obtains a continuous extension $\widetilde{H}(t, w) = g\left(|w|\right) \widetilde{G}(t, w)$ of H which may not be smooth in t but is smooth in w and has linearly bounded gradient. The Hamiltonian system associated with $\widetilde{H}$ can therefore be solved for all $t \in [0, 1]$, and the resulting time-1-map Φ_A is a symplectomorphism of $\mathbb{R}^{2n}$ such that $\Phi_A|_A = \varphi|_A$. We refer to [71] for the detailed construction of Φ_A. In order to find a *smooth* Hamiltonian function generating Φ_A one proceeds as in the proof of Proposition A.1: Applying Alexander's trick one obtains a smooth path Φ^t, $t \in \mathbb{R}$, of symplectomorphisms of $\mathbb{R}^{2n}$ such that $\Phi^0 = \mathrm{id}$ and $\Phi^1 = \Phi_A$, and since $\mathbb{R}^{2n}$ is contractible, there exists a smooth Hamiltonian function $H \colon \mathbb{R} \times \mathbb{R}^{2n} \to \mathbb{R}$ generating this path. $\qquad \square$

B Flexibility for volume preserving embeddings

We endow each open subset U of $\mathbb{R}^n$ with the volume form

$$\Omega_0 = dx_1 \wedge \cdots \wedge dx_n.$$

A *volume form* Ω on an arbitrary smooth n-dimensional manifold M is a nowhere vanishing differential n-form on M. A smooth embedding $\varphi \colon U \hookrightarrow M$ is called *volume preserving* if $\varphi^* \Omega = \Omega_0$. The volume forms Ω_0 and Ω orient each open subset of $\mathbb{R}^n$ and M. We write $\mathrm{Vol}(U, \Omega_0) = \int_U \Omega_0$ and $\mathrm{Vol}(M, \Omega) = \int_M \Omega$. This appendix, which is taken from [72], contains a proof of

Theorem B.1. *Consider an open subset U of $\mathbb{R}^n$ and a connected n-dimensional manifold M endowed with a volume form Ω. Then there exists a volume preserving embedding $\varphi\colon U \hookrightarrow M$ if and only if $\mathrm{Vol}(U, \Omega_0) \leq \mathrm{Vol}(M, \Omega)$.*

If U is a bounded subset whose boundary has zero measure and if $\mathrm{Vol}(U, \Omega_0) < \mathrm{Vol}(M, \Omega)$, Theorem B.1 is an easy consequence of Moser's deformation method, for which we refer to [66] or [39, Chapter 1.3]. Moreover, if U is a ball and M is compact, Theorem B.1 has been proved by Katok in [43]. Our point is that Theorem B.1 holds true for an arbitrary open subset of $\mathbb{R}^n$ and an arbitrary connected manifold even in case that the volumes are equal or infinite.

Proof of Theorem B.1.   Assume first that $\varphi\colon U \hookrightarrow M$ is a smooth embedding such that $\varphi^*\Omega = \Omega_0$. Then

$$\mathrm{Vol}(U, \Omega_0) = \int_U \Omega_0 = \int_U \varphi^*\Omega = \int_{\varphi(U)} \Omega \leq \int_M \Omega = \mathrm{Vol}(M, \Omega).$$

Assume now that $\mathrm{Vol}(U, \Omega_0) \leq \mathrm{Vol}(M, \Omega)$. We are going to construct a smooth embedding $\varphi\colon U \hookrightarrow M$ such that $\varphi^*\Omega = \Omega_0$.

We abbreviate the Lebesgue measure of a measurable subset V of $\mathbb{R}^n$ by $|V|$, so that $|V| = \mathrm{Vol}(V, \Omega_0)$ if V is open, and we write $\overline{V}$ for the closure of V in $\mathbb{R}^n$. Moreover, we denote by B_r the open ball in $\mathbb{R}^n$ of radius r centred at the origin.

Proposition B.2. *Assume that V is a non-empty open subset of $\mathbb{R}^n$. Then there exists a smooth embedding $\sigma\colon V \hookrightarrow \mathbb{R}^n$ such that $|\mathbb{R}^n \setminus \sigma(V)| = 0$.*

Proof. We choose an increasing sequence

$$V_1 \subset V_2 \subset \cdots \subset V_k \subset V_{k+1} \subset \cdots$$

of non-empty open subsets of V such that $\overline{V_k} \subset V_{k+1}$ for $k \geq 1$ and $\bigcup_{k=1}^{\infty} V_k = V$. To fix the ideas, we assume that the V_k have smooth boundaries.

Let $\sigma_1\colon V_2 \hookrightarrow \mathbb{R}^n$ be a smooth embedding such that $\sigma_1(V_1) \subset B_1$ and

$$|B_1 \setminus \sigma_1(V_1)| \leq 2^{-1}.$$

Since $\overline{V_1} \subset V_2$ and $\overline{\sigma_1(V_1)} \subset \overline{B_1} \subset B_2$, we find a smooth embedding $\sigma_2\colon V_3 \hookrightarrow \mathbb{R}^n$ such that $\sigma_2|_{V_1} = \sigma_1|_{V_1}$ and $\sigma_2(V_2) \subset B_2$ and

$$|B_2 \setminus \sigma_2(V_2)| \leq 2^{-2}.$$

Arguing by induction we find smooth embeddings $\sigma_k\colon V_{k+1} \hookrightarrow \mathbb{R}^n$ such that $\sigma_k|_{V_{k-1}} = \sigma_{k-1}|_{V_{k-1}}$ and $\sigma_k(V_k) \subset B_k$ and

$$|B_k \setminus \sigma_k(V_k)| \leq 2^{-k}, \tag{B.1}$$

$k \geq 1$. The map $\sigma : V \to \mathbb{R}^n$ defined by $\sigma|_{V_k} = \sigma_k|_{V_k}$ is a well defined smooth embedding of V into $\mathbb{R}^n$. Moreover, the inclusions $\sigma_k(V_k) \subset \sigma(V)$ and the estimates (B.1) imply that

$$|B_k \setminus \sigma(V)| \leq |B_k \setminus \sigma_k(V_k)| \leq 2^{-k},$$

and so

$$\left| \mathbb{R}^n \setminus \sigma(V) \right| = \lim_{k \to \infty} |B_k \setminus \sigma(V)| = 0.$$

This completes the proof of Proposition B.2.   □

Our next goal is to construct a smooth embedding of $\mathbb{R}^n$ into the connected n-dimensional manifold M such that the complement of the image has measure zero. If M is compact, such an embedding has been obtained by Ozols [68] and Katok [43, Proposition 1.3]. While Ozols combines an engulfing method with tools from Riemannian geometry, Katok successively exhausts a smooth triangulation of M. Both approaches can be generalized to the case of an arbitrary connected manifold M, and we shall follow Ozols.

We abbreviate $\mathbb{R}_{>0} = \{r \in \mathbb{R} \mid r > 0\}$ and $\overline{\mathbb{R}}_{>0} = \mathbb{R}_{>0} \cup \{\infty\}$. We endow $\overline{\mathbb{R}}_{>0}$ with the topology whose base of open sets consists of the intervals $]a, b[\subset \mathbb{R}_{>0}$ and the subsets of the form $]a, \infty] =]a, \infty[\cup \{\infty\}$. We denote the Euclidean norm on $\mathbb{R}^n$ by $\| \cdot \|$ and the unit sphere in $\mathbb{R}^n$ by S_1.

Proposition B.3. *Endow $\mathbb{R}^n$ with its standard smooth structure, let $\mu : S_1 \to \overline{\mathbb{R}}_{>0}$ be a continuous function and let*

$$S = \left\{ x \in \mathbb{R}^n \,\Big|\, 0 \leq \|x\| < \mu \left(\frac{x}{\|x\|} \right) \right\}$$

be the starshaped domain associated with μ. Then S is diffeomorphic to $\mathbb{R}^n$.

Remark B.4. The diffeomorphism guaranteed by Proposition B.3 may be chosen such that the rays emanating from the origin are preserved.   ◇

Proof of Proposition B.3. If $\mu(S_1) = \{\infty\}$, there is nothing to prove. In the case that μ is bounded, Proposition B.3 has been proved by Ozols [68]. In the case that neither $\mu(S_1) = \{\infty\}$ nor μ is bounded, Ozols's proof readily extends to this situation. Using his notation, the only modifications needed are: Require in addition that $r_0 < 1$ and that $\epsilon_1 < 2$, and define continuous functions $\tilde{\mu}_i : S_1 \to \mathbb{R}_{>0}$ by

$$\tilde{\mu}_i = \min \left\{ i, \, \mu - \epsilon_i + \tfrac{\delta_i}{2} \right\}.$$

With these minor adaptations the proof in [68] applies word-by-word.   □

In the following we shall use some basic Riemannian geometry. We refer to [44] for basic notions and results in Riemannian geometry. Consider an n-dimensional complete Riemannian manifold (N, g). We denote the cut locus of a point $p \in N$ by $C(p)$.

Corollary B.5. *The maximal normal neighbourhood $N \setminus C(p)$ of any point p in an n-dimensional complete Riemannian manifold (N, g) is diffeomorphic to $\mathbb{R}^n$ endowed with its standard smooth structure.*

Proof. Fix $p \in N$. We identify the tangent space $(T_p N, g(p))$ with Euclidean space $\mathbb{R}^n$ by a (linear) isometry. Let $\exp_p \colon \mathbb{R}^n \to N$ be the exponential map at p with respect to g, and let S_1 be the unit sphere in $\mathbb{R}^n$. We define the function $\mu \colon S_1 \to \overline{\mathbb{R}}_{>0}$ by

$$\mu(x) = \inf\{t > 0 \mid \exp_p(tx) \in C(p)\}. \tag{B.2}$$

Since the Riemannian metric g is complete, the function μ is continuous [44, VIII, Theorem 7.3]. Let $S \subset \mathbb{R}^n$ be the starshaped domain associated with μ. In view of Proposition B.3 the set S is diffeomorphic to $\mathbb{R}^n$, and in view of [44, VIII, Theorem 7.4 (3)] we have $\exp_p(S) = N \setminus C(p)$. Therefore, $N \setminus C(p)$ is diffeomorphic to $\mathbb{R}^n$. $\square$

A main ingredient of our proof of Theorem B.1 are the following two special cases of a theorem of Greene and Shiohama [29].

Proposition B.6. (i) *Assume that Ω_1 is a volume form on the connected open subset U of $\mathbb{R}^n$ such that $\mathrm{Vol}(U, \Omega_1) = |U| < \infty$. Then there exists a diffeomorphism ψ of U such that $\psi^* \Omega_1 = \Omega_0$.*

(ii) *Assume that Ω_1 is a volume form on $\mathbb{R}^n$ such that $\mathrm{Vol}(\mathbb{R}^n, \Omega_1) = \infty$. Then there exists a diffeomorphism ψ of $\mathbb{R}^n$ such that $\psi^* \Omega_1 = \Omega_0$.*

End of the proof of Theorem B.1. Let $U \subset \mathbb{R}^n$ and (M, Ω) be as in Theorem B.1. After enlarging U, if necessary, we can assume that $|U| = \mathrm{Vol}(M, \Omega)$. We set $N = M \setminus \partial M$. Then

$$|U| = \mathrm{Vol}(M, \Omega) = \mathrm{Vol}(N, \Omega). \tag{B.3}$$

Since N is a connected manifold without boundary, there exists a complete Riemannian metric g on N. Indeed, according to a theorem of Whitney [86], N can be embedded as a closed submanifold in some $\mathbb{R}^m$. We can then take the induced Riemannian metric. A direct and elementary proof of the existence of a complete Riemannian metric was given by Nomizu and Ozeki in [67].

Fix a point $p \in N$. As in the proof of Corollary B.5 we identify $(T_p N, g(p))$ with $\mathbb{R}^n$ and define the function $\mu \colon S_1 \to \overline{\mathbb{R}}_{>0}$ as in (B.2). Using polar coordinates on $\mathbb{R}^n$ we see from Fubini's Theorem that the set

$$\widetilde{C}(p) = \{\mu(x)x \mid x \in S_1\} \subset \mathbb{R}^n$$

has measure zero, and so $C(p) = \exp_p\left(\widetilde{C}(p)\right)$ also has measure zero (see [13, VI, Corollary 1.14]). It follows that

$$\mathrm{Vol}(N \setminus C(p), \Omega) = \mathrm{Vol}(N, \Omega). \tag{B.4}$$

According to Corollary B.5 there exists a diffeomorphism

$$\delta \colon \mathbb{R}^n \to N \setminus C(p).$$

After composing δ with a reflection of $\mathbb{R}^n$, if necessary, we can assume that δ is orientation preserving. In view of (B.3) and (B.4) we then have

$$|U| = \mathrm{Vol}(\mathbb{R}^n, \delta^* \Omega). \tag{B.5}$$

Case 1. $|U| < \infty$. Let $U_1, U_2, \ldots$ be the countably many components of U. Then $0 < |U_i| < \infty$ for each i. Given numbers a and b with $-\infty \le a < b \le \infty$ we abbreviate the "open strip"

$$S_{a,b} = \{(x_1, \ldots, x_n) \in \mathbb{R}^n \mid a < x_1 < b\}.$$

In view of the identity (B.5) we have

$$\sum_{i \ge 1} |U_i| = |U| = \mathrm{Vol}(\mathbb{R}^n, \delta^* \Omega).$$

We can therefore inductively define $a_0 = -\infty$ and $a_i \in \,] - \infty, \infty]$ by

$$\mathrm{Vol}(S_{a_{i-1}, a_i}, \delta^* \Omega) = |U_i|.$$

Abbreviating $S_i = S_{a_{i-1}, a_i}$ we then have $\mathbb{R}^n = \bigcup_{i \ge 1} \overline{S_i}$.

For each $i \ge 1$ we choose an orientation preserving diffeomorphism $\tau_i \colon \mathbb{R}^n \to S_i$. In view of Proposition B.2 we find a smooth embedding $\sigma_i \colon U_i \hookrightarrow \mathbb{R}^n$ such that $\mathbb{R}^n \setminus \sigma_i(U_i)$ has measure zero. After composing σ_i with a reflection of $\mathbb{R}^n$, if necessary, we can assume that σ_i is orientation preserving. Using the definition of the volume, we can now conclude that

$$\mathrm{Vol}(U_i, \sigma_i^* \tau_i^* \delta^* \Omega) = \mathrm{Vol}(\sigma_i(U_i), \tau_i^* \delta^* \Omega) = \mathrm{Vol}(\mathbb{R}^n, \tau_i^* \delta^* \Omega) = \mathrm{Vol}(S_i, \delta^* \Omega) = |U_i|.$$

In view of Proposition B.6 (i) we therefore find a diffeomorphism ψ_i of U_i such that

$$\psi_i^* \left(\sigma_i^* \tau_i^* \delta^* \Omega \right) = \Omega_0. \tag{B.6}$$

We define $\varphi_i \colon U_i \hookrightarrow M$ to be the composition of diffeomorphisms and smooth embeddings

$$U_i \xrightarrow{\psi_i} U_i \xrightarrow{\sigma_i} \mathbb{R}^n \xrightarrow{\tau_i} S_i \subset \mathbb{R}^n \xrightarrow{\delta} N \setminus C(p) \subset M.$$

The identity (B.6) implies that $\varphi_i^* \Omega = \Omega_0$. The smooth embedding

$$\varphi = \coprod \varphi_i \colon U = \coprod U_i \hookrightarrow M$$

therefore satisfies $\varphi^* \Omega = \Omega_0$.

Case 2. $|U| = \infty$. In view of (B.5) we have $\mathrm{Vol}(\mathbb{R}^n, \delta^*\Omega) = \infty$. Proposition B.6 (ii) shows that there exists a diffeomorphism ψ of $\mathbb{R}^n$ such that

$$\psi^*\delta^*\Omega = \Omega_0. \tag{B.7}$$

We define $\varphi \colon U \hookrightarrow M$ to be the composition of inclusions and diffeomorphisms

$$U \subset \mathbb{R}^n \overset{\psi}{\to} \mathbb{R}^n \overset{\delta}{\to} N \setminus C(p) \subset M.$$

The identity (B.7) implies that $\varphi^*\Omega = \Omega_0$. The proof of Theorem B.1 is complete. $\square$

C   Symplectic capacities and the invariants c_B and c_C

In the first two sections of this appendix we collect and prove those results on symplectic capacities which are relevant for this book. We refer to [4], [16], [20], [21], [36], [37], [38], [39], [48], [62], [77], [78], [82] for more information on symplectic capacities. In Section C.3 we compare the invariants c_B and c_C studied in Chapters 2 to 7 with symplectic capacities. In the last two sections we compare c_B with the symplectic diameter and interpret c_C as a symplectic projection invariant.

C.1   Intrinsic and extrinsic symplectic capacities on $\mathbb{R}^{2n}$.   We say that a subset $S \subset \mathbb{R}^{2n}$ symplectically embeds into $T \subset \mathbb{R}^{2n}$ if there exists a symplectic embedding φ of an open neighbourhood of S into $\mathbb{R}^{2n}$ such that $\varphi(S) \subset T$. We again write $Z^{2n}(a)$ for the symplectic cylinder $D(a) \times \mathbb{R}^{2n-2}$. Recall the

Definition C.1. An *intrinsic symplectic capacity on* $(\mathbb{R}^{2n}, \omega_0)$ is a map c associating with each subset S of $\mathbb{R}^{2n}$ a number $c(S) \in [0, \infty]$ in such a way that the following axioms are satisfied.

A1.   Monotonicity: $c(S) \leq c(T)$ if S symplectically embeds into T.

A2.   Conformality: $c(\lambda S) = \lambda^2 c(S)$ for all $\lambda \in \mathbb{R} \setminus \{0\}$.

A3.   Nontriviality: $0 < c(B^{2n}(\pi))$ and $c(Z^{2n}(\pi)) < \infty$.

An intrinsic symplectic capacity c on $\mathbb{R}^{2n}$ is *normalized* if

A3′.   Normalization: $c(B^{2n}(\pi)) = c(Z^{2n}(\pi)) = \pi$.

Remark C.2. In view of the monotonicity axiom, symplectic embedding results lead to constraints on *the space of* intrinsic symplectic capacities on $\mathbb{R}^{2n}$. We refer to [16] for more information on this application of symplectic embedding results. $\diamond$

For any subset S of $\mathbb{R}^{2n}$ we define the Gromov width $w_G(S)$ and the cylindrical capacity $z(S)$ by

$$w_G(S) := \sup\{a \mid B^{2n}(a) \text{ symplectically embeds into } S\},$$

$$z(S) := \inf\{a \mid S \text{ symplectically embeds into } Z^{2n}(a)\}.$$

It follows from Gromov's Nonsqueezing Theorem stated in Section 1.2.2 that both w_G and z are normalized intrinsic symplectic capacities on $\mathbb{R}^{2n}$. Another example of a normalized intrinsic symplectic capacity on $\mathbb{R}^{2n}$ is the Hofer–Zehnder capacity, [39].

Proposition C.3. *For any subset S of $\mathbb{R}^{2n}$ and any normalized intrinsic symplectic capacity c on $\mathbb{R}^{2n}$ we have*

$$w_G(S) \le c(S) \le z(S).$$

Proof. If $B^{2n}(a) \hookrightarrow S$ is a symplectic embedding, we conclude from monotonicity, conformality and normalization that $a = B^{2n}(a) \le c(S)$. Taking the supremum, we find $w_G(S) \le c(S)$. Similarly, if $S \hookrightarrow Z^{2n}(a)$ is a symplectic embedding, we conclude from the axioms that $c(S) \le c(Z^{2n}(a)) = a$. Taking the infimum we find $c(S) \le z(S)$. $\qquad\square$

If S is an open connected subset of $\mathbb{R}^2$, the inequalities in Proposition C.3 are equalities, see Corollary C.10 (i). For $n \ge 2$, however, the Symplectic Hedgehog Theorem stated in Section 1.2.2 shows that there exist bounded starshaped domains $S \subset \mathbb{R}^{2n}$ with arbitrarily small Gromov width and $z(S) = 1$. Still, for an ellipsoid and, more generally, a convex Reinhardt domain all normalized symplectic capacities on $\mathbb{R}^{2n}$ coincide, [37]. It is conjectured that this holds for any bounded convex domain, [85]. The subsequent proposition due to Viterbo, [85], confirms this conjecture up to a constant.

Proposition C.4 (Viterbo). *For any two normalized intrinsic symplectic capacities c and c' on $(\mathbb{R}^{2n}, \omega_0)$ and any bounded convex domain $K \subset \mathbb{R}^{2n}$ we have*

$$c'(K) \le n^2 c(K).$$

Moreover, $c'(K) \le n\, c(K)$ if K is centrally symmetric.

Proof. The proof is similar to the proof of Proposition 8.2.2. Let E be John's ellipsoid of K satisfying $\frac{1}{n}E \subset K \subset E$. According to [39, p. 40], there exists a linear symplectomorphism of $\mathbb{R}^{2n}$ mapping E to an ellipsoid $E' = E(a_1, \ldots, a_n)$, and Gromov's Nonsqueezing Theorem and Proposition C.3 show that $c(E') = c'(E')$, so that $c(E) = c'(E)$. Together with the monotonicity and conformality axioms we conclude

$$c'(K) \le c'(E) = c(E) = n^2 c\left(\tfrac{1}{n}E\right) \le n^2 c(K).$$

If K is centrally symmetric, then $\frac{1}{\sqrt{n}} E \subset K \subset E$, so that $c'(K) \leq n\,c(K)$.   □

As in Chapter 2 we denote by $\mathcal{D}(n)$ the group of symplectomorphisms of $(\mathbb{R}^{2n}, \omega_0)$. We also recall the

Definition C.5. An *extrinsic symplectic capacity on* $(\mathbb{R}^{2n}, \omega_0)$ is a map c associating with each subset S of $\mathbb{R}^{2n}$ a number $c(S) \in [0, \infty]$ in such a way that the following axioms are satisfied.

A1. Monotonicity: $c(S) \leq c(T)$ if there exists $\varphi \in \mathcal{D}(n)$ such that $\varphi(S) \subset T$.

A2. Conformality: $c(\lambda S) = \lambda^2 c(S)$ for all $\lambda \in \mathbb{R} \setminus \{0\}$.

A3. Nontriviality: $0 < c(B^{2n}(\pi))$ and $c(Z^{2n}(\pi)) < \infty$.

An extrinsic symplectic capacity c on $\mathbb{R}^{2n}$ is *normalized* if

A3'. Normalization: $c(B^{2n}(\pi)) = c(Z^{2n}(\pi)) = \pi$.

Comparing Definitions C.1 and C.5 we see that any intrinsic symplectic capacity on $\mathbb{R}^{2n}$ is an extrinsic symplectic capacity on $\mathbb{R}^{2n}$. The converse is not true as we shall see in Proposition C.7.

For any subset S of $\mathbb{R}^{2n}$ we define

$$\hat{w}_G(S) := \sup\{a \mid \text{there exists } \varphi \in \mathcal{D}(n) \text{ such that } \varphi(B^{2n}(a)) \subset S\},$$

$$\hat{z}(S) := \inf\{a \mid \text{there exists } \varphi \in \mathcal{D}(n) \text{ such that } \varphi(S) \subset Z^{2n}(a)\}.$$

It follows again from Gromov's Nonsqueezing Theorem that both $\hat{w}_G$ and $\hat{z}$ are normalized extrinsic symplectic capacities on $\mathbb{R}^{2n}$. Other examples of normalized extrinsic symplectic capacities on $\mathbb{R}^{2n}$ are the first Ekeland–Hofer capacity c_1, which was introduced in [20] and used in the proof of Theorem 2.2.4, and the displacement energy e introduced in [38]. The simply connected hull $\hat{S}$ of a bounded subset S of $\mathbb{R}^{2n}$ is the union of its closure $\overline{S}$ and the bounded components of $\mathbb{R}^{2n} \setminus \overline{S}$.

Proposition C.6. *For any subset S of $\mathbb{R}^{2n}$ and any normalized extrinsic symplectic capacity c on $\mathbb{R}^{2n}$ we have*

$$\hat{w}_G(S) \leq c(S) \leq \hat{z}(S).$$

Moreover, $\hat{w}_G(S) = w_G(S)$ and $\hat{z}(S) = \hat{z}(\hat{S})$.

Proof.   The inequalities $\hat{w}_G(S) \leq c(S) \leq \hat{z}(S)$ are proved in the same way as Proposition C.3.

The inequality $\hat{w}_G(S) \leq w_G(S)$ is obvious. In order to prove $\hat{w}_G(S) \geq w_G(S)$ we fix $\epsilon > 0$ and assume that φ symplectically embeds $B^{2n}(a)$ into S. According to Proposition A.1 there exists $\Phi \in \mathcal{D}(n)$ such that $\Phi|_{B^{2n}(a-\epsilon)} = \varphi|_{B^{2n}(a-\epsilon)}$. In particular, $\Phi\left(B^{2n}(a-\epsilon)\right) \subset S$, and so $\hat{w}_G(S) \geq a - \epsilon$. Since $\epsilon > 0$ was arbitrary, we conclude that $\hat{w}_G(S) \geq a$. Taking the supremum over a we find $\hat{w}_G(S) \geq w_G(S)$.

In view of the monotonicity of $\hat{z}$ we have $\hat{z}(S) \leq \hat{z}(\hat{S})$. In order to prove $\hat{z}(\hat{S}) \leq \hat{z}(S)$ we assume that $\varphi \in \mathcal{D}(n)$ embeds S into $Z^{2n}(a)$. Then $\varphi(\overline{S}) \subset \overline{Z^{2n}(a)}$. Moreover, if U is a bounded component of $\mathbb{R}^{2n} \setminus \overline{S}$, then $\varphi(U)$ is a bounded component of $\mathbb{R}^{2n} \setminus \varphi(\overline{S})$, and so $\varphi(U) \subset Z^{2n}(a)$. It follows that $\varphi(\hat{S}) \subset \overline{Z^{2n}(a)}$, and hence $\hat{z}(\hat{S}) \leq a$. Taking the infimum over a we conclude that $\hat{z}(\hat{S}) \leq \hat{z}(S)$. The proof of Proposition C.6 is complete. $\qquad\square$

The next proposition shows that neither the first Ekeland–Hofer capacity c_1 nor the displacement energy e nor the outer cylindrical capacity $\hat{z}$ are intrinsic symplectic capacities on $\mathbb{R}^{2n}$. Let $S^1 = \left\{(u, v) \mid u^2 + v^2 = 1\right\}$ be the unit circle.

Proposition C.7. *For the standard Lagrangian torus $T^n = S^1 \times \cdots \times S^1$ in $\mathbb{R}^{2n}$ we have*

$$c_1(T^n) = e(T^n) = \hat{z}(T^n) = \pi$$

while $c(T^n) = 0$ for any intrinsic symplectic capacity c on $\mathbb{R}^{2n}$.

Proof. According to Théorème (b) on page 43 of [78] or [79] we have $c_1(T^n) = \pi$, Theorem 1.6 (i) in [38] implies $c_1(T^n) \leq e(T^n)$, and Proposition C.6 shows that $e(T^n) \leq \hat{z}(T^n)$. Since clearly $\hat{z}(T^n) \leq \pi$, we conclude that

$$\pi = c_1(T^n) \leq e(T^n) \leq \hat{z}(T^n) \leq \pi$$

and so the first assertion follows. In order to prove the second assertion we assume that c is an intrinsic symplectic capacity on $\mathbb{R}^{2n}$. Fix $\epsilon > 0$. In view of Theorem C.8 (ii) below there exists a symplectic embedding $\varphi \colon S^1 \hookrightarrow D(\epsilon)$. The product

$$\varphi \times \cdots \times \varphi \colon T^n \hookrightarrow D(\epsilon) \times \cdots \times D(\epsilon) \subset Z^{2n}(\epsilon)$$

symplectically embeds T^n into $Z^{2n}(\epsilon)$, and so

$$c(T^n) \leq c\left(Z^{2n}(\epsilon)\right) = \frac{\epsilon}{\pi} c\left(Z^{2n}(\pi)\right).$$

Since $\epsilon > 0$ was arbitrary and since $c\left(Z^{2n}(\pi)\right) < \infty$ it follows that $c(T^n) = 0$ as claimed. $\qquad\square$

Notice that for $n \geq 2$, Proposition C.7 implies the remarkable Nonsqueezing Theorem mentioned in Section 1.2.2 and implying the Symplectic Hedgehog Theorem: Even though the volume of T^n in $\mathbb{R}^{2n}$ vanishes and T^n does not bound an open set, there is no symplectomorphism of $\mathbb{R}^{2n}$ mapping T^n into $Z^{2n}(\pi)$.

C.2 Symplectic capacities on $\mathbb{R}^2$. If $n \geq 2$ the computation of a capacity of a subset of $\mathbb{R}^{2n}$ is usually a difficult problem. In this section we show that for subsets of $\mathbb{R}^2$ the situation is different.

Since $B^2(\pi) = Z^2(\pi)$ we can assume that any (intrinsic or extrinsic) symplectic capacity c on $\mathbb{R}^2$ is normalized. Given a subset S of $\mathbb{R}^2$ we denote by Int S and $\bar{S}$ its interior and its closure and by S_i the components of Int S. If S is bounded, $\hat{S}$ denotes again the simply connected hull of S. The outer Lebesgue measure of S is denoted by $\bar{\mu}(S)$, and if S is measurable, $\mu(S)$ denotes its Lebesgue measure.

Theorem C.8. *Consider a subset S of $\mathbb{R}^2$.*

  (i) $w_G(S) = \sup_i \mu(S_i)$.

  (ii) $z(S) = \bar{\mu}(S)$.

  (iii) *If S is bounded, $\hat{z}(S) = \mu(\hat{S})$. If S is unbounded, $\hat{z}(S) = \infty$.*

Proof. (i) We abbreviate $s := \sup_i \mu(S_i)$. If φ symplectically embeds $D(a)$ into S, then $\varphi(D(a)) \subset S_i$ for some i, and so $w_G(S) \leq s$. In order to prove the reverse inequality we can assume that $s \in \,]0, \infty]$. If $s \in \,]0, \infty[$ we fix $\epsilon \in \,]0, s[$. Then there exists S_i such that $\mu(S_i) > s - \epsilon$. According to Proposition 1 in Section 1.2.2 we find a symplectic embedding of $D(s - \epsilon)$ into S_i, and so $w_G(S) \geq s - \epsilon$. Since $\epsilon \in \,]0, s[$ was arbitrary we conclude that $w_G(S) \geq s$. A similar argument shows that $w_G(S) = \infty$ if $s = \infty$. Assertion (i) thus follows.

  (ii) If φ symplectically embeds S into $D(a)$, then

$$\bar{\mu}(S) = \bar{\mu}(\varphi(S)) \leq \bar{\mu}(D(a)) = a.$$

Taking the infimum over a we find that $\bar{\mu}(S) \leq z(S)$. In order to prove the reverse inequality we can assume that $\bar{\mu}(S) < \infty$. Fix $\epsilon > 0$. In view of the definition of $\bar{\mu}(S)$ there exists an open neighbourhood U of S such that $\mu(U) \leq \bar{\mu}(S) + \epsilon$. By Proposition 1 stated in Section 1.2.2 we find a symplectic embedding φ of U into $D(\bar{\mu}(S) + \epsilon)$. Therefore $z(S) \leq \bar{\mu}(S) + \epsilon$. Since $\epsilon > 0$ was arbitrary we conclude that $z(S) \leq \bar{\mu}(S)$. Assertion (ii) thus follows.

  (iii) Using Proposition C.6, the identity in (ii) and the fact that $\hat{S}$ is measurable we find

$$\hat{z}(S) = \hat{z}(\hat{S}) \geq z(\hat{S}) = \bar{\mu}(\hat{S}) = \mu(\hat{S}).$$

In order to prove $\hat{z}(S) \leq \mu(\hat{S})$ we therefore need to show $\hat{z}(\hat{S}) \leq \mu(\hat{S})$. By construction, $\hat{S}$ is compact and $\mathbb{R}^2 \setminus \hat{S}$ is connected. Therefore, each path-component of $\hat{S}$ is simply connected. Since $\hat{S}$ is compact, $\mu(\hat{S})$ is finite. Fix $\epsilon > 0$.

Lemma C.9. *There exists a simply connected open neighbourhood W of $\hat{S}$ such that $\mu(W) \leq \mu(\hat{S}) + 2\epsilon$.*

Proof. We say that a subset of $\mathbb{R}^2$ is *elementary* if it is the union of finitely many open rectangles in $\mathbb{R}^2$. By definition of μ and since $\hat{S}$ is compact we find an elementary subset R of $\mathbb{R}^2$ such that $\hat{S} \subset R$ and $\mu(R) \leq \mu(\hat{S}) + \epsilon$. Let U be one of the finitely many components of R. The elementary set U is diffeomorphic to an open disc from which k closed discs have been removed. Since $\hat{S} \cap U$ is a compact subset of U all of whose path-components are simply connected, the set $U \setminus \hat{S}$ is open and connected. We therefore find k curves γ_i in $\overline{U} \setminus \hat{S}$ and elementary neighbourhoods N_i of γ_i such that the elementary subset $V := U \setminus \bigcup \overline{N_i}$ is a simply connected neighbourhood of $U \cap \hat{S}$. Applying this construction to each component of R we obtain disjoint simply connected elementary subsets $V_1, \ldots, V_l$ of R whose union contains $\hat{S}$. We finally choose $l - 1$ curves γ^i in $\mathbb{R}^2 \setminus \bigcup V_i$ and elementary neighbourhoods N^i of γ^i such that the elementary set

$$W := \bigcup_{i=1}^{l} V_i \cup \bigcup_{i=1}^{l-1} N^i$$

is simply connected and such that $\mu(\bigcup N^i) \leq \epsilon$. Then W is a simply connected neighbourhood of $\hat{S}$ and

$$\mu(W) \leq \mu\left(\bigcup V_i\right) + \mu\left(\bigcup N^i\right) \leq \mu(R) + \epsilon \leq \mu(\hat{S}) + 2\epsilon.$$

Lemma C.9 thus follows.   □

Let W be a simply connected open neighbourhood of $\hat{S}$ as guaranteed by Lemma C.9. Then W is diffeomorphic to $D(\mu(W))$. According to Proposition B.6 (i) we therefore find a symplectomorphism $\varphi \colon W \to D(\mu(W))$. Since $\varphi(\hat{S})$ is a compact subset of $D(\mu(W))$, we find $a \in \,]0, \mu(W)[$ such that $\varphi(S) \subset D(a)$. According to Proposition A.1 we find a symplectomorphism Φ of $\mathbb{R}^2$ such that $\Phi|_{D(a)} = \varphi^{-1}|_{D(a)}$. The map Φ^{-1} is then a symplectomorphism of $\mathbb{R}^2$ which embeds $\hat{S}$ into $D(a)$. It follows that $\hat{z}(\hat{S}) \leq a \leq \mu(W) \leq \mu(\hat{S}) + 2\epsilon$. Since $\epsilon > 0$ was arbitrary we conclude that $\hat{z}(\hat{S}) \leq \mu(\hat{S})$. This completes the proof of the first claim in assertion (iii).

If S is unbounded, then $\varphi(S)$ is unbounded for any diffeomorphism φ of $\mathbb{R}^2$, and so $\hat{z}(S) = \infty$. Assertion (iii) thus follows, and so the proof of Theorem C.8 is complete. □

Corollary C.10. (i) *Assume that S is a subset of $\mathbb{R}^2$ such that* Int S *is connected and* $\mu\,(\text{Int } S) = \bar{\mu}(S)$. *Then* $c(S) = \bar{\mu}(S)$ *for any normalized intrinsic symplectic capacity* c *on* $\mathbb{R}^2$. *If in addition* $\mu\,(\text{Int } S)$ *is finite, then S is measurable and so* $c(S) = \mu(S)$.

(ii) *Assume that S is a bounded subset of $\mathbb{R}^2$ such that* Int S *is connected and* $\mu\,(\text{Int } S) = \mu(\hat{S})$. *Then S is measurable and* $c(S) = \mu(S)$ *for any normalized extrinsic symplectic capacity* c *on* $\mathbb{R}^2$.

Proof. Let c be any normalized intrinsic symplectic capacity on $\mathbb{R}^2$. In view of Proposition C.3, (i) and (ii) in Theorem C.8 and the assumptions on S we have

$$w_G(S) \leq c(S) \leq z(S) = \bar{\mu}(S) = \mu(\text{Int } S) = w_G(S)$$

and so $c(S) = \bar{\mu}(S)$. If $\mu(\text{Int } S)$ is finite, the identities

$$\mu(\text{Int } S) = \bar{\mu}(S) = \inf\{\mu(U) \mid S \subset U, \ U \text{ open}\},$$

the monotonicity of $\bar{\mu}$ and the additivity of μ imply that $\bar{\mu}(S \setminus \text{Int } S) = 0$. Therefore $S = \text{Int } S \cup S \setminus \text{Int } S$ is measurable, and so $c(S) = \mu(S)$.

(ii) Since S is bounded, $\mu(\text{Int } S) = \mu(\hat{S})$ is finite. This and the inclusion $S \subset \hat{S}$ yield

$$\bar{\mu}(S \setminus \text{Int } S) \leq \bar{\mu}(\hat{S} \setminus \text{Int } S) = \mu(\hat{S}) - \mu(\text{Int } S) = 0.$$

It follows that S is measurable and that $\mu(S) = \mu(\text{Int } S)$. Let now c be any normalized extrinsic symplectic capacity on $\mathbb{R}^2$. In view of Proposition C.6, (i) and (iii) in Theorem C.8 and the assumptions on S we have

$$w_G(S) = \hat{w}_G(S) \leq c(S) \leq \hat{z}(S) = \mu(\hat{S}) = \mu(\text{Int } S) = w_G(S)$$

and so $c(S) = \mu(\text{Int } S) = \mu(S)$ as claimed. $\square$

We recall that Gromov's Nonsqueezing Theorem states that a ball $B^{2n}(a)$ symplectically embeds into a cylinder $Z^{2n}(A)$ only if $A \geq a$.

Corollary C.11. *Fix $a \in]0, \pi]$ and a normalized extrinsic symplectic capacity c on $\mathbb{R}^2$. Then the Nonsqueezing Theorem is equivalent to the identity*

$$\inf_{\varphi} c\big(p\big(\varphi(B^{2n}(a))\big)\big) = a \tag{C.1}$$

where φ varies over all symplectomorphisms of $\mathbb{R}^{2n}$ which embed $B^{2n}(a)$ into $Z^{2n}(\pi)$, and where $p \colon Z^{2n}(\pi) \to B^2(\pi)$ is the projection.

Proof. Assume first that the Nonsqueezing Theorem does not hold. Then there exist $a' > 0$ and $A < a'$ and a symplectic embedding $\varphi \colon B^{2n}(a') \hookrightarrow Z^{2n}(A)$. According to Proposition A.1 we find $a'' \in]A, a'[$ and a symplectomorphism Φ of $\mathbb{R}^{2n}$ such that

$$\Phi\big(B^{2n}(a'')\big) \subset Z^{2n}(A).$$

Notice that $\frac{a}{a''}A < a \leq \pi$. We denote the dilatation $z \mapsto \frac{a''}{a}z$ of $\mathbb{R}^{2n}$ by α. The composition $\alpha^{-1} \circ \Phi \circ \alpha$ is a symplectomorphism of $\mathbb{R}^{2n}$ which embeds $B^{2n}(a)$ into the subset $Z^{2n}\left(\frac{a}{a''}A\right)$ of $Z^{2n}(\pi)$, i.e.,

$$p\big((\alpha^{-1} \circ \Phi \circ \alpha)\big(B^{2n}(a)\big)\big) \subset D\left(\frac{a}{a''}A\right).$$

The axioms for the normalized symplectic capacity c on $\mathbb{R}^2$ imply that

$$c\big(p\big((\alpha^{-1} \circ \Phi \circ \alpha)\big(B^{2n}(a)\big)\big)\big) \le \frac{a}{a''} A.$$

Since $\frac{a}{a''} A < a$ we conclude that (C.1) does not hold.

Assume now that (C.1) does not hold. Since $c\left(p\left(B^{2n}(a)\right)\right) = c(D(a)) = a$ we then find a symplectomorphism φ of $\mathbb{R}^{2n}$ and $\epsilon > 0$ such that

$$c\big(p\big(\varphi\big(B^{2n}(a)\big)\big)\big) \le a - 2\epsilon. \tag{C.2}$$

We abbreviate $U := p\left(\varphi\left(B^{2n}(a)\right)\right)$. Since U is open and connected, from Corollary C.10 (i), Proposition C.6 and the estimate (C.2) it follows that

$$z(U) = \mu(U) = w_G(U) = \hat{w}_G(U) \le c(U) \le a - 2\epsilon.$$

We therefore find a symplectic embedding $\psi : U \hookrightarrow D(a - \epsilon)$. The composition $(\psi \times \mathrm{id}_{2n-2}) \circ \varphi$ symplectically embeds $B^{2n}(a)$ into $Z^{2n}(a - \epsilon)$ and so the Nonsqueezing Theorem does not hold. $\qquad\square$

According to Proposition C.7 we have $c(S^1) = 0$ for any intrinsic symplectic capacity c on $\mathbb{R}^2$ and $c(S^1) = \pi$ for the first Ekeland–Hofer capacity, the displacement energy and the outer cylindrical capacity. The following result, which was pointed out to me by David Hermann, shows that there exist other normalized extrinsic symplectic capacities on $\mathbb{R}^2$.

Proposition C.12 (D. Hermann). *For any $a \in [0, \pi]$ there exists a normalized extrinsic symplectic capacity c on $\mathbb{R}^2$ such that $c(S^1) = a$.*

Proof. Fix $a \in [0, \pi]$. For $r \ge 0$ we set

$$c\left(rD(\pi)\right) := r^2\pi \quad \text{and} \quad c(rS^1) := r^2 a,$$

and given an arbitrary subset $S \subset \mathbb{R}^2$ we set

$$c(S) := \max\left\{\hat{w}_G(S), \sigma(S)\right\}$$

where

$$\hat{w}_G(S) := \sup\left\{r^2\pi \mid \text{there exists } \varphi \in \mathcal{D}(2) \text{ such that } \varphi\left(rD(\pi)\right) \subset S\right\},$$
$$\sigma(S) := \sup\left\{r^2 a \mid \text{there exists } \varphi \in \mathcal{D}(2) \text{ such that } \varphi(rS^1) \subset S\right\}.$$

Then c is a well-defined normalized extrinsic symplectic capacity on $\mathbb{R}^2$, and $c(S^1) = a$. $\qquad\square$

C.3 Comparison of c_B and c_C with symplectic capacities. For any subset S of $\mathbb{R}^{2n}$ we define the symplectic invariants c_B and c_C by

$$c_B(S) := \inf\{a \mid S \text{ symplectically embeds into } B^{2n}(a)\}, \tag{C.3}$$

$$c_C(S) := \inf\{a \mid S \text{ symplectically embeds into } C^{2n}(a)\}. \tag{C.4}$$

For $n = 1$, the invariants c_B and c_C coincide with the cylindrical capacity z. For $n \geq 2$, the invariants c_B and c_C fulfil all axioms of a normalized intrinsic symplectic capacity on $\mathbb{R}^{2n}$ except that they are infinite for the symplectic cylinder. Recall from Proposition C.3 that the cylindrical capacity z is the largest normalized intrinsic symplectic capacity on $\mathbb{R}^{2n}$.

Proposition C.13. *For any subset S of $\mathbb{R}^{2n}$ we have*

$$z(S) \leq c_C(S) \leq c_B(S).$$

Proof. The claim follows from the inclusions $B^{2n}(a) \subset C^{2n}(a) \subset Z^{2n}(a)$. $\qquad\square$

Remark C.14. If $n \geq 2$, the inequalities in Proposition C.13 are in general not equalities: For a polydisc $P = P^{2n}(\pi, \ldots, \pi, a)$ with $a > \pi$ we have $z(P) = \pi < c_C(P)$. Moreover, (2.2.3) and (2.3.1) imply $c_C(C^{2n}(\pi)) = \pi < c_B(C^{2n}(\pi)) = n\pi$. (This example is extremal in the sense that for any $S \subset \mathbb{R}^{2n}$ we have $c_B(S) \leq n\, c_C(S)$.) $\diamond$

C.4 Comparison of c_B with the symplectic diameter. Recall that for any subset S of $\mathbb{R}^{2n}$ the *diameter* $d(S)$ is defined by

$$d(S) := \sup\left\{|z - z'| \mid z, z' \in S\right\}.$$

Symplectifying the Euclidean invariant $d(S)$ we obtain the symplectic invariant $d_s(S)$ defined by

$$d_s(S) := \inf\left\{d\big(\varphi(S)\big) \mid \varphi \text{ symplectically embeds } S \text{ into } \mathbb{R}^{2n}\right\}.$$

For convenience we shall work with the symplectic invariant $\delta(S)$ defined by

$$\delta(S) := \pi\left(\frac{d_s(S)}{2}\right)^2 \tag{C.5}$$

rather than with $d_s(S)$.

Theorem C.15. *For any subset S of $\mathbb{R}^{2n}$ we have*

$$\delta(S) \leq c_B(S) \leq \frac{4n}{2n + 1}\, \delta(S) \tag{C.6}$$

and if $n = 1$, then $\delta(S) = c_B(S) = \bar{\mu}(S)$.

Proof. Fix $S \subset \mathbb{R}^{2n}$. The *circumradius* $R(S)$ of S is the radius of the smallest ball containing S, i.e.,

$$R(S) := \inf \left\{ R \mid \text{there exists } w \in \mathbb{R}^{2n} \text{ such that } \tau_w(S) \subset B^{2n}(\pi R^2) \right\}$$

where τ_w denotes the translation $z \mapsto z + w$ of $\mathbb{R}^{2n}$. Symplectifying the Euclidean invariant $R(S)$ we obtain the symplectic invariant

$$R_s(S) := \inf \left\{ R(\varphi(S)) \mid \varphi \text{ symplectically embeds } S \text{ into } \mathbb{R}^{2n} \right\}.$$

Notice that each translation τ_w of $\mathbb{R}^{2n}$ is symplectic. Comparing the definition (C.3) of the invariant $c_B(S)$ with the definitions of $R(S)$ and $R_s(S)$ we therefore find that

$$c_B(S) = \pi \, (R_s(S))^2. \tag{C.7}$$

The main ingredient of the proof of Theorem C.15 are the inequalities

$$\frac{1}{2} d(T) \le R(T) \le \sqrt{\frac{n}{2n+1}} \, d(T) \tag{C.8}$$

valid for any subset T of $\mathbb{R}^{2n}$. While the left inequality in (C.8) is obvious, the right inequality is the main content of Jung's Theorem [14, Chapter 2, Theorem 11.1.1]. Applying (C.8) to all symplectic images $\varphi(S)$ of S in $\mathbb{R}^{2n}$ and taking the infimum we find that

$$\frac{1}{2} d_s(S) \le R_s(S) \le \sqrt{\frac{n}{2n+1}} \, d_s(S).$$

These inequalities, the definition (C.5) and the identity (C.7) imply (C.6).

Assume now $S \subset \mathbb{R}^2$. The left inequality in (C.6) and Theorem C.8 (ii) show that

$$\delta(S) \le c_B(S) = z(S) = \bar{\mu}(S).$$

In order to show that $\delta(S) \ge \bar{\mu}(S)$ we assume that φ symplectically embeds S into $\mathbb{R}^2$ and denote the convex hull of $\varphi(S)$ by conv $\varphi(S)$. Applying the Bieberbach inequality [14, Chapter 2, Theorem 11.2.1] to $\varphi(S)$ we find

$$\pi \left(\frac{d\,(\varphi(S))}{2} \right)^2 \ge \mu \, (\text{conv } \varphi(S)) \ge \bar{\mu} \, (\varphi(S)) = \bar{\mu}(S).$$

Since φ was arbitrary it follows that $\delta(S) \ge \bar{\mu}(S)$. We conclude that $\delta(S) = c_B(S) = \bar{\mu}(S)$. The proof of Theorem C.15 is complete. $\square$

Remarks C.16. 1. Assume that $n \ge 2$. It follows from the Bieberbach inequality [14, Chapter 2, Theorem 11.2.1] that $\delta \left(B^{2n}(\pi) \right) = c_B \left(B^{2n}(\pi) \right)$ and so the left inequality in (C.6) is sharp. We do not know, however, whether the right inequality in (C.6) is sharp.

2. The invariants $\delta = \pi \left(\frac{d_s}{2} \right)^2$ and $c_B = \pi (R_s)^2$ are examples of "Euclidean invariants for symplectic domains" as studied in [76]. $\diamond$

C.5   The invariant c_C as a symplectic projection invariant.   As was pointed out in [26, p. 580], symplectic capacities on $\mathbb{R}^{2n}$ measure to some extent the area of two dimensional symplectic projections of a set. We are now going to make this point precise for the invariants z and c_C.

For $i = 1, \ldots, n$ we set

$$E_i = \left\{ z \in \mathbb{C}^n \mid z_j = 0 \text{ for } j \neq i \right\},$$

and we denote the orthogonal projection $\mathbb{R}^{2n} \to E_i$ by p_i. The Lebesgue measure on E_i is denoted μ_i, and the outer Lebesgue measure on E_i is denoted $\bar{\mu}_i$. Given any subset S of $\mathbb{R}^{2n}$ we set

$$s_m(S) = \inf_{\varphi} \min_{1 \leq i \leq n} \bar{\mu}_i \left(p_i(\varphi(S)) \right),$$

$$s_M(S) = \inf_{\varphi} \max_{1 \leq i \leq n} \bar{\mu}_i \left(p_i(\varphi(S)) \right),$$

where φ varies over all symplectic embeddings of S into $\mathbb{R}^{2n}$.

Remarks C.17.   1.   If $S \subset \mathbb{R}^{2n}$ is Lebesgue measurable, then so are the sets $p_i(\varphi(S)) \subset E_i$, $i = 1, \ldots, n$, and so

$$s_m(S) = \inf_{\varphi} \min_{1 \leq i \leq n} \text{area} \left(p_i(\varphi(S)) \right),$$

$$s_M(S) = \inf_{\varphi} \max_{1 \leq i \leq n} \text{area} \left(p_i(\varphi(S)) \right).$$

2.   Composing an "infimal" embedding $\varphi \colon S \hookrightarrow \mathbb{R}^{2n}$ in the definition of $s_m(S)$ with the linear symplectomorphism

$$(z_1, z_2, \ldots, z_{i-1}, z_i, z_{i+1}, \ldots, z_n) \mapsto (z_i, z_2, \ldots, z_{i-1}, z_1, z_{i+1}, \ldots, z_n)$$

we find that for all subsets S of $\mathbb{R}^{2n}$,

$$s_m(S) = \inf_{\varphi} \bar{\mu}_1 \left(p_1(\varphi(S)) \right).$$

In view of the Extension after Restriction Principle Proposition A.1, the restriction of s_m to $\mathcal{K}_s^{2n}$ thus agrees with the invariant π_s considered in Section 8.2.   ◇

Theorem C.18.   *For any subset S of $\mathbb{R}^{2n}$ we have*

$$z(S) = s_m(S) \quad \text{and} \quad c_C(S) = s_M(S).$$

Proof.   We show $z(S) \leq s_m(S)$. We may assume that $s_m(S)$ is finite. Fix $\epsilon > 0$. By Remark C.17.2 there exists a symplectic embedding $\varphi \colon S \hookrightarrow \mathbb{R}^{2n}$ such that $\bar{\mu}_1 \left(p_1(\varphi(S)) \right) \leq s_m(S) + \epsilon$. According to Theorem C.8 (ii) we find a symplectic embedding ψ of $p_1\,(\varphi(S))$ into $D\,(s_m(S)+2\epsilon)$. The composition $(\psi\times\mathrm{id}_{2n-2})\circ\varphi$ symplectically embeds S into $Z^{2n}\,(s_m(S)+2\epsilon)$. Since $\epsilon>0$ was arbitrary, we conclude $z(S)\le s_m(S)$. The reverse inequality $s_m(S)\le z(S)$ is obvious.

The equality $c_C(S)=s_M(S)$ is proved in the same way as $z(S)=s_m(S)$. $\square$

Remark C.19. The invariants $s_m=z$ and $s_M=c_C$ are special symplectic projection invariants as studied in [76]. $\diamond$

D Computer programs

This appendix contains computer programs computing the functions f_{EB} and f_{EC} defined in 4.3.1 and 4.4.1 and drawn in Figures 1.1 and 7.1 and in Figure 4.13. The Mathematica programs below can be found under

ftp://ftp.math.ethz.ch/pub/papers/schlenk/folding.m

For convenience, in the programs (but not in the text) both the u-axis and the capacity-axis are rescaled by the factor $1/\pi$.

D.1 The estimate f_{EB}. We fix a and u_1 and try to embed the ellipsoid $E(\pi,a)$ into the ball $B^4(2\pi+(1-2\pi/a)u_1)$ by the multiple folding procedure specified in 4.3.1. If u_1 is admissible, we set $A(a,u_1)=2\pi+(1-2\pi/a)u_1$ and $A(a,u_1)=a$ otherwise.

```
A[a_, u1_] :=
  Block[{A=2+(1-2/a)u1},
     i  = 2;
     ui = (a+1)/(a-1)u1-a/(a-1);
     ri = a-u1-ui;
     li = ri/a;
     While[True,
          Which[EvenQ[i],
                If[ri < ui,
                   Return[A],
                   If[ui <= 2li,
                      Return[a],
                      i++;
                      ui = a/(a-2)(ui-2li);
                      ri = ri-ui;
                      lj = li;
                      li = ri/a
                   ]
```

```
           ],
       OddQ[i],
         If[ri < ui+lj,
            Return[A],
            i++;
            ui = (a+1)/(a-1)ui;
            ri = ri-ui;
            li = ri/a
            ]
         ]
       ]
     ]
```

This program just does what we proposed to do in 4.3.1 in order to decide whether u_1 is admissible or not. Note, however, that in the Oddq[i]-part we do not check whether the stairs S_{i+1} are contained in $T^4(A)$. This negligence does not cause troubles, since if $S_{i+1} \not\subset T^4(A)$, then u_1 will be recognized to be non-admissible in the subsequent EvenQ[i+1]-part. Indeed, recall that $S_{i+1} \not\subset T^4(A)$ is equivalent to $l_{i+1} \geq u_{i+1}$; hence $r_{i+1} = (a/\pi)l_{i+1} > u_{i+1}$ and $u_{i+1} \leq 2\,l_{i+1}$.

We denote the infimum of the admissible u_1's again by $u_0 = u_0(a)$. Then $A(a, u_1) = a$ for $u_1 < u_0$, and $A(a, u_1) = 2\pi + (1 - 2\pi/a)u_1$ is a linear increasing function for $u_1 > u_0$. Since, by (4.3.7), we can assume that $u_0 \leq a/2$, we have $A(a, u_0) \leq A(a, a/2) = \pi + a/2 < a$. Therefore, u_0 is found up to accuracy $acc/2$ by the following bisectional algorithm.

```
u0[a_, acc_] :=
  Block[{},
    b = a/(a+1);
    c = a/2;
    u1 = (b+c)/2;
    While[(c-b)/2 > acc/2,
      If[A[a,u1] < a, c=u1, b=u1];
      u1 = (b+c)/2
          ];
    Return[u1]
    ]
```

Here the choice $b = a\pi/(a + \pi)$ is also based on (4.3.7). Up to accuracy acc, the resulting estimate $f_{\mathrm{EB}}(a)$ is given by

```
fEB[a_, acc_] := 2 + (1-2/a)u0[a,acc].
```

D.2   The estimate f_{EC}. We fix $a > \pi$. As we have seen in Remark 4.4.1.1, folding only once does not yield an optimal embedding. We therefore fold at least twice and hence choose $u_1 \in \left]\frac{a\pi}{a+\pi}, \frac{a}{2}\right[$. We first calculate the height of the image of $T(a, \pi)$ determined by a and u_1. The following program is best understood by looking at Figure 4.12.

```
h[a_, u1_] :=
  Block[{l1=1-u1/a},
     i  = 2;
     ui = (a+1)/(a-1)u1-a/(a-1);
     ri = a-u1-ui;
     li = ri/a;
     hi = 2l1;
     While[ri > u1+l1 - li,
             i++;
             ui = (a+1)/(a-1)ui;
             ri = ri-ui;
             lj = li;
             li = ri/a;
             If[EvenQ[i], hi = hi+2lj]
           ];
     Which[EvenQ[i],
             hi = hi+li,
             OddQ[i],
             hi = hi+Max[lj,2li]
           ];
     Return[hi]
        ]
```

As explained in 4.4.1, the optimal folding point $u_0(a)$ is the u-coordinate of the unique intersection point of the graphs of $h(a, u_1)$ and $w(a, u_1)$. It can thus be found by a bisectional algorithm.

```
u0[a_, acc_] :=
  Block[{},
     b = a/(a+1);
     c = a/2;
     u1 = (b+c)/2;
     While[(c-b)/2 > acc/2,
       If[h[a,u1] > 1+(1-1/a)u1, b=u1, c=u1];
       u1 = (b+c)/2
          ];
     Return[u1]
       ]
```

Up to accuracy *acc*, the resulting estimate $f_{EC}(a)$ is given by

```
fEC[a_, acc_] := 1+(1-1/a)u0[a,acc].
```

E   Some other symplectic embedding problems

In this book we studied only some few symplectic embedding problems. In this appendix we briefly describe some other embedding problems in symplectic geometry and topology.

1. Local Nonsqueezing. Recall from the introduction that $Z^{2n}(\pi)$ denotes the open cylinder $D(\pi) \times \mathbb{R}^{2n-2}$ in $(\mathbb{R}^{2n}, \omega_0)$, and let $\overline{Z^{2n}(\pi)}$ be its closure. Assume that $K \subset \overline{Z^{2n}(\pi)}$ is a compact subset with smooth boundary ∂K such that $\partial K \cap \partial \overline{Z^{2n}(\pi)}$ is non-empty.

Can K be mapped into $Z^{2n}(\pi)$ via a symplectic isotopy?

This problem was almost fully solved in [50], and interesting variants were studied in [69].

2. Embeddings of Lagrangian submanifolds. Let K be a closed n-dimensional manifold, and let (M, ω) be a $2n$-dimensional symplectic manifold. By the Whitney Embedding Theorem, K smoothly embeds into M. An n-dimensional submanifold L of (M, ω) is called *Lagrangian* if ω vanishes on TL.

Does K embed as a Lagrangian submanifold into (M, ω)?

This is a famous problem in symplectic topology, and for the state of the art we refer to [1], [11], [63, Section 9.3] and the references therein.

3. Neighbourhoods of Lagrangian submanifolds. Consider a closed Lagrangian submanifold L of a $2n$-dimensional symplectic manifold (M, ω). By a theorem of Weinstein, there exists a symplectomorphism from a neighbourhood of L in (M, ω) to a neighbourhood of the zero-section 0_L of the cotangent bundle (T^*L, ω_{can}) mapping L to 0_L. Here, $\omega_{can} = \sum_{i=1}^{n} dp_i \wedge dq_i$ is the canonical symplectic form on T^*L.

How large can this neighbourhood be chosen?

Here, the size of a neighbourhood can be measured in many ways.

(i) If L comes with a Riemannian metric g, one can look at the ball bundles $T_r^*L = \{(q, p) \in T^*L \mid g_q(p, p) < r\}$ and study

$$\text{size}_g(M, L) = \sup \left\{ \pi r^2 \mid (T_r^*L, 0_L) \overset{s}{\hookrightarrow} (M, L) \right\}.$$

(ii) The size of a neighbourhood of L as above can be measured by any symplectic capacity as defined in [39].

(iii) A more local size of a neighbourhood of L in M is given by the relative Gromov width

$$w_G(M, L) = \sup \left\{ a \mid \left(B^{2n}(a), B^{2n}(a) \cap \mathbb{R}^n(x) \right) \overset{s}{\hookrightarrow} (M, L) \right\}$$

considered recently by Barraud and Cornea in [3].

4. The isotopy problem. Given a natural number k, a compact subset K of $(\mathbb{R}^{2n}, \omega_0)$ homeomorphic to a closed $2n$-dimensional ball, and a $2n$-dimensional connected symplectic manifold (M, ω), let $\mathrm{Emb}_k(K, M, \omega)$ be the space of symplectic embeddings of k disjoint copies of K into (M, ω) endowed with the C^∞-topology.

Is $\mathrm{Emb}_k(K, M, \omega)$ connected?

(i) We say that K is starshaped if its interior is starshaped. The following proposition for $k = 1$ is due to Banyaga, [2], and for arbitrary k was pointed out to me together with the proof by Paul Biran.

Proposition E.1. *If K is starshaped, then $\mathrm{Emb}_k(K, \mathbb{R}^{2n}, \omega_0)$ is connected for every k.*

The proof is given below.

(ii) The *camel space* in $\mathbb{R}^{2n}$ with eye of width a is the subset

$$\mathcal{C}^{2n}(a) = \{x_1 < 0\} \cup \{x_1 > 0\} \cup B^{2n}(a)$$

of $(\mathbb{R}^{2n}, \omega_0)$. If $b > a$, the "symplectic camel" $\overline{B^{2n}(b)}$ cannot pass the eye of width a, so that $\mathrm{Emb}_k\big(B^{2n}(b), \mathcal{C}^{2n}(a), \omega_0\big)$ has at least 2^k components, see [22], [65], [84].

(iii) For some closed symplectic 4-manifolds such as $\mathbb{CP}^2$ and ruled symplectic 4-manifolds and for the 4-ball it is known that $\mathrm{Emb}_k\big(B^4(a), M, \omega\big)$ is connected for all k, see [58].

(iv) Denote by $\overline{P(a, b)}$ the closure of the polydisc $P(a, b) = D(a) \times D(b)$. If $a + b > \pi$ and $a, b < \pi$, then $\mathrm{Emb}_1\big(\overline{P(a, b)}, \overline{P(\pi, \pi)}, \omega_0\big)$ has at least two components, see [26]. No similar result is known in higher dimensions.

(v) Nothing is known for ellipsoids if (M, ω) is closed or a bounded subset of $(\mathbb{R}^{2n}, \omega_0)$, nor is anything known for any K if (M, ω) is a closed symplectic manifold of dimension ≥ 6.

Proof of Proposition E.1. Let

$$\varphi = \coprod_{i=1}^{k} \varphi_i : \coprod_{i=1}^{k} K \hookrightarrow \mathbb{R}^{2n}$$

be a symplectic embedding of k copies of the starshaped set K. By definition, the embeddings φ_i extend to disjoint symplectic embeddings of an open neighbourhood of K. As is easy to see, this neighbourhood can be chosen starshaped. We can assume that 0 is a star point of K. Since the linear symplectic group $\mathrm{Sp}(n; \mathbb{R})$ is connected, we can also assume that $d\varphi_1(0) = \mathrm{id}$. Using Alexander's trick as in Step 1 of the proof of Proposition A.1 and extending the thereby obtained symplectic isotopy to $\mathbb{R}^{2n}$ as in Step 2 of that proof, we obtain a (compactly supported) symplectic isotopy $\Phi^t, t \in [0, 1]$, of $\mathbb{R}^{2n}$ such that $\Phi^0 = \mathrm{id}$ and

$$\phi_1 := \Phi^1 \circ \varphi_1 = \mathrm{id}_K. \tag{E.1}$$

If $k = 1$ we are done. If $k \geq 2$, set $\phi_i = \Phi^1 \circ \varphi_i$ and $p_i = \phi_i(0)$, $K_i = \phi_i(K)$. After slightly translating K_i for $i = 3, \dots, k$, if necessary, we can assume that the open rays $\mathcal{C}(p_i) = \{t\, p_i \mid t > 0\}$ are disjoint for $i = 2, \dots, k$. For $i = 1, \dots, k$ we consider now the symplectic isotopy

$$\phi_i^t(z) = \frac{1}{1-t}\, \phi_i\big((1-t)z\big), \quad z \in K, \ t \in [0, 1[,$$

starting at ϕ_i. Since the embeddings φ_i are disjoint, the embeddings ϕ_i^t are disjoint for each t. In view of (E.1) the isotopy ϕ_1^t is constant. To describe the other isotopies, we set $K_i^t = \phi_i^t(K)$ and consider the punctured cones

$$\mathcal{C}\left(K_i^t\right) := \left\{tz \mid t > 0,\ z \in K_i^t\right\}.$$

For $t \to 1$, the cone $\mathcal{C}\left(K_i^t\right)$ "converges" to the ray $\mathcal{C}(p_i)$. Since these rays are disjoint, we find $t_0 \in [0, 1[$ such that the cones $\mathcal{C}_i := \mathcal{C}\left(K_i^{t_0}\right)$ are mutually disjoint for $i = 2, \dots, k$. Notice that the cones $\mathcal{C}_i$ have non-empty interior since K was assumed to be homeomorphic to a $2n$-ball. Set $q_i = \phi_i^{t_0}(0) = \frac{1}{1-t_0}\, p_i$. Using again that $\mathrm{Sp}(n; \mathbb{R})$ is connected and applying Alexander's trick, we find for each $i = 2, \dots, k$ a symplectic isotopy $\psi_i \colon [0, 1] \times K \to \mathbb{R}^{2n}$ such that $\psi_i^t(0) = q_i$ for all $t \in [0, 1]$ and

$$\psi_i^0(z) = \phi_i^{t_0}(z) \quad \text{and} \quad \psi_i^1(z) = z + q_i \quad \text{for all } z \in K.$$

Choose ρ_i so large that the image $\psi_i\left([0, 1] \times K\right)$ is contained in the ball $B_{\rho_i}(q_i)$ of radius ρ_i centred at q_i. Since $\mathcal{C}_i$ has non-empty interior, we find $s_i > 0$ such that the translate $B_{\rho_i}\left(q_i + s_i q_i\right)$ of $B_{\rho_i}(q_i)$ is contained in $\mathcal{C}_i$ and is disjoint from K. Translating $K_i^{t_0}$ by $s_i q_i$ and finally applying the symplectic isotopy

$$[0, 1] \times K \to \mathbb{R}^{2n}, \quad (t, z) \mapsto \psi_i^t(z) + s_i q_i,$$

whose image lies in $B_{\rho_i}\left(q_i + s_i q_i\right)$, we end up with a symplectic isotopy

$$\varphi^t = \coprod_{i=1}^{k} \varphi_i^t \colon [0, 1] \times \coprod_{i=1}^{k} K \ \to \ \mathbb{R}^{2n}$$

starting at $\varphi^0 = \varphi$ and ending at φ^1 with $\varphi_1^1 = \mathrm{id}$ and $\varphi_i^1(z) = z + q_i + s_i q_i$ for $i = 2, \dots, k$. Since $\varphi_1^1(K) = K$ and since the sets $\varphi_i^1(K)$ for $i = 2, \dots, k$ lie in the mutually disjoint cones $\mathcal{C}_i$, it is clear that a symplectic embedding ψ^1 obtained in this way from another symplectic embedding $\psi \colon \coprod_{i=1}^{k} K \hookrightarrow \mathbb{R}^{2n}$ is symplectically isotopic to φ^1. The proof of Proposition E.1 is thus complete. $\square$

Bibliography

[1] M. Audin, F. Lalonde and L. Polterovich. Symplectic rigidity: Lagrangian sub-manifolds. In *Holomorphic curves in symplectic geometry*, Progr. Math. 117, Birkhäuser, Basel, 1994, 271–321.

[2] A. Banyaga. Sur la structure du groupe des difféomorphismes qui préservent une forme symplectique. *Comment. Math. Helv.* **53** (1978), 174–227.

[3] J. F. Barraud and O. Cornea. Lagrangian Intersections and the Serre Spectral Sequence. math.DG/0401094.

[4] S. M. Bates. Some simple continuity properties of symplectic capacities. In *The Floer memorial volume*, Progr. Math. 133, Birkhäuser, Basel, 1995, 185–193.

[5] U. Betke and M. Henk. Finite packings of spheres. *Discrete Comput. Geom.* **19** (1998), 197–227.

[6] P. Biran. Connectedness of spaces of symplectic embeddings. *Internat. Math. Res. Notices* **1996**, 487–491.

[7] P. Biran. Symplectic packing in dimension 4. *Geom. Funct. Anal.* **7** (1997), 420–437.

[8] P. Biran. A stability property of symplectic packing. *Invent. Math.* **136** (1999), 123–155.

[9] P. Biran. Constructing new ample divisors out of old ones. *Duke Math. J.* **98** (1999), 113–135.

[10] P. Biran. From symplectic packing to algebraic geometry and back. In *European Congress of Mathematics, Vol. II (Barcelona, 2000)*, Progr. Math. 202, Birkhäuser, Basel, 2001, 507–524.

[11] P. Biran. Geometry of symplectic intersections. In *Proceedings of the International Congress of Mathematicians, Vol. II (Beijing, 2002)*, Higher Ed. Press, Beijing, 2002, 241–255.

[12] P. Biran and K. Cieliebak. Symplectic topology on subcritical manifolds. *Comment. Math. Helv.* **76** (2001), 712–753.

[13] W. Boothby. *An introduction to differentiable manifolds and Riemannian geometry*. Second edition, Pure and Applied Mathematics 120, Academic Press, Orlando, 1986.

[14] Yu. D. Burago and V. A. Zalgaller. *Geometric Inequalities*. Grundlehren Math. Wiss. 285, Springer-Verlag, Berlin, 1988.

[15] K. Cieliebak. Symplectic boundaries: creating and destroying closed characteristics. *Geom. Funct. Anal.* **7** (1997), 269–321.

[16] K. Cieliebak, H. Hofer, J. Latschev and F. Schlenk. Quantitative symplectic geometry. To appear in the AMS volume in honour of Anatole Katok's 60'th birthday.

[17] H. Cohn and N. Elkies. New upper bounds on sphere packings. I. *Ann. of Math.* **157** (2003), 689–714.

[18] C. Conley and E. Zehnder. The Birkhoff-Lewis fixed point theorem and a conjecture of V. I. Arnol'd. *Invent. Math.* **73** (1983), 33–49.

[19] J. Conway and N. Sloane. *Sphere packings, lattices and groups*. Third edition, Grundlehren Math. Wiss. 290, Springer-Verlag, New York, 1999.

[20] I. Ekeland and H. Hofer. Symplectic topology and Hamiltonian dynamics. *Math. Z.* **200** (1989), 355–378.

[21] I. Ekeland and H. Hofer. Symplectic topology and Hamiltonian dynamics II. *Math. Z.* **203** (1990), 553–567.

[22] Y. Eliashberg and M. Gromov. Convex symplectic manifolds. In *Several Complex Variables and Complex Geometry, Proceedings, Summer Research Institute, Santa Cruz, 1989, Part 2*, ed. by E. Bedford et al., Proc. Sympos. Pure Math. **52**, Amer. Math. Soc., Providence, 1991, 135–162.

[23] Y. Eliashberg and H. Hofer. Unseen symplectic boundaries. In *Manifolds and geometry* (Pisa, 1993), Sympos. Math. XXXVI, Cambridge University Press, Cambridge, 1996, 178–189.

[24] Y. Eliashberg and N. Mishachev. *Introduction to the h-principle*. Grad. Stud. Math. 48, Amer. Math. Soc., Providence, RI, 2002.

[25] A. Floer and H. Hofer. Symplectic Homology I: Open sets in $\mathbb{C}^n$. *Math. Z.* **215** (1994), 37–88.

[26] A. Floer, H. Hofer and K. Wysocki. Applications of symplectic homology I. *Math. Z.* **217** (1994), 577–606.

[27] U. Frauenfelder, V. Ginzburg and F. Schlenk. Energy capacity inequalities via an action selector. To appear in *Proceedings on Geometry, Groups, Dynamics and Spectral Theory in memory of Robert Brooks*.

[28] R. Gardner. *Geometric tomography*. Encyclopedia Math. Appl. 58, Cambridge University Press, Cambridge, 1995.

[29] R. Greene and K. Shiohama. Diffeomorphisms and volume preserving embeddings of non-compact manifolds. *Trans. Amer. Math. Soc.* **255** (1979), 403–414.

[30] P. Gritzmann. Finite packing of equal balls. *J. London Math. Soc.* **33** (1986), 543–553.

[31] M. Gromov. Pseudo-holomorphic curves in symplectic manifolds. *Invent. Math.* **82** (1985), 307–347.

[32] M. Gromov. *Partial Differential Relations*. Ergeb. Math. Grenzgeb. (3) 9, Springer-Verlag, Berlin, 1986.

[33] T. Hales. The Kepler conjecture. math.MG/9811078.

[34] *Handbook of convex geometry*. Vol. B, edited by P. M. Gruber and J. M. Wills, North-Holland Publishing Co., Amsterdam, 1993.

[35] B. Hasselblatt and A. Katok. *Introduction to the modern theory of dynamical systems*. Encyclopedia Math. Appl. 54, Cambridge University Press, Cambridge, 1995.

[36] D. Hermann. Holomorphic curves and Hamiltonian systems in an open set with restricted contact-type boundary. *Duke Math. J.* **103** (2000), 335–374.

[37] D. Hermann. Non-equivalence of symplectic capacities for open sets with restricted contact type boundary.
http://www.math.u-psud.fr/~biblio/pub/1998/abs/ppo1998_32.html

[38] H. Hofer. On the topological properties of symplectic maps. *Proc. Roy. Soc. Edinburgh Sect. A* **115** (1990), 25–38.

[39] H. Hofer and E. Zehnder. *Symplectic Invariants and Hamiltonian Dynamics*. Birkhäuser, Basel, 1994.

[40] M.-Y. Jiang. Symplectic embeddings from $\mathbf{R}^{2n}$ into some manifolds. *Proc. Roy. Soc. Edinburgh Sect. A* **130** (2000), 53–61.

[41] Y. Karshon. Appendix to [61]. *Invent. Math.* **115** (1994), 431–434.

[42] Y. Karshon and S. Tolman. The Gromov width of complex Grassmannians. math.SG/0405391.

[43] A. Katok. Bernoulli diffeomorphisms on surfaces. *Ann. of Math.* **110** (1979), 529–547.

[44] S. Kobayashi and K. Nomizu. *Foundations of Differential Geometry.* Volume II, Interscience, New York, 1969.

[45] B. Kruglikov. A remark on symplectic packings. *Dokl. Akad. Nauk* **350** (1996), 730–734.

[46] W. Kuperberg. An extremum property characterizing the n-dimensional regular cross-polytope. math.MG/0112290.

[47] F. Lalonde. Isotopy of symplectic balls, Gromov's radius and the structure of ruled symplectic 4-manifolds. *Math. Ann.* **300** (1994), 273–296.

[48] F. Lalonde and D. McDuff. The geometry of symplectic energy. *Ann. of Math.* **141** (1995), 349–371.

[49] F. Lalonde and D. McDuff. Hofer's L^∞-geometry: energy and stability of Hamiltonian flows, part II. *Invent. Math.* **122** (1995), 35–69.

[50] F. Lalonde and D. McDuff. Local non-squeezing theorems and stability. *Geom. Funct. Anal.* **5** (1995), 364–386.

[51] F. Lalonde and D. McDuff. The classification of ruled symplectic 4-manifolds. *Math. Res. Lett.* **3** (1996), 769–778.

[52] R. Lazarsfeld. Lengths of periods and Seshadri constants of abelian varieties. *Math. Res. Lett.* **3** (1996), 439–447.

[53] E. Makai and H. Martini. On bodies associated with a given convex body. *Canad. Math. Bull.* **39** (1996), 448–459.

[54] F. M. Maley, J. Mastrangeli and L. Traynor. Symplectic packings in cotangent bundles of tori. *Experiment. Math.* **9** (2000), 435–455.

[55] D. McDuff. Blowing up and symplectic embeddings in dimension 4. *Topology* **30** (1991), 409–421.

[56] D. McDuff. Symplectic manifolds with contact type boundaries. *Invent. Math.* **103** (1991), 651–671.

[57] D. McDuff. An irrational ruled symplectic 4-manifold. In *The Floer memorial volume*, Progr. Math. 133, Birkhäuser, Basel, 1995, 545–554.

[58] D. McDuff. From symplectic deformation to isotopy. In *Topics in symplectic 4-manifolds* (Irvine, CA, 1996), First Int. Press Lect. Ser., I, Internat. Press, Cambridge, MA, 1998, 85–99.

[59] D. McDuff. Fibrations in symplectic topology. In *Proceedings of the International Congress of Mathematicians*, Vol. I (Berlin, 1998), *Doc. Math.* 1998, Extra Vol. I, 339–357.

[60] D. McDuff. Geometric variants of the Hofer norm. *J. Symplectic Geom.* **1** (2002), 197–252.

[61] D. McDuff and L. Polterovich. Symplectic packings and algebraic geometry. *Invent. Math.* **115** (1994), 405–429.

[62] D. McDuff and D. Salamon. *Introduction to Symplectic Topology*. Second edition, Oxford Mathematical Monographs, The Clarendon Press, Oxford University Press, New York, 1998.

[63] D. McDuff and D. Salamon. *J-holomorphic curves and symplectic topology*. Amer. Math. Soc. Colloq. Pub. 52, Amer. Math. Soc., Providence, RI, 2004.

[64] D. McDuff and J. Slimowitz. Hofer-Zehnder capacity and length minimizing Hamiltonian paths. *Geom. Topol.* **5** (2001) 799–830.

[65] D. McDuff and L. Traynor. The 4-dimensional symplectic camel and related results. In *Symplectic geometry*, London Math. Soc. Lecture Note Ser. 192, Cambridge University Press, Cambridge, 1993, 169–182.

[66] J. Moser. On the volume elements on a manifold. *Trans. Amer. Math. Soc.* **120** (1965), 286-294.

[67] K. Nomizu and H. Ozeki. The existence of complete Riemannian metrics. *Proc. Amer. Math. Soc.* **12** (1961), 889–891.

[68] V. Ozols. Largest normal neighborhoods. *Proc. Amer. Math. Soc.* **61** (1976), 99–101.

[69] G. Paternain, L. Polterovich and K. F. Siburg. Boundary rigidity for Lagrangian submanifolds, non-removable intersections, and Aubry-Mather theory. Dedicated to Vladimir I. Arnold on the occasion of his 65th birthday. *Moscow Math. J.* **3** (2003), 593–619.

[70] L. Polterovich. *The geometry of the group of symplectic diffeomorphisms*. Lectures in Mathematics ETH Zürich, Birkhäuser, Basel, 2001.

[71] F. Schlenk. An extension theorem in symplectic geometry. *Manuscripta Math.* **109** (2002), 329–348.

[72] F. Schlenk. Volume preserving embeddings of open subsets of $\mathbb{R}^n$ into manifolds. *Proc. Amer. Math. Soc.* **131** (2003), 1925–1929.

[73] F. Schlenk. Symplectic embeddings of ellipsoids. *Israel J. Math.* **138** (2003), 215–252.

[74] F. Schlenk. On a question of Dusa McDuff. *Internat. Math. Res. Notices* **2003**, 77–107.

[75] F. Schlenk. Packing symplectic manifolds by hand. math.SG/0409568.

[76] F. Schlenk. Euclidean geometry of symplectic domains. In preparation.

[77] K. F. Siburg. Symplectic capacities in two dimensions. *Manuscripta Math.* **78** (1993), 149–163.

[78] J.-C. Sikorav. *Systèmes Hamiltoniens et topologie symplectique.* Dipartimento di Matematica dell' Università di Pisa, ETS Editrice, Pisa, 1990.

[79] J.-C. Sikorav. Quelques propriétés des plongements lagrangiens. Analyse globale et physique mathématique (Lyon, 1989), *Mém. Soc. Math. France* **46** (1991), 151–167.

[80] C. Taubes. $SW \Rightarrow Gr$: from the Seiberg-Witten equations to pseudo-holomorphic curves. *J. Amer. Math. Soc.* **9** (1996), 845–918.

[81] L. Traynor. Symplectic packing constructions. *J. Differential Geom.* **42** (1995), 411–429.

[82] C. Viterbo. Capacités symplectiques et applications (d'après Ekeland–Hofer, Gromov). Séminaire Bourbaki, Vol. 1988-89. *Astérisque* **177-178** (1989), Exp. No. 714, 345–362.

[83] C. Viterbo. Plongements lagrangiens et capacités symplectiques de tores dans $\mathbb{R}^{2n}$. *C. R. Acad. Sci. Paris, Sér. I Math.* **311** (1990), 487–490.

[84] C. Viterbo. Symplectic topology as the geometry of generating functions. *Math. Ann.* **292** (1992), 685–710.

[85] C. Viterbo. Metric and isoperimetric problems in symplectic geometry. *J. Amer. Math. Soc.* **13** (2000), 411–431.

[86] H. Whitney. Differentiable manifolds. *Ann. of Math.* **37** (1936), 645–680.

[87] G. Xu. Curves in P^2 and symplectic packings. *Math. Ann.* **299** (1994), 609–613.

[88] C. Zong. *Sphere packings.* Universitext. Springer-Verlag, New York, 1999.

Index